Moving Back From Midnight
Working together to save our planet

Roman Bystrianyk
Kathryn Schmutter, BSc

© 2023 Roman Bystrianyk and Kathryn Schmutter
All rights reserved.

Cover design and book formatting by Salome Mumford
www.salomemumford.com
contact@salomemumford.com

Editing by Maria Purcell-Herald
mariaj.herald@gmail.com
ISBN-13: 9798790077425

www.movingbackfrommidnight.com

*To the two most amazing women I have had the honor and
privilege to know –
Meryl Barr and Michelle Heltay.
Their love and inspiration made this work possible.
Thank you from the depths of my heart and soul.*

*To Willow and all the world's children,
I hope we leave you a living and vibrant planet.*

– Roman

*To my incredible husband, Daniel Schmutter,
whose love supports me in everything this life has to offer.
Thank you.*

*To my beautiful daughter Hayah Joy Schmutter, you are the reason
for this work.
Your presence makes me strive to be a better person.
I hope to leave a lighter Earth for you.*

To our Mother Earth and future generations, we are trying.

– Kate

Acknowledgments

Many thanks to those that helped make this book a reality – Meryl Barr, Jessica Shiever, Paula Gomez, Bryan Bystrianyk, Krista Ryan, George William, Suzanna Louise Symington, Florin-Constantin Mihai, Joanna Meyer, Michelle Heltay, Rich Purcell, Lesia Amato, and many others.

To Wasyl, Daria, Orest, Lesia, Natalia, and Nadia. Thank you for being the most incredible parents and siblings anyone could wish for. To Bryan, Kyle, and Dylan, thanks for being the best kids who grew up to be outstanding men. I'm grateful for having such a wonderful family that supported and inspired me throughout my life.

To Salome, thank you for the fantastic art and layout work that made this book visually incredible. To Maria for the great editing and many bits of wisdom.

To the thousands of scientists, historians, researchers, reporters, and others whose efforts made this work possible.

Foreword

There is increasing evidence that the adverse changes being observed and predicted to worsen in the world arounds us, are leading to a different sort of malaise. Psychiatrists are becoming increasingly concerned about the mental wellbeing of people facing a continuing barrage of bad news stories concerning our natural world. Therapists are now recognising the broader social implications of collective trauma arising from the perceived environmental damage being caused and from deep concerns over the type of existence being left to their children.

Patrick McGorry AO is Professor of Youth Mental Health at the University of Melbourne and in an editorial in the Australian Medical Journal[1], reported that currently in Australia, the prevalence of mental health problems among children aged 4–12 years lies between 7% and 14%, rises to 19% among adolescents aged 13–17 years and increases again to 27% among young adults aged 18–24. This means that up to one in four young people in Australia are likely to be suffering from a mental health problem.

Richard Louv, in his international best-selling book 'Last Child in the Woods: Saving Our Children From Nature-Deficit Disorder,'[2] provides new and updated research confirming that direct exposure to nature is essential for the physical and emotional health of children and adults. Louv also provides 100 actions that can be taken to create change in your community school and family, 35 discussion points to inspire people of all ages to talk about the importance of nature in their lives.

Glenn Albrecht[3] coined in 2007 the term 'Solastalgia' to give greater meaning and clarity to environmentally induced distress produced by environmental change impacting on people while they are directly connected to their home environment. A dictionary definition of Solastalgia describes it as a neologism formed by the combination of the Latin words sōlācium (comfort) and the Greek root -algia (pain, suffering, grief), that describes a form of emotional or existential distress caused by environmental change. Albrecht further concluded that:

> *the defeat of solastalgia and non-sustainability will require that all of our emotional, intellectual and practical efforts be redirected towards healing the rift that has occurred between ecosystem and human health, both broadly defined.*

Thomas Hübl's 'Global social witnessing[4]' and Joanna Macey's 'Active hope[5]' are two examples of powerful responses to the systemic eco-trauma currently arising throughout society. Both initially guide participants through a process of truth-telling about what is actually happening to most global ecosystems. This turning toward, rather than turning away from global events as a group, has been found to build a sense of social cohesion that translates into intentionality, activation, and informed action. Slowing the current pace of environmental damage that is occurring on so many fronts will provide much-needed time for communities to adapt and respond.

Roman Bystrianyk and Kate Schmutter's book is an important clarion call, that lays out in all its brutality, the damage that is being wreaked on our life-support systems and our mental wellbeing. Few learned articles available today provide such a comprehensive portrayal of the state and trajectory of planetary systems as a result of uninformed human activity.

This work is an important contribution to the environmental challenges dialogue and constitutes an earnest invitation to dive deep into truth-telling as part of global social witnessing and active hope.

Dr. David M Deeley Landscape Ecologist - Bushfire Ecologist

References

1 McGorry, P.D., Purcell, R., Hickie, I.B. and Jorm, A.F. (2007) Investing in youth mental health is a best buy. The logic and plan for achieving early intervention in youth mental health in Australia. The Medical Journal of Australia, Volume 187, Number 7 pp. S5-S7.

2 Louv, R. (2010). Last Child in the Woods. Atlantic Books, http://

	richardlouv.com/books/last-child
3	Glenn Albrecht coined in 2007 Solistalgia
4	Thomas Hübl's 'Global Social Witnessing.'
5	Joanna Macey's 'Active Hope.'

Table of Contents

Dedications iii
Acknowledgements v
Foreword vii
Awakenings 1
Plastic Oceans 11
Fished Out 49
Coral Reef Carnage 79
Dead Zones 101
Acid Seas 125
Glaciers Going Going Gone 147
Flattened Forests 165
Exterminate 195
Consumerism Gone Mad 231
Toxic World 261
Air Pollution Catch-22 297
Lights Out 311
Turning Back 329
Glossary 341
References 359
Images 433
Resources 445
Index 459
Authors 467

AWAKENINGS

Though it might be nice to imagine there once was a time when man lived in harmony with nature, it's not clear that he ever really did.
– Elizabeth Kolbert, The Sixth Extinction: An Unnatural History

And we really should be considering the moral implications of what we're doing. What kind of a species are we that we treat the rest of life so cheaply? There are those who think that's the destiny of Earth: We arrived, we're humanizing the Earth, and it will be the destiny of Earth for us to wipe humans out and most of the rest of biodiversity. But I think the great majority of thoughtful people consider that a morally wrong position to take, and a very dangerous one.
– Edward Osborne Wilson

The Earth is a very small stage in a vast cosmic arena. Think of the rivers of blood spilled by all those generals and emperors so that in glory and in triumph they could become the momentary masters of a fraction of a dot. Think of the endless cruelties visited by the inhabitants of one corner of the dot on scarcely distinguishable inhabitants of some other corner of the dot. How frequent their misunderstandings, how eager they are to kill one another, how fervent their hatreds. Our posturings, our imagined self-importance, the delusion that we have some privileged position in the universe, are challenged by this point of pale light.
– Carl Sagan, Speech at Cornell University, 1994

The blue dot

Our home, the Earth, is a pale blue dot that orbits an average-sized star in an enormous sea of four hundred billion or more stars, making up the Milky Way galaxy.[1] There could be as many as six billion stars with Earth-like planets in our galaxy, with our galaxy being only a speck among the vast universe of a possible two trillion other galaxies.[2] With so many likely habitable planets, it would seem conceivable that humans could eventually venture to and even colonize some of those worlds.

Although there are such a massive number of potentially habitable destinations, the distances between stars are almost beyond imagination. Proxima Centauri, our nearest neighboring star, is 40 trillion kilometers (25 trillion miles) away. It takes light emitted by that star traveling at 299,792 kilometers a second (186,282 miles per second) 4.25 years to reach the Earth.[3] The Voyager 1 spacecraft, launched in 1977, has been traveling at 17 kilometers per second (38,000 miles per hour) for over 40 years.[4] Even at this incredible speed, it has only covered a tiny fraction of the distance to our neighbor, taking around another 16,000 years to arrive there.[5] A spaceship with living humans on board reaching even the closest star that may or may not have a habitable planet in orbit will not happen any time soon.

Terraforming literally means Earth-shaping. Most discussions about terraforming are often centered around using technology to transform Mars into an Earth-like paradise.[6] Such a massive undertaking would take decades or centuries and ultimately may not even be feasible. Like traveling to a nearby Earth-like planet, this notion is really more in the realm of science fiction than anything seriously attainable, at least in the foreseeable future.

Yet, while our imaginations often focus on possible Earth-like planets out there, we seem to forget we already inhabit an incredibly beautiful and bountiful world. The Earth has vast quantities of liquid water and other foundations for life, an electromagnetic shield protecting the surface from deadly solar radiation, the proper distance from a stable, long-lived star, and other factors making it a perfect place for an incredible variety of life.[7] The Earth is an ideal place for humans and other inhabitants we share it with to live and grow. It is truly a paradise. Or it was.

Demolishing Eden

On a daily or even yearly time scale, the world may seem basically the same, but our planet is rapidly changing and is now in an extreme environmental crisis. The ecological deterioration is almost imperceptible from a human perspective, but it is incredibly rapid from a geological or even an entire human history standpoint. For centuries humans have been altering the environment and, in effect, de-terraforming the planet, making it less habitable for life.

Humans are certainly one of the most prolific species in history, dramatically altering the planet over the last few hundred years and significantly since the Industrial Revolution. Since the mid-twentieth century, the transformation has accelerated with massive social and technological changes lifting millions out of the struggle of existence to enjoy the fruits of incredible human ingenuity and marvelous innovations. Yet, others on our planet live in poverty and appalling slavery or near-slavery conditions to supply the needs of these fortunate and wealthier inhabitants. Human brilliance has also helped fuel an enormous human population explosion, an immense, unsustainable increase in natural resource consumption, and worldwide environmental degradation and destruction.

Once vast numbers of buffalo, antelope, wolves, and many other creatures filled the North American countryside. Huge flocks of birds like the passenger pigeon endlessly passed overhead, sharing the landscape. The oceans were brimming with sea life beyond imagination, filled with fish like cod that were so bountiful there seemed no reason to believe their numbers would ever diminish. Beautiful corals hugged the coasts and were teeming with life. Awe-inspiring glaciers were a natural feature on parts of the continent. The air, water, and land were largely pristine and uncontaminated. It was a natural paradise. Other parts of the world were equally unspoiled, bountiful, and stunningly beautiful.

Today, countless wild animals have now declined to a fraction of the populations that once existed. Those that remain are rapidly vanishing in the wild, with a paltry few finding sanctuary in zoos and wildlife parks. Some animals have been driven to extinction, while others are now on the precipice of oblivion. The massive flocks of birds and schools of fish that once flourished are now not even a distant memory. The beauty of coral reefs has, in many instances, been degraded or entirely reduced to ocean rubble. The once-massive glaciers have retreated and are rapidly vanishing. Tropical rainforests, such as the Amazon, are all being cut down to supply products to fill the shelves of insatiable consumers. The clean air, water, and land are now often filled with trash and toxic chemicals. The landscape is now crisscrossed with shopping malls, fast-food chains, roads crammed with vehicles pumping out toxins, gas stations, motels, movie theaters, and other forms of distractions. An extensive aging power grid, transportation system, and massive industrial farms support and feed the ever-growing mass of people.

Discarded plastics and other debris of modern conveniences dot the landscape, with much of it ending up in our oceans due to humanity's bottomless, mad need for more possessions to fill their homes and junk foods and drinks to satiate their cravings. Similar environmental decimation is occurring in almost every corner of the planet. Consumption and convenience are driving factors of the human footprint that is truly pervasive and ever-expanding. We would need multiple Earths to satisfy this worldwide devouring of resources. Ironically, rampant consumerism has also fueled a global obesity and health epidemic while simultaneously leading people further away from the wellbeing and happiness they ultimately desire.

In just over a half-century of plastic production, our consuming, throwaway societies have contaminated every corner of the world with this virtually indestructible material. Not only are vast amounts of life directly decimated by plastics, but this substance and associated toxins can be found everywhere, from the Arctic to the deepest parts of the oceans. Plastic particles are present in our water, soil, and even the air. We are now eating, drinking, and breathing in particles from synthetic clothing, bottles, plastic wraps, cosmetics, as well as an endless number of products, and more recently, facemasks, thoughtlessly discarded to join the millions of tons of other trash that contaminates the world. Plastic is everywhere, and as the situation has magnified, there aren't many places in the world where it can't be found.

As these calamities continue to unfold, they seem to be hardly noticed. Most people are distracted by the latest triviality, political scandal, or real and imagined crises that keep them from addressing these truly horrific realities. And any notice taken of these life-destroying disasters is usually followed by optimistic blindness to believe that technology will somehow solve our problems. Plastic-eating bacteria, carbon capture, or other technology-profit-centered notions become the solution so that blind consumption can continue unimpeded. Thoughts on reigning in consumerism and focusing on these problems aren't seriously considered as economic growth is essentially the only barometer of success used by the political and elite classes.

While many Westerners sometimes reflect on their own past of slavery and other abuses that people have suffered, the same conditions still exist for millions who toil and suffer daily to supply the products they consume and enjoy. Occasionally the plight of the wretched slaves throughout the world momentarily appears to the forefront of consciousness only to quickly dissipate as many believe that these stories are simply deviations from our equitable, ethical, modern world. The reality is that cheap, disposable human labor is a fundamental component of our present economic system to provide inexpensive products and enormous profits, especially for a small percentage at the top of the global economic structure. Minerals, food, clothing, electronics, and other consumables are often supplied by the poor men, women, and children, who later clean up these discarded products and associated product packaging.

Despite the warnings from various scientists and groups, this environmental devastation continues, with potentially catastrophic consequences for the planet's future. In 2021, seventeen prominent international scientists warned that critical environmental issues, such as biodiversity loss, climate change, human overpopulation, and overconsumption, coupled with ignorance and inaction, drive the world towards a "ghastly future."[8] The problems will only worsen in the coming years, with ramifications spanning over the coming decades and centuries. That environmentally destructive change is accelerating, and as a result, the Earth is rapidly approaching a series of devastating tipping points. It's the "Perfect Storm" as the world will need to deal with cascading, interlocking catastrophes, all of which have considerable momentum.

Ticking towards midnight

In 1947, the Bulletin of the Atomic Scientists created the Doomsday Clock. The clock is a metaphor to warn the world about how close we are to destroying the planet through humanity's actions. Reaching the midnight hour signifies the destruction of civilization, and over the years, scientists have set the clock to varying numbers of minutes before midnight. In 2020, they set the clock to only 100 seconds to midnight, focusing on two simultaneous existential dangers—nuclear war and climate change.

Yet, not generally appreciated, other dangers exist that make the situation even more alarming. Giant fishing trawlers circling the globe are rapidly decimating the oceans, leaving some seas barren and devoid of life. Coral reefs, which house critical concentrations of aquatic life, have been disappearing for decades and are plunging towards extinction. The oceans are also being inundated with massive amounts of plastic continually breaking down into microplastics, contaminating the entire food chain. Driven by modern agriculture, dead zones – ocean areas devoid of oxygen – have appeared and expanded worldwide. With increasing water temperatures, oceans are experiencing a decline in oxygen, while ever-growing atmospheric CO_2 levels acidify the same oceans.

Tropical rainforests are not immune to humanity's ingenious destruction to obtain resources as they are leveled at a rate of forty football fields per minute. Mountain glaciers are melting and vanishing, threatening a key source of freshwater for the world. As creatures across the globe plummet toward extinction, they become more unique and become more lucrative through the lens of human greed. This only accelerates their demise as hunters seek to make ever-increasing profits by acquiring them or their body parts.

Even if the world's inhabitants suddenly decided to seriously try to end our plastic addiction, this material has infiltrated every corner of modern existence, which most can't conceive of ever living without this wonder of modern human ingenuity. There are virtually endless products weighing millions of tons made of this material already, with more being created every second. The throw-away culture is so ingrained that while hundreds of millions of tons of plastic already sit in our environment, the amount adding to this massive garbage pile continues to escalate. The plastic addiction problem is similar to other monumental worldwide environmental disasters unfolding to satisfy humanity's needs and wants.

We're not just excessively polluting, we're not just wiping species out across the planet, we're not just hacking down all the tropical rainforests, we're not just changing the climate, we're not just destroying life in the oceans, but we're doing it all simultaneously. And virtually all human activities exist on an electrical grid on which all of modern society depends. The massively complex grid is old, decaying, and vulnerable. It would only take a single severe solar storm to seriously damage and destroy large portions of it, throwing our modern world into an apocalyptic chaos few can even imagine.

To date, as a whole, humanity seems to be only mildly concerned by the potential for global catastrophe. Most choose to ignore these problems instead of being proactive. However, a growing number of people are recognizing and calling attention to the present and oncoming storms.

Turning back

For a long time, humanity has been almost solely focusing on profits and ignoring essentially anything else of importance. Greed, convenience, and consumption have been virtually the sole societal mantras for decades. This has been not only to the detriment of our planet but also to many of its inhabitants. But, as awareness and understanding of these issues increases, our perceptions of what is essential also shift. More and more people are taking action to make changes. Sometimes these changes are seemingly small, such as avoiding plastic water bottles in favor of a reusable container. Some are larger, such as organizations working to stop the destruction of endangered species. Each may seem insignificant on its own, but every revolution starts with individuals that want change. All positive change matters as the energy of these actions ripple out and multiply.

We all need to stop waiting for others to act and be willing to do everything we can to mitigate the destruction of our only home, regardless of what others are doing. We all need to become active doers and leaders. Once we become educated on an issue, then we have the power to enact change.

The following chapters detail the problems we all face. But with every issue, there are multiple solutions we can all implement. At the end of each chapter, there are actions listed that we can all take on an individual basis to make a difference—to change course and create a sustainable world. And what may be surprising is that many of these changes will lead you to be healthier and happier—something everyone wants.

Some of these suggestions you can immediately act on. Others may take more time, but they are all constructive ways to work on these predicaments. Human ingenuity is boundless when focused on an issue. Some brilliant solutions already exist and are waiting to become acted on and become part of the societal consciousness. There are no doubt new clever solutions just waiting for someone to focus their mind, time, and attention on these problems. That innovative solution might be yours. The time for change and action is now. This crisis is an opportunity for each of us to transform and make it a better world. The midnight hour is at hand, and we need to alter course before nature does it for us.

Awakenings

PLASTIC OCEANS

*Mr. McGuire: I want to say one word to you. Just one word.
Benjamin: Yes, sir.
Mr. McGuire: Are you listening?
Benjamin: Yes, I am.
Mr. McGuire: Plastics.
Benjamin: Exactly, how do you mean?
Mr. McGuire: There's a great future in plastics.
Think about it. Will you think about it?
Benjamin: Yes, I will.
– The Graduate, 1967*

*Marine debris – trash in our oceans –
is a symptom of our throw-away society and our approach
to how we use our natural resources.
– Achim Steiner*

You're enjoying your day at the beach when suddenly a gust of wind blows your plastic bags and plates off of your blanket. You try and catch them as the various pieces lurch erratically across the sand. You recover a few items, but some of them get away and are carried by the breeze down the coast. At the end of the day, a plastic water bottle, a spoon, or two, maybe a shovel or Frisbee, get left behind in the sand.

Plastic products can escape from overflowing trash cans, piles of garbage at municipal and illegal dumps, and through many other circumstances. They make their way through streams, lakes, rivers, and stormwater to the ocean.[9] Styrofoam food containers, sunglasses, bottle caps, drinking straws, beach coolers, fishing lines, and a wide assortment of other consumer products gradually make their way to the ocean, where they are worn down by time and the elements. That bobbing plastic bottle will eventually disappear from view, but its ever smaller long-lived particles will still be in the environment.

Floating garbage off the shore of Manila Bay in the Philippines.

Some of these plastic objects may be picked up and thrown out, while some remain exposed to the environment. The sun, wind, and sand slowly do their work, making the plastics more brittle, breaking them down into smaller pieces. Some of these items and smaller fragmented parts make their way into the ocean, where sunlight and wave action continue to break them into smaller and smaller pieces. Through ocean currents, some of this plastic waste collects on distant shores. Many end up in one of the enormous ocean whirlpools where they accumulate and continue to break apart.

Rise of the plastics

Plastics are a man-made substance that has been incorporated into modern life over the last century. Plastic's characteristics of flexibility, durability, strength, versatility, light weight, and low production cost have contributed to its entering all aspects of everyday life. Many types of products use this synthetic material, which comes in many shapes, sizes, and colors. Some common types of plastic are polystyrene (PS, aka Styrofoam) used for take-out food containers; polyethylene terephthalate (PET) used for soda bottles; polyethylene (PE) used for plastic bags; high-density polyethylene (HDPE) used for detergent bottles; polyvinyl chloride (PVC) used for plumbing pipes; polypropylene (PP) used for drinking straws; polyamide (PA, aka nylon) used for toothbrushes; and polyester (PES) used for clothing.[10]

> Plastics have become increasingly dominant in the consumer marketplace since their commercial development in the 1930s and 1940s... The largest market sector for plastic resins is packaging; that is, materials designed for immediate disposal. In 1960, plastics made up less than 1% of municipal solid waste by mass in the United States; by 2000, this proportion increased by an order of magnitude. By 2005, plastic made up at least 10% of solid waste.[11]

Worldwide plastic production began just after World War II, increasing from 2.3 million tons in 1950 to 162 million tons in 1993[12] and 359 million tons by 2015.[13] Since the 1940s, a total of about 9.2 billion tons of plastic have been produced.[14] Of that amount, 6.9 billion tons have become waste, with only about 9% of discarded plastic having been recycled, leaving 6.3 billion tons sitting in landfills and in the environment[15,16] equal to the weight of over 17,200 Empire State Buildings.[17]

> The Empire State Building, located in New York City, has a height of 443 meters (1,454 feet) and weighs 331,000 metric tons (365,000 tons), equal to over 220,000 midsize sedans.

Flooding our oceans

In 2010, the 6.4 billion people living in countries within 50 kilometers (31 miles) of an ocean coast produced an estimated 2.5 billion metric tons of garbage: Approximately 11% or 275 million metric tons of plastic. An estimated 1.7 to 4.6% (or 4.8 to 12.7 million metric tons) of plastic waste generated by those countries entered the ocean in 2010.[18] To put things into perspective, the largest living animal in the world is the blue whale, with an average adult weight of about 115 metric tons,[19] meaning that as much as 110,000 blue whales' weight equivalent of plastic enters our oceans each year. Another way to look at this, if the average midsize sedan weighs 3,300 pounds,[20] that plastic entering the ocean each year would equal the weight of over 8 million cars.

> It's hard to imagine the blue whale's size, which is the largest living animal in the world. They average 21 to 27 meters (70 to 90 feet) and weigh 90 to 136 metric tons (100 to 150 tons), equal to the weight of 75 midsize sedans. The blue whale is as long as three school buses, and their heart alone is as large as a small car.

According to the global risk consulting firm Verisk Maplecroft, Americans recycle just 35% of their municipal waste, while the most efficient country, Germany, recycles 68%. The firm estimates the United States produces about 106 kilograms (234 pounds) of plastic waste per person per year.

"The US is the only developed nation whose waste generation outstrips its ability to recycle, underscoring a shortage of political will and investment in infrastructure," the firm said. Will Nichols, the firm's head of environment, said the US had better recycling abilities than much of the world, "but the sheer amount of waste that is being generated is not being dealt with as well."[21]

A 2020 study in the journal Science Advances shows that in 2016 the United States generated 46.3 million tons

> In the United States, the equivalent of 1,300 plastic grocery bags per person end up in places such as oceans and roadways.

of plastic waste, by far the most in the world.[22] Of that waste, between 2.7% and 5.3% (1.2 million and 2.5 million tons) was mismanaged. According to study co-author Jenna Jambeck, an environmental engineering professor at the University of Georgia,

> *If you took nearly 2.5 million tons of mismanaged plastic waste — bottles, wrappers, grocery bags and the like — and dumped it on the White House lawn, it would pile as high as the Empire State Building.*[23]

The plastic was dumped on land, rivers, lakes, or shipped abroad, where it was not properly disposed of. An estimated 560,000 to 1.6 million tons of United States plastic waste likely went into oceans. This makes the United States the largest plastic pollution generator globally and the third-worst ocean plastic polluter.[24]

Diagram showing the main sources and movement pathways for plastics in the marine environment, with sinks occurring (1) on beaches, (2) in coastal waters and their sediments and (3) in the open ocean. Curved arrows depict wind-blown litter, grey arrows waterborne litter, stippled arrows vertical movement through the water column (including burial in sediments) and black arrows ingestion by marine organisms.

Recent work by researchers at The Ocean Cleanup, a Dutch foundation developing new technologies for ridding the oceans of plastic, found that two-thirds of oceanic plastic debris comes from the 20 most polluting rivers. The overwhelming majority of these rivers are in Asia. In China, the Yangtze River is the most significant culprit, dumping some 330,000 tons (the weight of over 200,000 cars) of plastic into the East China Sea every year.[25] It is perhaps unsurprising, with the United States and China leading the world in Gross Domestic Product (GDP), that they also lead in plastic production and pollution.

Shore cleanups have been organized by the Ocean Conservancy since 1986. The organization has rallied communities together with the common goal of collecting and documenting the trash littering their coastlines. In 25 years, a total of 166,144,420 items were collected in 152 different countries and locations. Plastic items, such as plastic bags and bottles and six-pack holders, accounted for 11% of the total amount of collected waste. Over those 25 years, 957,975 six-pack holders alone were collected.[26]

By far, the largest single item collected was cigarette butts, at nearly 53,000,000. In fact, as many as 5.6 trillion cigarette butts or 766,571 metric tons (844,000 tons) of butts (the weight of over 6,500 blue whales or nearly half a million cars) are deposited into the environment worldwide every year.[27] Cigarette butts are made of compressed cellulose acetate, a plastic product wrapped in an external paper layer, and because of this, they degrade very slowly.[28] A typical cigarette butt can take 18 months to 10 years or more to decompose, slowly releasing trace amounts of toxic substances such as cadmium, arsenic, and lead into the environment.[29]

> "Many people, even smokers, are not aware that the cigarette filter is comprised of thousands of little particles of plastic," says Nicolas Mallos, director of Trash Free Seas Program at the Ocean Conservancy in Washington DC. "One solid filter ends up being thousands of tiny fibres that can be released into the marine environment."[30]

Cigarette filters are composed of about 12,000 fibers, and this material's fragments may be released during the inhalation of a cigarette. Threads can be inhaled and ingested, with these filter fibers reportedly found in the lung tissue of patients with lung cancer.[31] Ironically, a monograph published by the Public Health Service, National Institutes of Health, and the National Cancer Institute, concluded that filtered and low tar cigarettes have not provided any real benefit to public health, including any reduction in death from lung cancer.[32]

This enormous amount of plastic entering the environment is a considerable problem as plastics don't decompose like natural substances. Wood, grass, and food leftovers all undergo a process known as biodegradation. This means these natural substances are decomposed by bacteria and fungi into environmentally beneficial compounds. These same biological processes don't act on plastics or work very slowly, and so the net effect is that plastic products remain in the environment for a very long time.

While there is no agreed figure for the time plastic takes to fully degrade, it could be hundreds of years.[33] In the marine environment, depending upon water conditions, ultraviolet light penetration, and the level of physical abrasion, plastics can last up to 600 years.[34] Plastic degeneration happens far faster in a hot, abrasive environment, like a beach, more so than in the ocean's colder water. However, objects eventually split up into tiny pieces of plastic. The main ways that plastics reach the sea are from beaches and land-based sources like rivers, stormwater runoff, wastewater discharges, or the transport of land litter by the wind.[35]

All of this plastic waste continues to accumulate in our oceans. By 2025 there will be an estimated 100 to 250 million metric tons of cumulative plastic debris.[36] Given that there are 10,000 to 25,000 blue whales in the world's oceans,[37] by 2025, there will be as much as 217 times more plastic by weight than blue whales. This is equal to as much as 750 times the weight of the Empire State Building or over 165 million cars.

Unless human behavior changes radically, plastic waste will continue to grow with increased population and increased per capita consumption associated with the current forms of economic growth, especially in urban areas and developing countries, with "peak waste" not expected to be reached before the year 2100.[38] According to a group of scientists who created a new computer model to track the flow of global plastic pollution, without widespread intervention, more than 1.3 billion tons of plastic waste will flow into the world's oceans and land from 2020 to 2040.[39] This increase in plastic will equal the weight of over 3,500 Empire State Buildings or more than 780 million cars.

A great deal of the recent explosion in plastic production is due to a technology known as ethane cracking. This byproduct of fracking is used to create the sorts of plastics used in packaging, often single-use plastic packaging. A new ethane cracking plant being built by petrochemical company Shell is expected to produce 1.6 million tons of polyethylene plastic each year.[40]

A million plastic bottles are bought around the world every minute. That number is projected to jump another 20% by 2021. This is all due to an insatiable desire for bottled water and the spread of a Western "on-the-go" culture. More than 480 billion plastic drinking bottles were sold in 2016 across the world, up about 60% from just 10 years earlier. If placed end-to-end, they would extend more than halfway to the sun.[41] Major drink brands produce the highest number of plastic bottles. Coca-Cola alone generates an estimated 110 billion throwaway plastic bottles every year, equating to an astounding 3,490 per second.[42]

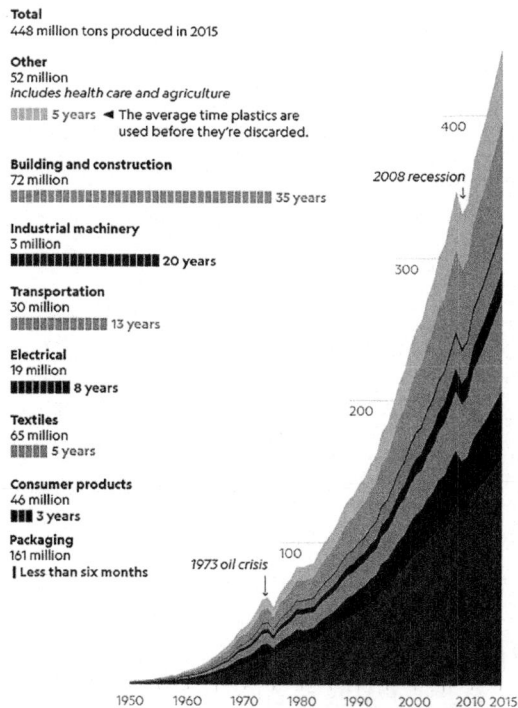

Global plastic production by industry in millions of tons.

Other consumer products are also flooding the oceans. Scientists estimate that 437 million to 8.3 billion plastic straws are slowly disintegrating on all of the world's coastlines.[43] Each year, an estimated 500 billion to 1 trillion plastic bags are used worldwide. That is equal to as many as 32,000 plastic bags being used per second, with only about 1 in 200 being recycled.[44] Single-use coffee pods also create a staggering amount of waste. In 2014, the discarded K-Cups for the Keurig Single Serve Coffee Maker were enough to circle the earth more than 10 times.[45] Sachet style, tear-off packets that once held a single serving of shampoo, toothpaste, coffee, condiments, or other products are sold by the millions. Roughly 40% of the now more than 406 million metric tons (448 million tons) of plastic produced every year are disposable.[46]

According to a Global Alliance for Incinerator Alternatives (GAIA) study, in the Philippines alone, the population uses more than 163 million plastic sachet packets, 48 million shopping bags, and 45 million thin-film bags daily.

> *GAIA revealed that for plastic sachet packets alone, Filipinos produce enough waste each year to cover the entire land area of Metro Manila with one foot of plastic. And that's not counting the 17.5 billion pieces of plastic shopping bags and 16.5 billion pieces of plastic labo bags Filipinos use each year.*[47]

Once in the ocean, plastic products' environmental fate primarily depends on the plastic density, which influences buoyancy and its position in the water column. Plastics that are denser than seawater, like PVC, will sink, while those with a lower density like polyethylene (e.g., plastic bags) and polypropylene (e.g., drinking straws) will tend to float in the water column. Dr. Shoichi Oshima of the Japan Hydrographic Association noted an example of plastics floating in the water column by observing;

> *...a fleet of flimsy white plastic, supermarket shopping bags, up-ended and suspended at depths of 2000 metres [1.25 miles], and drifting like an assembly of ghosts.*[48]

One of the responses to the declared COVID-19 pandemic (a flu-like pandemic that began in 2019) was the use of facemasks in an attempt to reduce viral transmission. Despite several scientific articles showing a lack of or uncertainty of effectiveness,[49][50][51] many locations worldwide mandated the public wearing of facemasks, which helped create a massive increase in production and use.

In 2020, facemask manufacturing increased to an estimated 52 billion masks.[52] According to a report by marine conservation NGO Oceans Asia, this massive increase in the number of masks being used resulted in an estimated 1.56 billion of them ending up in our oceans in 2020. This amounts to between 4,680 and 6,240 metric tons or the weight of as much as 4,100 cars. Disposable facemasks are now another major contributor to our oceans' ongoing plastic pollution crisis. Facemasks take as long as 450 years to break down, providing a source of microplastics for centuries. Dr. Teale Phelps Bondaroff, OceansAsia's director of research, noted,

> *The fact that we are starting to find masks that are breaking up indicates that this is a real problem that microplastics are being produced by masks.*[53]

Also, because of COVID-19, single-use plastic use has escalated, and many places rolled back efforts to reduce the use of single-use plastics. Some locations banned reusable options, and people turned to items wrapped in more plastic under the impression that it is safer. Companies removed bulk bins placing those items in plastic containers. People increased their takeaway food consumption where the packaging is often in styrofoam or other plastic containers. As a result, this has caused an estimated increase of 30% in plastic waste entering the oceans in 2020 than in 2019.[54] In 2021 a study found that 8 million tons (more than 4.8 million cars worth) of pandemic-related plastic waste were generated globally, of which 25,000 tons (equal to over 15,000 cars) entered the ocean.[55]

Many of these facemasks and other plastic items made it into our rivers and oceans, causing further damage to the environment and sealife. Early in

2020, facemasks were already washing up on the shores of beaches, adding to the already existing staggering amount flooding our oceans.

> "The 1.56 billion face masks that will likely enter our oceans in 2020 are just the tip of the iceberg," says Dr. Teale Phelps Bondaroff, Director of Research for OceansAsia, and lead author of the report. "The 4,680 to 6,240 metric tonnes of face masks are just a small fraction of the estimated 8 to 12 million metric tonnes of plastic that enter our oceans each year."[56]

Jason Ulset with the Chattahoochee Riverkeeper has seen an increase in Personal Protective Equipment (PPE), gloves, masks, and little plastic bottles of hand sanitizer, some of which end up in the Chattahoochee River.[57] That's a worry as the river supplies drinking water for 5 million people and eventually empties into the Gulf of Mexico. Kim Cobb, a Georgia Tech professor who studies oceans and climate, indicated this massive influx of COVID-19 related waste could be disastrous.

> "The fact that these kinds of influxes are headed the way of these ecosystems that are already so stressed by existing failures in our handling of plastics globally, really, could be the nail in the coffin."[58]

Microplastics

Because plastics don't decompose, they simply get smashed into smaller and smaller pieces. Pieces shorter than 5 millimeters (just under two-tenths of an inch) or about a pencil eraser's size have been termed microplastics.

A 2016 study estimates that there are already 245,000 metric tons (270,000 tons) of microplastics made of 5.25 trillion particles in our oceans.[59] Another study in the journal Nature estimated that there could be up to 51 trillion pieces of microplastics floating in the oceans.[60] That's roughly 7,000 plastic particles for every person on the planet. While most microplastics are due to the wearing down of larger products like water bottles, there are direct microplastic pollution sources.

> *The majority of microplastics in the oceans are secondary products derived from degradation and fragmentation of mesoplastics or larger fragments; primary microplastics, introduced directly into the oceans via runoff, are manufactured as micron-sized particles typically used as exfoliants for cosmetic formulations, in industrial abrasives and 'sand-blasting' media, in textile applications and synthetic clothes.[61]*

Microbeads made of various plastic particles are used in hundreds of products designed to be discarded down the drain. They are often used as abrasive scrubbers in face washes, body washes, cosmetics, cleaning supplies, and even in kinds of toothpaste.[62] A single bottle of facial cleanser can contain 350,000 microbeads.[63] The massive scope of the microbead problem has led some countries to outlaw the manufacture of products containing microbeads.[64]

An estimated 800+ trillion of these particles are to be washed down US pipes every day.[65] Many of these are recovered in wastewater treatment, ending up as sediment that accumulates at the bottom of settling tanks. These wastes are often applied to land as mulches or fertilizers that may eventually enter aquatic habitats via runoff or become airborne and distributed throughout the environment as the sludge dries out and decomposes. The airborne microplastics can then be inhaled by anitmals and people, accumulating in and potentially delivering chemicals to the lower parts of their lungs. These toxins may even cross into the circulatory system.[66]

More than 10 million metric tons (11 million tons) of sewage sludge was produced in wastewater treatment plants in the European Union (EU) in 2010. Every kilogram of this sludge has been found to contain thousands of microplastic particles, most of which are plastic fibers. The microplastic-laden sewage sludge that is then spread on fields and forests as mulch or fertilizer will undoubtedly cause further accumulation of microplastics in the environment.[67] Not all microplastics are separated via wastewater treatment. As many as 8 trillion of these particles escape into the Earth's waterways daily just in the United States. That's enough to cover more than 300 tennis courts daily or over 21.4 square kilometers (8.3 square miles) yearly.

In 2001, a high concentration of plastic debris was first observed in the North Pacific central gyre or whirlpool. It was christened an "ocean garbage patch." There are currently five ocean garbage patches that have been identified in the North Atlantic, South Atlantic, South Indian, North Pacific, and South Pacific. The total estimated combined surface size of these patches is 15,916,000 square kilometers (or 6,145,000 square miles), or roughly double the size of Australia.[68] The Great Pacific Garbage Patch alone is comprised up of an estimated 1.8 trillion pieces of plastic, with 94% being microplastics.[69] Julia Reisser, a researcher based at The University of Western Australia, noted that traversing the giant rubbish-strewn whirlpools in a boat was like sailing through "plastic soup."

Exposure to wind, waves and sun degrades plastic trash into tiny plastic particles that soak up pollutants. These 'microplastics' made up 80% of total plastic samples collected in a recent survey of Lake Erie.

> *"You put a net through it for half an hour, and there's more plastic than marine life there," she said. "It's hard to visualise the sheer amount, but the weight of it is more than the entire biomass of humans. It's quite an alarming problem that's likely to get worse. Bigger fish eat the little fish, and then they end up on our plates. It's hard to tell how much pollution is being ingested, but certainly, plastics are providing some of it."[70]*

Even the remote and the once pristine Arctic Ocean is being infiltrated by plastic. The first survey of this region in 2013 found roughly 300 billion pieces of floating plastic. Most of these pieces are tiny but visible to the naked eye.

> *"Our data demonstrate that the marine plastic pollution has reached a global scale after only a few decades using plastic materials," said Andrés Cózar Cabañas, a biologist at the University of Cádiz [Spain]. It is, he said, "clear evidence of the human capacity to change our planet. This plastic accumulation is likely to grow further."[71]*

Besides floating particles, microplastics also accumulate on the seafloor, posing additional risk to those ecosystems. Some plastic flakes drift like "marine snow" down the water column where fish can consume them. Other bits fall farther to the muddy bottom, where they are gobbled up by grass shrimp and other sediment feeders. Other plastic pieces wash up onto beaches and salt marshes to become food for burrowing worms and filter-feeding oysters. On some beaches on the Big Island of Hawaii, as much as 15% of the sand is composed of microplastics.[72]

Some may find it surprising because of plastic's buoyancy, but a significant amount reaches the deep seafloor, which is the largest marine habitat on the planet. Once in the deep sea, plastic can persist for thousands of years.[73] Microplastics in the form of microfibers made from modified acrylic, polypropylene, viscose (rayon), and polyester have been found in deep-sea organisms.[74] A recent study "shocked" scientists when they found up to 1.9 million pieces of microplastics in just one square meter of the seafloor.[75]

It has been shown that microplastics are ingested by large marine organisms, such as whales.[76] Plastic debris and fibers from textiles have also been found in hundreds of species globally, including many fish species.[77] These include swordfish and tuna, and bivalves,[78] such as mussels and oysters.

In 2015 scientists examining rivers found as many as 111 microscopic pieces of plastic in a single fish.[79] That result would later seem minuscule when a 2018 study of the River Tame, the main river of the West Midlands of England, found more than 500,000 microplastic particles per square meter in the top 10 centimeters of the river bed.[80] That is a concerning 5,000,000 particles per cubic meter (6,000,000 per cubic yard) buried in the river's bottom.

> *More than 1,000 small pieces of plastic per litre were found in the River Tame... The River Thames in London was found to have about 80 microplastic particles per litre, as was the River Cegin in North Wales. The Blackwater River in Essex had 15. Ullswater has 30 and the Llyn Cefni reservoir on Anglesey 40.[81]*

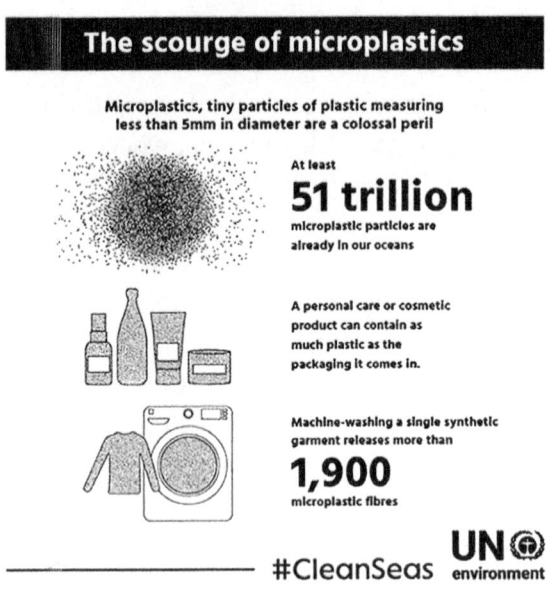

The scourge of microplastics.

A 2018 study in Frontiers in Marine Science found microplastics, particularly plastic fibers such as those used in textiles, in the stomachs of 73% of mesopelagic fish caught in the Northwest Atlantic.[82] The mesopelagic is an ocean zone at a depth of 200 to 1000 meters (660 to 3,300 feet), with the fish from that zone serving as a food source for a large variety of marine animals, including tuna, swordfish, dolphins, seals, and seabirds.

Studies have demonstrated that quantities of microscopic plastic fragments found in surface water plankton have significantly increased since the 1960s.[83] Their widely distributed occurrences suggest environmental impacts whose magnitude and significance have yet to be fully understood. Through environmental exposure, fragmentation turns larger plastics into fine powders that pass from view and are soaked up by the environment. More microplastics and their associated toxins continue to accumulate in the oceans, and ultimately in the entire environment causing irreversible harm. Their tiny size makes it exceptionally difficult, if not impossible, to remove them from the vast open ocean and other environments.

Killing life

Over 630 species have been recorded interacting with plastic debris.[84] Seabirds, sea snakes, sea turtles, penguins, seals, sea lions, manatees, sea otters, fish, crustaceans, and half of all marine mammals are the most impacted by macro debris.[85] They can choke on grocery bags and become entangled in six-pack rings and fishing nets. Studies have shown that fish and other marine life, such as birds, eat these plastics, damaging their digestive systems. Plastics that stay in the gut make the animal feel full, leading to malnutrition or even starvation. In one experiment, grass shrimp were fed a brine shrimp diet mixed in with polypropylene plastic beads. After six days, all of the shrimp were dead. They stopped eating because their guts were blocked with plastic, which caused them to starve to death.[86]

Each year, an estimated 1 million seabirds, 100,000 sea mammals, and countless fish are killed due to plastic pollution.[87] An autopsy of a young male sperm whale that had washed up dead on the southeastern coast of Spain was found to have 29 kilograms (65 pounds)

> ...of plastic trash crammed into the dead whale's stomach and intestines, including dozens of plastic bags, chunks of mangled rope and glass, a large water container and several 'sacks of raffia [a fiber derived from palm trees].[88]

The whale died because its digestive system had become lethally impacted or infected. A young whale washed up in the Philippines, which died of dehydration and starvation after eating 40 kilograms (88 pounds) of plastic bags made up of "40 kilos of rice sacks, grocery bags, banana plantation bags, and general plastic bags. Sixteen rice sacks in total."[89] Around the world, sightings are being made of dead whales washed ashore killed by eating plastic.[90,91] Most of the endangered sperm whales that have been found dead in the eastern Mediterranean have been killed by plastic debris. According to an article in The Times,

Post-mortem examinations on nine of 24 [sperm whale] carcasses found in Greek waters revealed that the animals experienced slow and painful deaths after their stomachs were blocked by the large amounts of plastic they had ingested.[92]

By being entangled in or eating macro plastics, large organisms, such as birds, can be killed or experience sublethal effects. These include a compromised ability to capture and digest food, sense hunger and/or escape from predators, decrease body condition, and impaired locomotion, including migration. In 1960, plastic was found in the stomachs of fewer than 5% of seabirds. By 1980 that number had jumped to 80%, and by 2015 it was 90%.[93] According to one estimate, by 2050, almost all seabirds will have ingested plastic.[94]

A baby albatross chick can have an ounce of plastic in its belly and remain healthy; the dead chicks have twice as much.

Recent studies show that fish that hatched in waters with high quantities of microplastics were "smaller, slower, and more stupid" than those hatched in clean waters, making them easier targets for predator fish. Disturbingly, the fish actually choose to eat plastic instead of their typical food.

> "They all had access to zooplankton, and yet they decided to just eat plastic in that treatment. It seems to be a chemical or physical cue that the plastic has, that triggers a feeding response in fish," Dr. Lonnstedt told BBC News. "They are basically fooled into thinking it's a high-energy resource that they need to eat a lot of. I think of it as unhealthy fast food for teenagers, and they are just stuffing themselves."[95]

In another study, perch, still in their larval state, were shown to take in plastics and prefer it over their real food. Larval perch with access to microplastic particles ate only the plastics, ignoring their natural food source of plankton.

"This is the first time an animal has been found to preferentially feed on plastic particles and is cause for concern," said Peter Eklöv, co-author of the study. "Larvae exposed to microplastic particles during development also displayed changed behaviours and were much less active than fish that had been reared in water that contained no microplastic particles."[96]

A 2017 study published in the journal Marine Pollution Bulletin shows that corals are also gobbling up plastics.[97] Corals, which are sightless symbiotic animals and plant organisms, find something about plastics appealing enough that they want to eat it. This is a threat to corals because, like other animals, corals can't digest plastic. The ingested plastic can lead to intestinal blockages, which create a false sense of fullness, impacting coral health.

Some insects appear to be consuming microplastics from their environment as well. Research published in 2018 showed that mosquitoes that ate microplastics as mosquito larva developing in their aquatic environment still contained some plastic as they became airborne adults.[98] Later as they are eaten by spiders, bats, birds, and other creatures, they could be dispersing those plastic bits throughout the food chain. Other winged insects have similar life cycles, making them another possible way of spreading microplastics throughout the food web. Other work has already shown that mayflies and caddisfly larvae also contain microplastics.[99]

Ghost Fishing

Knotted masses of lost but relatively intact nets and fishing lines can retain the ability to trap fish and other species for long periods. This unintentional capture of sea life has been termed "ghost fishing." Hundreds of kilometers of nets and lines estimated at 580,000 metric tons (640,000 tons or the weight of over 375,000 cars) are lost in the ocean every year.[100]

> *Fishing gear can be lost by accident or abandoned at sea deliberately. Once there, nets and lines can pose a threat to wildlife for years or decades, ensnaring everything from small fish and crustaceans to endangered turtles, seabirds and even whales. Spreading throughout the ocean on tides and currents, lost and discarded fishing gear is now drifting to Arctic coastlines, washing up on remote Pacific Islands, entangled on coral reefs and littering the deep sea floor.[101]*

About half the weight of the Great Pacific Garbage Patch is made up of fishing nets and other fishing industry gear, including ropes, oyster spacers, eel traps, crates, and baskets.[102] Due to the resilient nature of the materials used to produce this fishing gear, they can and will keep ghost fishing for multiple decades or even centuries.

> *Prior to the 1950s, rope, and cordage used in all marine activities, including fisheries, was made of natural fibres—typically Indian or Manila hemp and cotton, and it was often strengthened with a coating of tar or strips of worn canvas. These materials lose their resilience in usage, and if lost or discarded at sea tend to disintegrate quickly... over the past 50+ years, these natural fibres have been replaced by nylon and other synthetic materials that are generally buoyant and far more endurable.[103]*

It's virtually impossible to know how many marine animals are killed each year by "ghost fishing." However, various reports suggest staggering numbers—many of them of commercially valuable or endangered species. Off the coast of Washington state in Puget Sound, lost fishing gear is thought to kill more than 3.5 million animals a year, including nearly 25 seals, porpoises, and other marine mammals every week.[104]

Toxins and microbes

The chemical ingredients in 50% of plastics are listed as hazardous by the United Nations Globally Harmonized System of Classification and Labeling of Chemicals.[105] Bisphenol-A, commonly known as BPA, and phthalates are called "everywhere chemicals" because they are widely used in making countless plastic products. Plasticizers are frequently added to increase plastic's flexibility, flame retardants to reduce the spread of combustion, and colorants and other materials to modify basic plastic properties. Plastics may contain ingredients such as oxybenzone used as an ultraviolet light absorber and stabilizer.[106]

BPA has been recognized since the 1940s as an endocrine-disrupting chemical that interferes with normal hormonal function.[107] Researchers have linked phthalates, which are used as plasticizers, to asthma, attention-deficit hyperactivity disorder, breast cancer, obesity and type II diabetes, low IQ, neurodevelopmental issues, behavioral issues, autism spectrum disorders, altered reproductive development, and male fertility issues.[108] These plastic additives and pollutants might be released when eaten by a wide variety of marine organisms. In 2018, "surprising levels" of phthalates were found in wild bottlenose dolphins, which are high up on the food chain.[109]

> Commonly used additives, including phthalates, bisphenol A (BPA), alkylphenols, [and] polybrominated diphenyl ethers are hazardous to biota [ocean animal and plant life] acting as endocrine-disrupting chemicals that can mimic, compete with, or disrupt the synthesis of endogenous [growing within the body] hormones. These compounds have been measured at high concentrations in plastic fragments sampled both at remote and urban beaches, as well as in those floating in the open ocean.[110]

PCBs (polychlorinated biphenyls) and PBDEs (polybrominated diphenyl ethers) were once commonly used as electrical insulators and flame retardants. Although they were banned in the United States in 1979, these substances' global production is estimated to have been about

1.3 million metric tons. They have been found in high concentrations in crustaceans in the deep ocean.[111] The surface area of microplastics allows them to absorb these pollutants from the surrounding ocean water. Plastic debris accumulates pollutants such as PCBs up to 100,000 to 1,000,000 times the levels found in seawater.[112]

Japanese scientist Yukie Mato demonstrated that plastics bind to chemicals in seawater and concentrates them. DDE (a breakdown product from the insecticide DDT), PCBs, and other endocrine-disrupting chemicals were each found to be one million times more concentrated on plastic beads than the seawater they were placed into.[113] Seabirds that ingested microplastics have been found to have elevated amounts of PCBs and other persistent organic contaminants.[114]

Bacteria and other microbes have been found to also live on microplastic particles. The plastic fragments help to disperse these organisms throughout the environment while the organisms simultaneously influence contaminants leaching from these plastics. A researcher from the National University of Singapore found more than 400 types of bacteria on 275 pieces of microplastic collected from local beaches. They included microbes linked to coral reefs' bleaching and that cause wound infections and gastroenteritis in humans.[115]

Contaminated food and drink

The dangerous cocktail of pollutants associated with plastics can be transferred to fish and other sea life. When mammals, such as seals or people, eat marine animals that have consumed this marine debris, there is the potential to increase their own burden of these hazardous chemicals.[116] This raises important questions regarding the bioaccumulation and biomagnification of chemicals and the consequences for human health. Recent studies show that humans are ingesting microplastics from seafood, with 100% of mussels sampled from UK coastlines and supermarkets containing microplastics or other debris like cotton and rayon.[117] Scientists believe microplastic consumption by people eating seafood in Britain was likely "common and widespread."

> *When consuming an average portion of mussels (250 g wet weight), one consumes around 90 particles. An average portion of 6 oysters (100 g wet weight) contains around 50 particles... European top consumers will ingest up to 11,000 microplastics per year, while minor mollusc consumers still have a dietary exposure of 1,800 microplastics per year. Once inside the human digestive tract, intestinal uptake of the ingested particles may occur.*[118]

> *"This is a wake-up call to the fact that our waste management systems are not as tight and advanced as they should be, and that might be coming back to haunt us through the food chain," said Chelsea Rochman, a postdoctoral fellow at the University of California, Davis.*[119]

As might be expected, plastic contamination is already being found in other places besides seafood. Analysis of a variety of consumer sea salt brands showed they also contain plastics.[120] While the particle amounts found are currently low, smaller particles, which probably will be in even higher concentrations, were not measured due to study limitations. Also, the low detected quantities of plastic are likely to increase significantly over time as ocean plastic loads continue to accumulate and plastics continue to splinter and spread. Microplastics have also been found in 90% of the table salt brands worldwide, with Asian brands having exceptionally high amounts.[121] Through salt alone, the average adult consumes approximately 2,000 microplastic particles per year. Erik van Sebille, an oceanographer at Utrecht University in the Netherlands who studies global ocean circulation and plastic pollution, noted that,

> *"Over the last few years, whenever scientists have gone out to look for plastic in the ocean, they have almost always found it. Whether on the remote ocean floor, in the ice in the Arctic, in the stomachs of seabirds and fish, or now in sea salt. Plastic in the ocean is an atrocity, a testament to humanity's filthy habits."*[122]

Sea salt was not only contaminated with plastics but was also tainted with pigments associated with plastics. Examples of pigments found in table salt include victoria blue, commonly used as a coloring agent in polyacrylic fibers, and lead chromate (yellow) dye, a toxic compound with extensive applications in paints and plastic industries. Lead chromate pigment has been associated with cancers, cerebrovascular (brain blood vessels) disease, and nephritis (inflammation of the kidneys) in humans. Researchers recently found microplastic particles in the placentas of unborn babies of healthy women who had normal pregnancies and births.[123] The researchers said was that this was "a matter of great concern." Elizabeth Salter Green, at the chemicals charity Chem Trust, noted,

> *Babies are being born pre-polluted. The study was very small but nevertheless flags a very worrying concern.*[124]

Dr. Sherri Mason collaborated with researchers at the University of Minnesota to examine microplastics in drinking water. Dr. Mason found that Americans could be ingesting upwards of 660 particles of plastic every year. It is thought that the majority of this plastic contamination comes from microfibers and single-use plastics such as water bottles.[125] A small study in 2018 found that microplastics have already been detected in human waste, suggesting they are widespread in the food chain.[126]

A 2017 study found 83% of the world's tap water is already contaminated with microscopic plastic fibers.[127] The United States had the highest level of contamination at 94% of samples collected at various sites.[128] Another study found that 93% of leading brands of bottled water were contaminated with plastic debris, including nylon, polyethylene terephthalate, and polypropylene, which are used to make bottle caps.[129] Other studies have shown that plastics have spread throughout the aquatic environment and end up in products such as honey[130] and beer.[131]

Other plastic pollution sources

Sources of environmental and aquatic plastic contamination have expanded as more products use plastics. Products like wet wipes, marketed as flushable, add to the load in sewer systems, costing billions of dollars in worldwide maintenance. Because they don't biodegrade quickly, they can end up among all of the accumulating refuse on beaches.[132] These wet wipes are also made from plastics, which once flushed, break apart into microplastics and add to the world's environmental plastic burden.[133]

Production of synthetic diapers began in the 1960s. They gained popularity over the following decades, replacing traditional cloth diapers. In 2018, Americans generated an estimated 4,100,000 tons of waste from disposable diapers, accounting for 1.4% of municipal waste.[134]

> *In the United States, there are about four million babies born every year. During their first year of life, the average newborn uses about 2500 diapers. This means that from babies under one year old, Americans dispose of around a trillion diapers a year.*[135]

Worldwide, disposable diapers represent about 4% of solid waste and are the third-largest single consumer item in landfills discarded after a single use.[136] These synthetic diapers take hundreds of years to break down, meaning that the diapers you wore as a baby are likely still intact, sitting in a landfill. In addition to the enormous waste generated, disposable diapers contain substances that are harmful to human health and the environment.

Diaper ingredients include tributyltin (TBT) – a biocide used to prevent the growth of bacteria, which is considered highly toxic and poisonous to marine life and humans; dioxins – a group of persistent, highly toxic organic pollutants that are carcinogenic and linked to long-term health problems.

> *Dioxin is carcinogenic, and had been listed by EPA as the most toxic of all cancer-linked chemicals. Dioxin, in very small quantities*

(parts per trillion), causes birth defects, skin diseases, liver disease, immune system suppression and genetic damage in laboratory animals consequently, dioxin was banned in many countries of the world... While one may believe that the tiny amount of dioxin exposure from diapers is insignificant, it is however of much concern that a substance reputed as the most carcinogenic chemical known, is found in a baby care product.[137]

Phthalates are synthetic plastic materials that act as liquid absorbents to improve the functionality and softness of diapers and many other products, such as women's sanitary pads. Some of these plastic materials release volatile organic compounds (VOCs) and endocrine-disrupting chemicals, potentially posing health risks to children who wear them. Phthalates and VOCs are absorbed directly through the skin, so there is an increased public health concern about whether these substances may adversely influence children's health.[138] Where phthalates are released into the environment, they can create reproductive toxicity in humans and animals.

Synthetic rubber, a variant of plastic, makes up about 60% of the rubber used in car tires. As tires wear down, they emit small dust particles into the air, landing on adjoining surfaces, with an unknown amount carried out to the sea.[139] The total amount of microplastics generated from the wear of automotive tires in the EU alone is estimated at 503,500 metric tons (555,000 tons) per year or over 4,300 blue whales' worth.[140] Microplastics from tires and roads were found to make up 89% of the ultra-fine particles found in the air around busy motorways.[141] A 2019 study found an estimated 7 trillion pieces of microplastics flow into the San Francisco Bay via stormwater drains alone.[142] Nearly half of these particles found in stormwater looked suspiciously like tiny fragments of car tires, which rainfall washes off the streets and into waterways.

Another primary source of plastic pollution is nurdles. Nurdles are tiny pellets of plastic resin that manufacturers use to create plastic packaging and products. Billions are lost every year, ending up in waterways. They are the second-largest source of microplastic pollution in water after the amount generated from vehicle tires.[143] A 2018 study estimated that between 3 million to 36 million pellets may escape every year from just one small industrial area in Sweden. When smaller particles were considered, the number of particles released is a hundred times greater.[144]

Synthetic textiles, such as polyester and acrylic, also slowly break down while washing and drying clothes. A Plymouth University study showed that more than 700,000 microscopic fibers could be released into wastewater during each use of a domestic washing machine. Many of these are likely to pass through sewage treatment and into the environment.[145] A study in California found that in 2019 an estimated 13.3 quadrillion fibers (4,000 metric tons equal to the weight of 2,600 cars) were released into California's natural environment.[146] The 13.3 quadrillion figure is 130,000 times as many fibers as stars in the Milky Way galaxy. These fibers are primarily shed when articles such as yoga pants, stretchy jeans, and fleece jackets are repeatedly washed. Up to 40% of these microfibers pass through wastewater treatment plants and end up in rivers, lakes, and oceans, with 85% of the world's plastic debris containing some percentage of this waste.[147]

Nanoplastics

As plastics break down into smaller pieces, their fate in the marine environment is poorly understood, but they potentially become the most hazardous form – nanoplastics. Nanoplastics are particles less than 100 nanometers (nm) or about the size of a typical virus such as the influenza A virus.[148] These ultra-small plastics could enter living cells, causing inflammation and possible disruption of cellular functions.[149] Rachel Hurley from the University of Manchester noted:

> "It is the really small stuff we get worried about, as they can get through the membranes in the gut and in the bloodstream – that is the real fear."[150]

Studies have shown that these particles can be transported through the aquatic food chain via algae into fish, affecting lipid metabolism and fish behavior.[151] A recent study found that these nano-sized particles can cross the blood-brain barrier accumulating inside fish brains, creating behavioral disorders through what researchers believe is brain damage.[152] The study also found that animal plankton (zooplankton) died when exposed to plastic nanoparticles, while larger plastic particles did not appear to affect them. According to Tommy Cedervall, a chemistry researcher at Lund University in Sweden,

> *It is important to study how plastics affect ecosystems and that nanoplastic particles likely have a more dangerous impact on aquatic ecosystems than larger pieces of plastics.*[153]

Breakdown of plastics into nanoplastics may actually take a long time on the order of decades or centuries. However, the direct use of nanoparticles in cosmetics, detergents, food, dental, and other commercial products is rapidly increasing despite very little knowledge of their effect on the environment, particularly on organism metabolism. One product, 3D printers, has been found to emit 200 billion ultra-fine particles (UFP) per minute, having potentially serious health consequences.

> *Inhaling UFPs is potentially harmful, as the particles' deposit efficiently in both the pulmonary and alveolar regions of the lung, as well as in head airways.' The particles could also enter the brain through the olfactory nerve. Symptoms of UFP inhalation include shortness of breath, stroke, cardiac arrest, and even death.*[154]

These nano-sized products can easily bypass any sewage treatment system, with these potentially potent particles ending up in freshwater and marine habitats. If the magnitude of adverse effects on wildlife is severe enough, such as population-level declines, world food security could be affected. Heidi Taylor, director of the marine debris organization Tangaroa Blue, recently noted that,

This is the next climate change, and nobody's thinking that it's going to be as bad as it is. If we start looking at communities like the islands here, that rely so heavily on seafood, and that [seafood] is contaminated by plastics and chemicals that are in the ocean, this is going to be not an issue about saving turtles, this is going to be a human health issue, and that will be a game changer.[155]

Greenhouse gases

As countries turn from fossil fuels to renewable energy, plastic production is a way for petrochemical companies to continue to make big profits. By 2030, plastic-linked emissions of production are expected to equal nearly 300 coal power plants.[156]

Once plastic waste enters the environment, it continues to affect the atmosphere. Dumping, incinerating, recycling, and composting of certain types of plastics all release carbon dioxide. The total emissions from plastics in 2015 were equivalent to nearly 1.8 billion metric tons of CO_2 (the weight of over 1.2 billion cars).[157]

Once the plastic waste enters the environment, it continues to affect the air, as most conventional plastics have also been found to discharge the greenhouse gases methane and ethylene when exposed to sunlight.[158] Polyethylene, used in shopping bags, was found to be the most prolific emitter of both gases. David Karl, the senior author of a study that examined this phenomenon, noted,

> *Plastic represents a source of climate-relevant trace gases that is expected to increase as more plastic is produced and accumulated in the environment. This source is not yet budgeted for when assessing global methane and ethylene cycles, and may be significant.*[159]

Recycling illusion

For decades the United States and other industrialized countries have counted plastic waste as "recycled" if exported. While this avoids disposal

> A 20-foot long shipping container has a volume of 33.2 cubic meters (42.3 cubic yards).

costs and local environmental impacts, this waste problem is often shifted to countries with poor waste management.[160] In 2018, the United States sent 157,000 twenty-foot-long shipping containers (430 per day) to such countries. This equals 5.2 million cubic meters (6.6 million cubic yards) or enough to fill over 2,000 Olympic-sized swimming pools.[161] Malaysia has become a "dumping ground" for Western plastics. In 2017, Malaysia imported approximately 550,000 metric tons of plastic (the weight of over 4,500 blue whales or over 330,000 cars). There the plastics are manually broken down by workers earning $10 a day. Large amounts of low-grade scrap, such as single-use plastic bags, end up in massive dump-sites where it eventually is buried in landfills or incinerated in the open, further releasing toxins into the environment.[162]

Exporting plastics to countries ill-equipped to manage it creates a comfortable illusion that predominantly Asian countries are to blame for the world's ocean plastic pollution. While this plastic pollution sleight of hand may make citizens and businesses in the Western world more comfortable with their plastic addiction, the end result is that the increasing plastic disaster is still impacting the planet as a whole. According to Kara Lavender Law, an oceanography professor at the Sea Education Association in Cape Cod, Massachusetts,

> *We're putting this in the blue bin and then it's getting trucked to Boston. And then it's getting put on a ship that's sailing most of the way around the world for somebody to unpack it and pick through it and cut labels off it in hopes that some portion of that material will be turned into (plastic) pellets and into a children's toy or whatever.*[163]

When recycling started in the 1980s, it was a noble and well-intentioned idea that provided a closed-loop for products and packaging. Unfortunately, the notion of recycling has given the manufacturers of disposable items the ability to essentially market overconsumption as environmentalism.[164] The system cannot keep up with massively escalating consumer consumption, and many things that consumers throw in the blue bins thinking they will be recycled are simply not.

> *Every year, reports come out touting rising recycling rates and neglecting to mention the soaring consumption that goes along with them. American consumers assuage any guilt they might feel about consuming mass quantities of unnecessary, disposable goods by dutifully tossing those items into their recycling bins and hauling them out to the curb each week.*[165]

Reports from September 2020 showed how the plastic industry-funded ads in the 1980s promoted recycling to solve our growing waste problem.[166] These makers of virgin plastics were the major proponents and financial sponsors of plastic recycling programs because they created the illusion of a sustainable, closed-cycle while actually promoting the continued use of raw materials for new single-use plastics. According to Judith Enck, a visiting professor at Bennington College in Vermont and president of Beyond Plastics, a nonprofit focused on ending plastics pollution,

> *The reason the public thinks recycling is the answer is that the plastic industry has spent 30 years on multimillion-dollar campaigns saying that. That was absolutely the wrong message. The message should have been: Don't use so much plastic.*[167]

> In the United States, about 76% of plastic garbage goes into landfills, 16% is burned at very high temperatures releasing greenhouse gases, 1% ends up in the oceans.

Unfortunately, most plastic can't and won't be recycled. For example, the EPA reported that plastic generated in 2018 was 35.7 million tons, accounting for 12.2 percent of municipal solid waste that year.[168] Of this total, only 3 million tons, less than 9%, were recycled. The vast majority – 27 million tons – ended up in landfills, and the rest was combusted. The environmental agency also estimated that less than 10% of plastic thrown in bins in the last 40 years has been recycled. According to David Biderman, CEO and executive director of the Solid Waste Association of North America,

> *Most people have the attitude that if they just put it in the blue bin, it will get taken away and somebody will figure out what to do with it, but putting something in the blue bin and actually recycling it are two very different things.*[169]

Why isn't more plastic recycled? Most products are made up of mixtures of various plastics and chemicals, which can make recycling impossible. The two recycling codes, 1 and 2, which are considered the most recyclable, are usually "downcycled," which means they're turned into lower-quality products that will eventually end up in a landfill because those materials can't be recycled again. Often consumer products utilize virgin plastic because the cost of new plastic is lower than recycled plastic.

Ultimately, recycling is mostly an illusion that allows for continued rampant consumerism. According to Professor Enck and others, recycling doesn't work if you continue to increase new plastic manufacturing. According to Enck,

> *We can't recycle our way out of the problem. The only solution is reducing the generation and use of plastic.*[170]

Plastic Armageddon

Although the plastic garbage patches and marine debris are not precisely quantifiable in all aspects, they are a symptom of a root problem, which is plastic end-of-life use. The microplastic endgame is not the ocean garbage patches, but ultimately it is the interaction with the entire ocean ecosystem. Los Angeles Captain Charles Moore, an environmental advocate, credited with bringing attention to the Great Pacific Garbage Patch, noted that,

> *The ocean is like a plastic soup, bulked up with the croutons of these larger items. It's like a toilet bowl that swirls but doesn't flush.*[171]

With global plastic production doubling every 11 years, during those 11 years, people will make as much plastic as has been produced since plastic was invented. [172] A 2017 Ellen MacArthur Foundation report shows that by 2050 plastics will consume 20% of all oil production, up from 5% today. One out of every five barrels of oil will not fuel or lubricate our machines but will be used to make plastic. The report states that at least 8 million metric tons of plastic waste enter the oceans each year. If action is not taken by 2050, there will be more plastic in the sea than fish, weighing 850 million metric tons (937 million tons)[173] or equal to over 2,500 Empire State Buildings or over 550 million cars. Those 550 million cars placed bumper to bumper would wrap around the Earth over 66 times.

> *Each year, at least 8 million tonnes of plastics leak into the ocean — which is equivalent to dumping the contents of one garbage truck into the ocean every minute. If no action is taken, this is expected to increase to two per minute by 2030 and four per minute by 2050. Estimates suggest that plastic packaging represents the major share of this leakage. In a business-as-usual scenario, the ocean is expected to contain 1 tonne of plastic for every 3 tonnes of fish by 2025, and by 2050, more plastics than fish (by weight).*[174]

We are facing an ever-swelling tsunami of plastic waste that is difficult to imagine. Measurements from the most contaminated regions of the world's oceans show that the mass of plastics already exceeds that of plankton sixfold.[175] The potential for biomagnification of plastic particulates in the environment is of significant concern for life all the way up on the food chain, biosecurity, and, ultimately, human health. Dr. Lisa Emelia Svensson, the former executive director of the oceans branch at UNEP, the UN Environment Programme, said plastics are "ruining the ecosystem of the ocean" and are nothing short of a "planetary crisis."[176] Erik Solheim, the former head of the UN's environment program, stated that "we're facing an ocean Armageddon."[177]

Despite numerous laws, regulations, and cleanup efforts, plastic-dominated marine debris appears to be ever expanding, and hence, so is the magnitude of the resulting problems. This plastic load of pollution not only reaches our oceans, but a large portion of this debris ends up on, or buried in, the seafloor. The potential is there for an unseen pervasive impact on deep-marine ecosystems.

The looming plastic catastrophe is something we need to tackle in a globally comprehensive way. Even if 100% of plastics in the Western world were truly recycled, the number of plastics flooding the oceans from the developing world would still be overwhelming. Globally, there are almost 2.8 billion people without access to waste collection services, of which 1.9 billion live in rural areas, generating large amounts of uncollected household waste ending up in surrounding water bodies, open dumps, or burnt in open-fire activities.[178] We need to work together as one human community to solve this massive mismanagement of waste materials across the planet. What is needed is a global plan to properly collect trash, recycle whatever can be, and keep it from inundating the environment.

With the sea covering over 73% of our world, it is not physically practical to remove all the existing plastic debris. With more than 90% of plastics in the ocean being less than 10 millimeters long,[179] it really becomes an impossible task. So instead, we must find ways to change our continuing impact because we can't significantly alter the enormous damage that has already been done.

Because of our modern obsession with convenient single-use items, we will be eating, drinking, and breathing in plastic product remnants and their toxins for decades and centuries to come. We must substantially decrease plastics use, especially these single-use types, which is nearly half of all plastics manufactured today. We must also think of better ways to intercept and capture all plastics before they infiltrate the marine environment. *If we don't take action, the problems will persist, continue to escalate, and become increasingly hopeless, with potentially disastrous consequences for all life on this planet.*

What you can do!

Plastic pollution is a worldwide crisis. If we work together, we can really make a difference and solve this problem. Here are some simple steps that you can take at a personal level to make a difference.

Switch from bottled water – Use glass or stainless steel bottles and refill from a filtered tap water source. According to a 2009 article in Environmental Research Letters, bottled water takes as much as 2,000 times the energy cost of producing tap water. In 2007, in the United States, about 33 billion liters of bottled water were consumed "equivalent to between 32 and 54 million barrels of oil or a third of a percent of total US primary energy consumption."[180]

Switch from other drinks that come in plastic bottles – Make your own drinks at home, such as fruit smoothies or juices, which are also much healthier alternatives than sodas, and place them into your reusable bottle. According to the Harvard School of Public Health, sugary drinks are calorie-rich and devoid of nutrition. "Beyond weight gain, routinely drinking these sugar-loaded beverages can increase the risk of type 2 diabetes, heart disease, and other chronic diseases. Furthermore, higher consumption of sugary beverages has been linked with an increased risk of premature death."[181]

Switch to a reusable travel mug for hot drinks – According to the Environmental Protection Agency (EPA), 25 billion Styrofoam cups are thrown out each year. Styrofoam cups are not biodegradable and cannot be recycled, and they will still be sitting in landfills in 500 years.[182] The plastic cup lids and stirrers are also not recycled. Paper cups are lined with a plastic coating, so they don't leak, which also means they can't be recycled. Billions of them ending up in landfills each year. Instead, use a stainless steel travel mug. Better yet, break free of the "on-the-go" culture and consider having your coffee or tea at home or at a sit-down café instead of taking it with you.

Switch from plastic straws – A plastic straw – It's something that comes with most beverages we order, from soft drinks to even a glass of water. In the United States, Americans use them at an average rate of 1.6 straws per person per day, which equals 175 billion straws a year used once and thrown out.[183] Most of us can simply do without this frill, but there are plenty of alternatives for those that need them – straws made of metal, paper, and even actual straw.

Use reusable utensils – An estimated 40 billion individual plastic utensils are produced each year. With low reuse and recycling rates, most of them end up in our landfills, beaches, and oceans.[184] Instead of plastic forks, spoons, and knives, carry one of the many sets of reusable utensils made from bamboo or stainless steel.

Use reusable shopping and produce bags – Plastic bags are an eyesore as they end up as litter in our parks, forests, and beaches. These single-use bags are made from fossil fuels and end up as deadly waste in landfills and the ocean. It only takes about 14 plastic bags to equal the amount of gas required to drive one mile.[185] Instead, use reusable hemp, cotton, or other cloth bags for groceries and other shopping. For produce, leave it loose for items like avocadoes and onions. For other things, use reusable mesh produce bags.

Purchase products in bulk – Many products such as fresh meat, grains, and condiments can be purchased from providers that offer their products in bulk. This takes a little more planning to provide your own containers and pantry for freezer storage space at home, but once the habit develops, it is remarkable how much plastic packaging can be avoided this way.

Use your own take-out food containers – Pack lunches and bring containers to restaurants so you can pack your own leftovers into reusable containers made from stainless steel or glass with snap-on lids.

Switch to bars of shampoo and conditioner – Shampoo and conditioner bars eliminate the need for plastic bottles. Most bars come wrapped in recycled paper or in paper boxes. Most bottled shampoos are 80% water, and conditioners can be up to 95% water. Shampoo and conditioner bars are concentrated and generally last longer than bottled versions. On average, a shampoo bar will outlast two to three bottles of liquid shampoo.[186]

Switch from using toothpaste tubes – Toothpaste tubes are, for the most part, not recycled, with more than a billion of them ending up in landfills every year. Use toothpaste tablets that come in glass jars, or you can also make your own tooth powder out of natural ingredients such as baking soda.

Switch to natural fibers for clothing – We often don't think of our clothing as plastic, but polyester, made from petroleum, is now used in about 60% of our clothes.[187] When synthetic fabrics are washed, they release tiny strands called microfibers, which end up everywhere in the environment. Switch to organic cotton, hemp, and linen, which are natural materials used to make clothing for thousands of years.

Avoid seafood – Over a ton of lost plastic fishing gear, also known as "ghost gear," ends up in the ocean every minute, which is destroying sea life on a massive scale. According to the Independent, "the world's biggest seafood firms are all contributing to the deaths of more than 100,000 whales, dolphins, seals, turtles, and seabirds that are killed in agony every year by discarded fishing equipment."[188] Moving away from eating seafood will reduce demand and lessen the damage caused by highly destructive industrial fishing practices.

Breastfeed – Plastic bottles release microplastics and nanoplastics. Although the impact on health isn't clear, to avoid any problems, breastfeed when possible. Breastfeeding provides the best nutrition for a baby and is the most widely recommended way to feed a newborn.[189] When breastfeeding is not possible, consider using glass bottles.

Use modern cloth diapers –Disposable baby diapers have primarily replaced older cloth diapers mainly due to perceived advantages, principally convenience. However, very little attention is given to the product after its use and its potential adverse health effects. Modern cloth diapers are washable and reusable. All diapers impact the environment; however, using cloth diapers and washing a full load and line drying is the most sustainable option. Cloth diapers can have flushable liners that can still add to landfills. Diaper liners are used to make diaper changes easier but are not essential. Cloth diapers are free of harmful chemicals and plastics for a healthier baby. If you need to use disposable diapers, there are biodegradable options available.

Quit smoking – This helps cut down on the massive amount of toxic cigarette butts entering the environment and eliminates a significant cause of lung cancer. Also, the discarded cigarette filters are of dubious benefit in actually reducing diseases in smokers. According to a 2009 article in the International Journal of Environmental Research and Public Health, "cigarette filters are primarily a marketing tool to help sell 'safe' cigarettes" which have little "scientific evidence, including patterns of mortality from smoking-caused diseases, does not indicate a benefit to public health from changes in cigarette design and manufacturing over the last fifty years." [190]

Innovate! – For every problem, there are many solutions. Human ingenuity and creativity are boundless when it comes to almost any situation. You may have a great idea to tackle this serious issue. Let your ideas blossom and work with others to make them a reality. Every positive change makes a difference.

A huge net is being dragged across the sea floor, destroying everything in its path. Ahead of it bloom undersea forests and their hundreds and thousands of living creatures, both plant, and animal; behind it is a desert. The net is pulled to the surface, and most of the dead and dying life forms in it are thrown out. A few marketable species are retained. This is like taking a front-end loader and scraping up your entire front garden and shredding it, keeping a few pebbles, and dumping the rest of it down the drain. Couple this with overfishing – really easy to do with megaships equipped with sonar for fast fish finding – and the eventual result is no fish.
– Margaret Atwood, Payback: Debt and the Shadow Side of Wealth

Fishery management is an endless argument about how many fish there are in the sea until all doubt has been removed – but so have all the fish.
– John Gulland

There is a sufficiency in the world for man's need but not for man's greed.
– Mahatma Gandhi

A brief history of overfishing

In the early sixteenth century, with the New World's discovery, later called the Americas, numerous stories were reported to Europe of an endless abundance of Atlantic cod. This seemingly never-ending supply of cod helped fuel the new American colonies for generations and even gave Massachusetts' famous cape its name, Cape Cod. At the turn of the sixteenth century, John Mason, an English fishing skipper working out of Newfoundland shore-station, noted,

> *Cods are so thick by the shore that we hardly have been able to row a boat through them... Three men going to Sea in a boat, with some on shore to dress them, in 30 days will commonly kill between 25 and thirty thousand...[191]*

At that time, cod were reported to be up to six and seven feet in length (roughly 2 meters) and weigh as much as 200 pounds (90 kilograms), with their numbers dense enough to slow down boats.[192] Hundreds of fishing vessels caught thousands of tons of cod off of the New England coast. Boats took in a daily catch of 15,000 to 30,000 fish, and it seemed the fish were an "inexhaustible manna."[193] Nearly 200 years later, by the end of the seventeenth century, travelers such as Baron De Lahontan still commented on the seemingly endless quantities of cod off Newfoundland, Canada:

> *You can scarce imagine what quantities of Cod-Fish were catch'd [caught] there by our Seamen... yet the Hook was no sooner at the bottom, than the Fish was catch'd; so that had nothing to do but throw in, and take up without interruption.* [194]

Cod fishing off the coast of New England was a huge business. In 1763 the English at Newfoundland's fisheries caught and cured

> A quintal is an antiquated unit of weight equivalent to 100 kilograms or 220 pounds.

386,274 quintals or about 42,500 tons of cod. That year, there were 106 fishing vessels with 265 ships transporting the salted cod to England and

British Colonies. In 1783 the amount had increased to 591,276 quintals or over 65,000 tons of fish.[195] Presumably, due to centuries of intense fishing, by 1827, the largest cod had decreased in size to just over four feet long (1.3 meters), weighing in at 46 pounds (20 kilograms). As many as 1,500 English, American, and French fishing vessels extracted thousands of tons of cod off Newfoundland's coast. A single fisherman was recorded to have caught about 12,000 cod annually, with the average fisherman catching about 7,000.[196]

Still, the iconic cod proved not to be inexhaustible. After about 400 years of a strong fishing industry, the population of cod collapsed in 1992. Cod catches decreased from 500,000 tons in the Grand Banks in the 1960s to 50,000 tons.[197] This led to a fishing moratorium that caused 40,000 people in Newfoundland to be thrown out of work.[198] Inshore fishermen, like 76-year-old Wilson Hayward, believed the seeds of the collapse were sown after the Second World War. Hundreds of factory trawlers had arrived on the Grand Banks, which stretch out more than 320 kilometers (200 miles) off Newfoundland's coast.

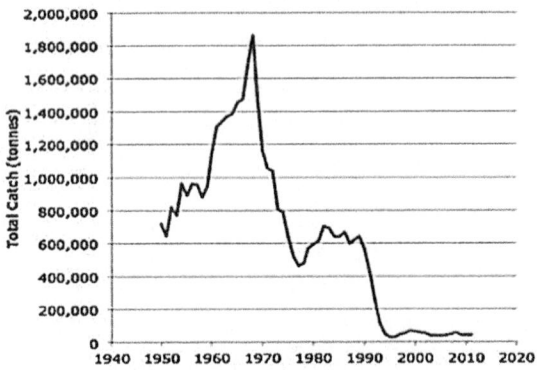

Collapse of Northwest Atlantic cod.

I remember going out on to the cape in the night, and all you could see were dragger (fishing trawler) lights as far as the eye could see, just like a city in the sea. We all knew it was wrong. They were taking the mother fish, which had been out there spawning over the years. They cleaned it all up; they dragged the ocean floor like the paved road. [199]

Trawling is a method of fishing that involves dragging a net through the water behind a boat. Bottom trawling is when a fishing net is pulled across the seafloor. Over the decades, fishing vessels have advanced significantly in terms of size, power, and technology. In 1954 the first new factory freezer trawlers appeared. These ships could sweep up as many cod in two 30-minute trawls as could be caught during the entire summer season in the sixteenth century.[200] Factory trawlers, mostly from Europe but some from as far as East Asia, began large fishing operations with massive increases in catches by the 1960s and early 1970s. By the 1960s, pair trawlers – two vessels that tow a single net – had tripled in size and engine power in only a decade.[201]

Atlantic cod, a relatively long-lived and slow-growing species, could not keep up with the increasing fishing rate. As the vast majority of spawning adults were packed into ships' freezers, catches began to decline until they collapsed in the early 1990s.[202] In 1492 the amount of fish off the coast of Nova Scotia was estimated to be 4,000,000 tons. By the early 2000s, that number had suffered a massive crash of nearly 99%, with only 50,000 tons remaining.[203]

The cod stocks off the coast of the United States still remain significantly below sustainable levels.[204] Off the coast of Newfoundland, a study indicated the start of a recovery. The authors attributed the recovery to improved environmental conditions, better fish management, and the increased availability of the small fish they feed on, capelin, whose populations also fell drastically in the early 1990s. Even so, today's cod, at an average length of about 2.2 feet (0.68 meters) and at a weight of 8 pounds (3.6 kilograms),[205] is at a meager less than 5% of the size of the legendary cod of the sixteenth century.

Other fish stocks experienced similar overfishing and collapse. In the mid-seventeenth century, Nicolas Denys described his time along the Miramichi River, which flows through New Brunswick, Canada. He experienced sleepless nights because of the noise made by the massive amounts of salmon swimming in that river:

So large a quantity of salmon enters the river at night one is unable to sleep, so great is the noise they make in falling upon the water after having thrown or darted themselves into the air passing over the river flats. I found a little river which I named Riviere au Saulmon... I made a cast of the seine net [a fishing net that hangs vertically in the water with floats at the top and weights at the bottom edge] at its entrance where it took so great a quantity of salmon that then men could not haul it to land and... had it not broken the salmon would have carried it off. We had a boat full of them, the smallest three feet long.[206]

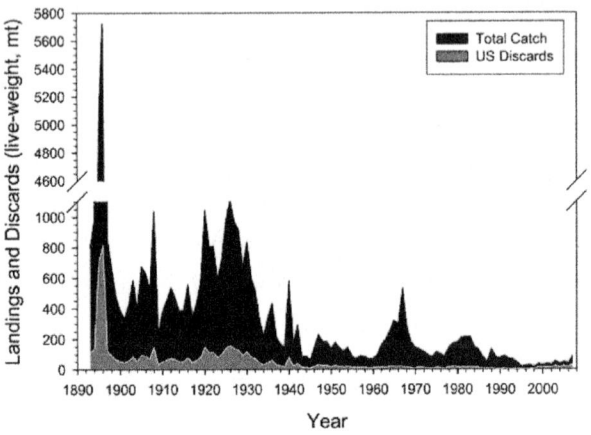

Atlantic halibut total catch in metric tons from the Gulf of Maine-Georges Bank region from 1893 to 2007.

George Cartwright fished salmon for export in the late 1700s. He noted that the salmon were so thick that he could not have fired a shot into the river without hitting one. In 1799 he said, "In the Eagle River [Labrador, Canada] we are killing 750 salmon a day, and we would have killed more had we had more nets... the fish averaging from 15 to 32 pounds apiece."[207]

Overfishing combined with lumber mills, damming of rivers, and pollution from tanneries and iron smelters caused the salmon populations to decline. John Rowan, an English visitor to Canada, noted as early as 1870 that the salmon fishing in Nova Scotia had already decreased due to these factors, and he predicted that eventually, the rivers would be "rendered barren."[208] Rowan's prescient forecast came to pass in just a few years. By 1900 salmon were virtually extinct in Connecticut and Massachusetts and in most New Hampshire and Maine rivers.[209,210] In the mid-1970s, the total population of Atlantic salmon in North America was close to 1.8 million. By 2013 the population had fallen to one-third of that level.[211] As of 2017, the number of Atlantic salmon has continued to decline.

As fishing greatly expanded off the coast of New England in the early years of the New World, cod was the fish of choice. Although halibut was also incredibly abundant, it was discarded mainly because the meat was considered a poor choice for salting, which was the method of preserving fish at the time. Later during the 1830s to 1840s, many fishing vessels shifted to using ice to store fish, which was ideal for preserving halibut. This new preservation technology, along with a preferential cultural shift to fresh from salted fish, led to the halibut fishery's height during the 1840s to the 1870s.

Atlantic halibut catches from the Gulf of Maine and Georges Bank peaked in 1896 with 5,725 metric tons, decreasing over the decades. In 1974 that number had fallen to 84 metric tons, peaking again in 1982 to 215 metric tons, and then reducing to a low of 18 metric tons in 1998.[212] The numbers remained well below 100 metric tons during the early 2000s, a tiny fraction of its once bountiful level during the late 1800s and early 1900s.

Ravaging the oceans

Overall the Northwest Atlantic has seen a significant decline in fish catch, down from about 4.2 million metric tons (4.6 million tons) in the early 1970s to 1.9 million metric tons (2.1 million tons) in 2013, more than a 50% drop. Strict regulations have been in place to allow for fish stock recovery, but some historically plentiful fish such as cod, flounder, and redfish are still showing little or no improvement.[213] Atlantic cod catches remain extremely low since their collapse in the early 1990s.

The Bay of Bengal lies off the coast of India. The Bay's fisheries have been severely depleted due to decades of overfishing. Many once-abundant species have all but vanished. Most unfavorably affected are the species at the top of the food chain. The man-eating sharks that were once feared by sailors are now rarely found in these waters. Other top predators like grouper, croaker, and rays have also been decimated. Fishing catches now are mainly made up of species like sardines, which are near the bottom of the marine food web.

> *Good intentions have played no small part in creating the current situation. In the 1960s, western aid agencies encouraged the growth of trawling in India, so that fishermen could profit from the demand for prawns in foreign markets. This led to a "pink gold rush," in which prawns were trawled with fine mesh nets that were dragged along the sea floor. But along with hauls of "pink gold," these nets also scooped up whole seafloor ecosystems as well as vulnerable species like turtles, dolphins, sea snakes, rays, and sharks. These were once called bycatch and were largely discarded. Today the collateral damage of the trawling industry is processed and sold to the fast-growing poultry and aquaculture industries of the region. In effect, the processes that sustain the Bay of Bengal's fisheries are being destroyed in order to produce dirt-cheap chicken feed and fish feed.* [214]

As many as 44,000 fishing boats and 147 trawlers skim the Bay of Bengal every day with sea-floor-scraping nets that rake up everything. Many of the fishing boats operate illegally, which makes the situation even worse. From 1985 to 2005, the amount of fish caught per net had dropped by about 80%, indicating the rapid decimation of fish stocks.[215] The result of this overexploitation may be disastrous for the region as millions of livelihoods are already endangered by ever-shrinking Bay's resources. A 2020 report on the state of fish in the Bay of Bengal indicates that most fish species are in decline, with some nearing extinction.[216]

> *"Some seas in the world, like the Gulf of Thailand, have run out of fish," one of the authors of the report, Sayedur Rahman Chowdhury, told BBC Bengali. Hundreds of large vessels are overfishing at an unsustainable rate, monitors suggest. Local fishermen say the government is turning a blind eye as the trawlers target key fish species they rely on.*

In the 1980s and 1990s, fisheries expanded into new areas and began to target different species. There was an increase in catches for some time, but catch rates began to decline in the late 1990s forcing trawlers to move increasingly further and further from their home waters.

> *The Mergui archipelago on the Thai-Myanmar border is one of the more secluded parts of the Bay. In the late 19th century, an English fisheries officer described this area as being "literally alive with fish." Today the archipelago's sparsely populated islands remain pristinely beautiful while some of its underwater landscapes present scenes of utter devastation. Fish stocks have been decimated by methods that include cyanide poisoning.*[217]

Historically, the oceans were considered limitless and thought to be able to supply enough fish to feed an ever-increasing number of people. However, the demands of this ever-growing world population now far outstrip the sustainable yield of the seas. Fish has historically been a vital source of food for humans, providing approximately 16% of the world's population's animal protein. One billion people rely on fish as their primary source of protein.[218] As fisheries became depleted and fish were harder to catch, many fishermen and governments responded with equipment and technology investments to fish longer and farther away from their home ports.

> *Consumer tastes in the First World have largely contributed to the problem. Increasing demand for top predators, such as swordfish or tuna, has put severe pressure on existing stocks... Long-line fishing for swordfish and other billfishes may significantly diminish the populations of many shark species, which are known to have slow reproductive rates and thereby slow recovery rates.*

Having exhausted the ocean of fish close to Chinese shores, vast numbers of Chinese trawlers sail beyond their local waters to exploit other countries' waters. These fishermen are subsidized by the Chinese government, which is not primarily concerned with the health of the world's oceans and the other countries that depend on them. Increasingly, China's expanding fleet of fishing vessels is heading to the waters of West Africa and off the coast of countries such as Senegal. Local fishermen cannot hope to compete with Chinese ships that are so massive that they drag up as many fish in one week as Senegalese boats catch in a year. [219]

> *In Senegal, an impoverished nation of 14 million, fishing stocks are plummeting. Local fishermen working out of hand-hewn canoes compete with megatrawlers whose mile-long nets sweep up virtually every living thing. Most of the fish they catch is sent abroad, with a lot ending up as fishmeal fodder for chickens and pigs in the United States and Europe.[220]*

Overfishing and habitat degradation have profoundly altered populations of marine animals, especially sharks and rays. Sharks are primarily caught to meet the demand for shark fin soup, a traditional and expensive Asian delicacy.[221] After a shark is caught, its fins are cut off and kept, and then in a final cruel and heartless act, the maimed animal is often thrown back into the ocean, where it sinks and dies. Millions of finless shark carcasses litter the ocean bottom every year.[222]

Queensland's shark control program, established in 1962 to "minimize the threat of shark attack on humans," has, to date, caught and killed 50,000 sharks. As a result, shark populations have fallen off Australia's east coast over the last 55 years. Hammerhead and great white sharks have plummeted by 92%, whaler sharks by 82%, and tiger sharks by 74%.[223]

An estimated 63 to 273 million sharks are killed each year, which is 6 to 8% of all sharks in the ocean.[224] This is equivalent to as many as 8 sharks being slaughtered per second. Because sharks are slow maturing and slow reproducing animals, their populations are being decimated. Elizabeth Wilson, Senior Director of Environmental Policy at The Pew Charitable Trusts, noted,

> *We are now the predators. Humans have mounted an unrelenting assault on sharks, and their numbers are crashing throughout the world's oceans.[225]*

A study published in the journal Nature in 2021 found that 18 shark and ray species populations have declined by 70% since 1970. The authors of that study caution that many of these species might disappear entirely in a decade or two.[226]

Worldwide, fish extraction from the oceans peaked in 1996 at just over 86 million metric tons (94 million tons) and has generally declined since.[227] However, that number in 1996 may have actually been 130 million metric tons (143 million tons) if discarded and illegally caught fish are included. Thus, the already severe decline in fish stocks since the mid-1990s becomes three times larger than initially thought by including unaccounted for catches. Since 1996 fish catches have dropped, on average, a massive 1.2 million metric tons (1.3 million tons) every year.[228] By 2010 fishers had a reduced yield of 109 million metric tons (120 million tons), a 16% decline since the peak in 1996 just 14 years earlier.[229]

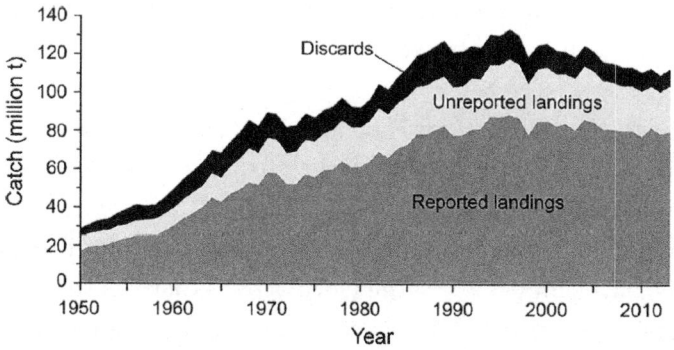

Total global catches (millions of tons), separated into reported landings, unreported landings and estimated discards from 1950 to 2014. Note these data exclude marine mammals, reptiles and plant material, as well as all freshwater catches.

The fish killed but tossed back overboard, called bycatch or "trash fish," and non-industrial levels of fishing accounts for one-fourth of fish catches. In the 1990s, between 20 and 27 million metric tons (22-30 million tons) of bycatch were discarded annually in the world's commercial fisheries.[230,231] For shrimpers, 80% of everything caught is bycatch and thrown back for dead.[232] Today, nearly 10 million metric tons are still discarded or about 10% of annual catches despite declining fish stocks worldwide.[233]

Fishers sometimes reject a portion of their catch because some caught fish are damaged and unmarketable, the fish are too small, the species is out of season, or the caught fish were not of interest. According to Dirk Zeller, a professor at the University of Western Australia,

> *Discards also happen because of a nasty practice known as high-grading where fishers continue fishing even after they've caught fish that they can sell. If they catch bigger fish, they throw away the smaller ones; they usually can't keep both loads because they run out of freezer space or go over their quota.*[234]

Still worse, a new phenomenon has emerged in Asia and elsewhere, where some trawlers are no longer targeting particular fish. Instead, they are scooping up any and all sea life that they find, turning this indiscriminate catch into fishmeal, fish oil, chicken feed, or surimi, the compressed white paste used to make fish cakes. Amanda Vincent, professor at the University of British Columbia, has termed this extremely destructive practice as "annihilation trawling."[235]

According to a United Nations Food and Agriculture Organization's (FAO) 2020 report on the state of the world's fisheries, 34.2% of the world's stocks are now overfished and unsustainable.[236] An additional 59.6% of fish stocks are the maximum limit and have no room for further expansion. This leaves only 6.2% of the world's fish stocks as so-called underfished. The over-exploitation of the planet's fish has more than tripled since the 1970s. Lasse Gustavssin, the director of Oceana, a marine conservation body, stated,

> *We now have a fifth more of global fish stocks at worrying levels than we did in 2000. The global environmental impact of overfishing is incalculable, and the knock-on impact for coastal economies is simply too great for this to be swept under the rug anymore.*[237]

The populations of all large predator fish in the oceans have declined by 90% in the 50 years since modern industrial fishing became widespread worldwide.[238] Dr. Maria Salta, a biological oceanographer and lecturer in environmental microbiology in the School of Biological Sciences at the University of Portsmouth, echoed this dire outlook on the state of the oceans:

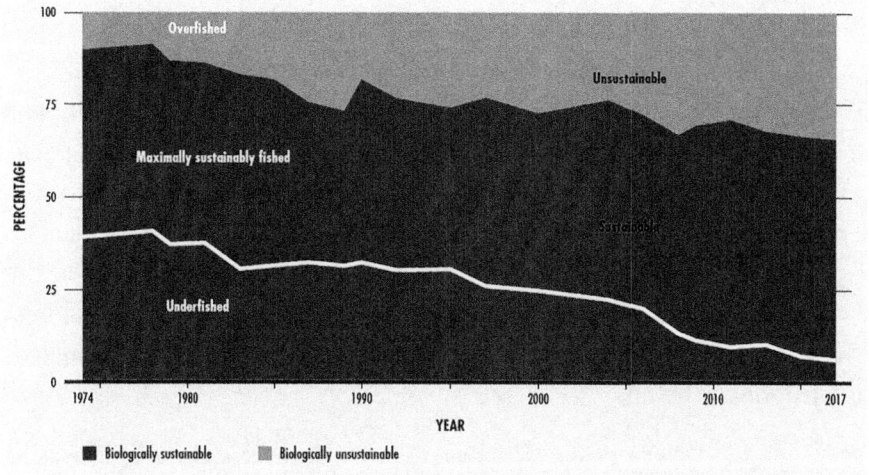

Global trends in the state of world marine fish stocks since, 1974 - 2017.

> *It is clear that if we continue like this, in a few years time, there is not going to be much left. We are losing species every day without ever knowing about them. Sometimes humans can be like a plague to the environment. The oceanic white-tip shark populations declined by 99 percent from 1950 to 1999, making it now an endangered species.*[239]

According to an NRDC (Natural Resources Defense Council) report on killing marine mammals in foreign fisheries, more than 650,000 marine mammals are killed or seriously injured every year in foreign fisheries after being hooked, entangled, or trapped in fishing gear.[240] Seals, sea lions, whales, dolphins, porpoises, and others are killed due to indiscriminate fishing practices, threatening the survival of numerous marine mammal populations.

Seabirds, such as terns and penguins, are starving because industrial fisheries are catching their food sources. The diminishing food supply for seabirds has caused a 70% population decline over the last 70 years. Since the 1970s and 80s, nearly 50% of terns and frigate birds and 25%

of penguins are gone. Not only are seabirds starving to death trying to compete with massive fishing vessels, but they are also getting tangled in fishing gear and dying from eating the plastic waste that has flooded into the oceans.[241]

Longline fishing alone is estimated to kill 160,000 to 320,000 seabirds every year.[242] Bristling with hooks, longlines, which can be up to 80 miles long, unintentionally trap, drown and harm seabirds, as well as turtles, dolphins, and other marine life. Industrial fishing vessels accidentally kill tens of thousands of albatrosses each year, bringing them ever closer to extinction.[243] The population of leatherback turtles has plummeted by 97% over the last 30 years in no small measure due to longline fishing. Every year, tens of thousands of sea turtles drown after getting snagged due to longline fishing.

> *"Outside national waters, in the high seas, it is essentially a no man's land when it comes to protecting sensitive environments and their inhabitants,"* says Paul Snelgrove, a deep-sea biologist at Memorial University in St John's, Canada. *"It is a highly unsatisfactory state of affairs."*[244]

In 2018 there were 4.56 million fishing vessels, with 67,800 of them 24 meters (78 feet) or longer.[245] Fishing operations have expanded to, quite literally, every corner of the ocean over the last 100 years. Technological advances have enabled humans to find and catch every single fish in the sea, no matter where they are located on the planet. A huge problem associated directly with overfishing is the widespread use of trawling. Doctor Maria Salta comments,

> *It's the equivalent of forest clear-cutting, but in the ocean, because when they trawl the entire bottom, whatever is there, is removed from the environment and changes the entire ecosystem. Biomass of the deep sea is in sharp decline because of trawling.*[246]

The rapid depletion of fish stocks on continental shelves helped create pressure to find alternative fishing grounds. At the end of the twentieth century, fleets of ships began fishing seamounts, which are mountains that rise from the ocean seafloor and do not quite reach the ocean's surface.

These more remote fishing areas are more easily prone to fish stock collapse as the fish targeted over seamounts are typically long-lived, slow-growing and slow-maturing. Many of these fisheries more closely resemble mining operations than sustainable fisheries, with targeted fish stocks showing signs of overexploitation within a short period from the start of fishing. This has been the case for the orange roughy fisheries off New Zealand, Australia, Namibia, and the North Atlantic.[247]

> *...on seamounts and on continental slopes, where virgin communities are fished, similar dynamics of extremely high catch rates are observed, which decline rapidly over the first 3–5 years of exploitation.*[248]

The problem is worsened by the dangers of trawling, which damages seamount surface communities. In addition, the fact that many seamounts are located in international waters makes proper monitoring nearly impossible.

> *It is widely accepted that seamounts are fragile habitats. Trawl gear is today being deployed across steeply irregular, and often boulder-strewn, sea floor surfaces at depths typically lying between 500 and 1000/2000m [1640 and 3280/6560 feet]. Netting caught during passage across the seabed can cause considerable damage to seabed environments (e.g., deep water coral reefs), and if not recovered, may remain there, out of sight, and continue ghost fishing almost indefinitely. The potential magnitude of disturbance to seabed environments can be likened 'forest clear cutting.'*[249]

It takes an average of only 4 years for seamount fisheries to collapse and 5 to 15 years after the collapse for recovery, making these seamount fisheries unsustainable.[250]

Expanding fishing areas, using bigger vessels, better nets, and new technology for spotting fish are not bringing the world's fleets more significant returns. These technological advances put immense pressure on fish stocks and leave fewer regions out of reach where fish can reproduce unmolested, thus worsening the effects of over-harvesting.
Fisheries experts trying to quantify illegal fishing along the African coast

say tens of thousands of tons of fish are stolen by foreign fishing vessels just along Senegal's coast alone.[251] This ongoing practice of fishing fleets moving their operations from depleted areas to new areas causes a long-term decline in global catches as overfishing spreads.[252]

Slavery on the high seas

When there are too few fish to catch in an area, fishing vessels venture farther from shore into deeper waters, staying out at sea longer. Investigative reports reveal how the fishing industry, especially in Southeast Asia, coerces or forces men against their will into modern-day slavery in vast fishing fleets. Men sometimes remain at sea for years on end. An exploited migrant worker from Myanmar describes the living conditions onboard illegal Thai trawlers:

> *When working, we invest our whole life. We're in their hands, and there is nothing we can do about it. When they give orders, we have to follow. It's quite dangerous. We have to eat when they tell us to, sleep when they tell us to. There's absolutely no excuse. When I make a mistake, they beat me with a metal rod.*[253]

Impoverished villagers are offered what seems to be a well-paying fishing job but then incur debt for food and lodging. According to Doctor Jessica Sparks, course director at the Department of Infectious Disease and Global Health at Tufts University, who researches the connections between fishing stock declines worldwide and forced labor on the open seas,

> *That's called debt bondage. There are stories of fisherman being out at sea for five to ten years, without ever setting foot on land, getting transferred from one vessel to another at sea. There's a lot of physical violence, sexual violence, mental violence—people getting thrown overboard to fish for sharks.*[254]

According to a study by the anti-trafficking group International Justice Mission (IJM), more than a third of migrant fishermen in Thailand have been victims of trafficking.[255] Thailand is the world's third-largest seafood exporter and, as of 2014, has more than 42,000 active fishing vessels with more than 172,000 people employed as fishermen. Three-quarters of migrants working on Thai fishing vessels are in debt bondage and struggle to pay that obligation. They are routinely underpaid and physically abused.

> *Thailand's multibillion-dollar seafood sector came under fire in recent years after investigations showed widespread slavery, trafficking and violence on fishing boats and in onshore food processing factories.*[256]

Most of these fishermen work at least 16 hours a day, and only 11% were paid the legal monthly minimum of $272 per month. Working 16 hours, 7 days a week, means these virtual slaves are paid just over $0.50 per hour. One indebted fisherman owed a debt to his supervisor fisherman's brother, "I fear for my life as he has killed in front of me before. I don't dare to run. He would kill my children."[257]

Aung Ye Tun was 17 years old when he was tricked and forced to work on a Thai fishing boat under slave-like conditions.[258] He described how they worked around the clock with only half an hour's rest a day, and anyone caught sleeping without permission would be beaten. Food was scarce, and some of them resorted to eating raw squid. For five years, he was exploited along with other trafficked youths. "When the situation was at its worst, we used to say it was hell," he said. The big boats work for three months without enough food supplies and let the workers starve. Some people take their own lives by jumping into the water.

According to the UK-based Business and Human Rights Resource Center (BHRRC), little has been done to combat slavery in the tuna supply chain.[259] As of 2019, about 80% of the world's biggest canned tuna brands do not know who caught their fish, putting workers in the industry at risk of exploitation and slavery. According to Amy Sinclair of BHRRC,

> *Modern slavery is endemic in the fishing industry, where the tuna supply chain is remote, complex and opaque. Yet despite years of shocking abuses being exposed, tuna companies are taking little action to protect workers.[260]*

In 2015, an investigation revealed the horrifying reality of men being held captive in cages for years, with many having no contact with their families for up to a decade. These modern-day slaves are trapped in perpetual debt bondage and are frequently subjected to physical abuse. They are forced to work excruciatingly long 22-hour days to pay for necessities such as food and shelter, perpetuating a never-ending cycle of debt.[261] Investigations found that Thai captains torture their slaves, and those that resist are often murdered. A former Cambodian slave shared a typical story,

> *I once saw a captain grab a metal spike used to mend nets and stab a fisherman in the chest. The crew pulled a sleeping bag over his corpse and rolled it overboard.[262]*

Reporters following the supply chain found it ended up under labels including Iams, Meow Mix, Fancy Feast, and other types of cat food.[263] The distributors in the United States who are receiving some of the seafood from these factories also sell to Wal-Mart, Kroger, Albertson's, Safeway, and others. As much as 70% of seafood for export markets is produced in developing countries where labor costs are relatively low as the risk of abuse and slavery is high. Cargo ships loaded with slave-caught seafood destined for Western markets are widespread in the industry. According to a study in Science Advances,

> *Seafood is made with a significant incidence of forced labor, child labor, or forced child labor in the seafood hub countries of Indonesia, Thailand, Vietnam, the Philippines, and Peru. In 2016, widespread forced labor in seafood work was reported in 47 countries, with incidents reported in additional countries, including New Zealand, Ireland, the United States, and Taiwan.[264]*

Warming waters

The Gulf of Maine extends from Cape Cod in Massachusetts to Cape Sable at the southern tip of Nova Scotia. In this area, years of strict fishing limits were instituted to repopulate cod stocks. Historically, the establishment of fishing quotas has helped stocks of various fish recover. Yet despite these limits, this wasn't happening with cod in the Gulf of Maine. In fact, populations were continuing to decline. As researchers investigated, they found that cod spawning and survival have been hampered by rapid, extraordinary ocean warming in the Gulf of Maine, where sea surface temperatures rose faster than anywhere else on the planet between 2003 and 2014.[265] According to Dr. Simon Boxall, an associate professor of oceanography at the University of Southampton,

> *Cod were overfished, but we also see climate change kicking in and warming the waters, and cod, which like a cooler climate, are being pushed further north. Our cod are migrating to Iceland.*[266]

Andy Pershing, chief scientific officer at the Gulf of Maine Research Institute in Portland, says, "We're really in the crosshairs of climate change right now."[267] The dramatic warming in the Gulf of Maine during the summer and early fall has extended "summer conditions" by 66 days over the last 33 years. Warming in the previous 10 years has occurred faster than 99% of the global oceans,[268] and in 2012 average water temperatures reached the highest level in the 150 years that humans have been recording them.[269] Andrew Thomas, a professor at the University of Maine School of Marine Sciences, noted,

> *There are going to be winners, and there are going to be losers. If you're a tourist and you want to swim on the beaches, a longer, warmer summer sounds fine. But there are definitely marine species that are going to have trouble adapting to that kind of change because it's happening very rapidly.*[270]

An update in 2020 by Andrew Pershing, Ph.D., shows the warming continues.[271] Since 2010, the Gulf of Maine's temperature has been above average 92% and at heatwave levels for 55% of the time. Dr. Pershing comments,

Whether it's the temperatures alone, or some of the species that come with them (e.g., recent reports of black sea bass and bonito), observations that would have surprised us a decade ago now feel expected. Like so many other parts of our lives this year, what used to be unimaginable is now a new normal.[272]

It's not only cod that is being affected by increasing water temperatures. According to National Oceanic and Atmospheric Administration (NOAA) researchers, about 50% of 36 fish stocks in the Northwest Atlantic Ocean, many of them commercially valuable species, have been shifting northward over the last four decades. Some stocks have nearly disappeared from United States waters as they move farther offshore. Janet Nye, a postdoctoral researcher at NOAA's Northeast Fisheries Science Center (NEFSC) laboratory in Woods Hole, Massachusetts, commented,

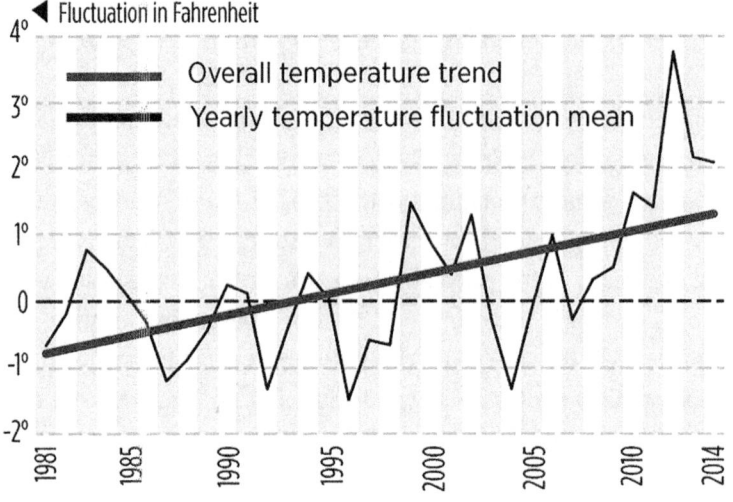

SOURCE: Andrew J. Pershing/Gulf of Maine Research Institute

Getting Warmer: Sea surface temperatures in the Gulf of Maine have been rising over the last 35 years, and at nearly the fastest rate on the planet over the last 10. 2012 had the warmest readings in the 150 years humans have been collecting them.

> *During the last 40 years, many familiar species have been shifting to the north where ocean waters are cooler, or staying in the same general area but moving into deeper waters than where they traditionally have been found. They all seem to be adapting to changing temperatures and finding places where their chances of survival as a population are greater.*[273]

A 2013 paper in the journal Nature showed that ocean warming has already affected global fisheries in the past four decades. Shifts in fish populations in most of the world's coastal and shelf areas have significantly and positively been related to regional changes in sea surface temperature.[274] The study's lead author William Cheung, an assistant professor at the University of British Columbia's Fisheries Centre, noted,

> *One way for marine animals to respond to ocean warming is by moving to cooler regions. As a result, places like New England on the northeast coast of the U.S. saw new species typically found in warmer waters, closer to the tropics. Meanwhile, in the tropics, climate change meant fewer marine species and reduced catches, with serious implications for food security. We've been talking about climate change as if it's something that's going to happen in the distant future -- our study shows that it has been affecting our fisheries and oceans for decades.*[275]

Farmed fish

Until recently, fish have been the only remaining important food source that is still mainly gathered from the wild rather than farmed. Yet, as the amount of global wild fish caught continues to decline, aquaculture, also known as fish or shellfish farming, has increasingly filled the gap. Because of this seafood alternative, fish consumption has continued without much public notice of wild fish stocks' devastation. Aquaculture has grown more than 20-fold since the 1980s.[276] By 2021 it was forecast to overtake wild-caught as the primary source of fish.[277] However, there are significant adverse effects of aquaculture on the environment and wild fish that are worth considering.

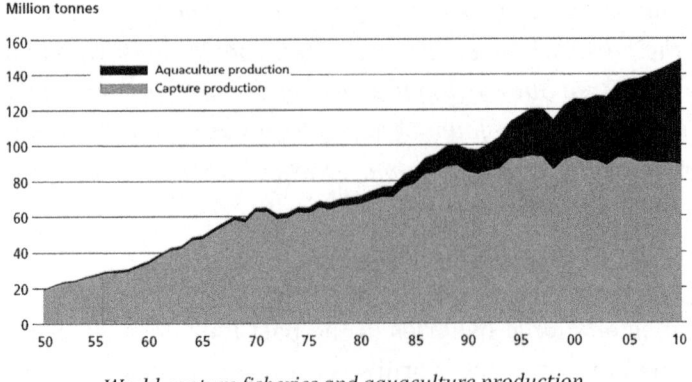

World capture fisheries and aquaculture production.

At the core of industrial food production is monoculture, which is growing single crops intensively on a large scale. Corn, wheat, soybeans, cotton, and rice are all commonly grown this way. Monoculture farming relies heavily on chemical inputs such as synthetic fertilizers and pesticides. Fertilizers are needed because growing the same plant in the same place year after year quickly depletes the plant's nutrients. These crops are also susceptible to disease and pests because a lack of variety in genetics makes them ecologically vulnerable—Farmers attempt to control these problems with chemicals, pesticides, and genetically modified (GMO) crops.

Aquaculture uses the same methods to achieve large-scale and profitable outputs as other industrial food production systems. When using these intense monoculture methods, aquaculture operations also experience similar types of negative consequences. The salmon farming industry uses open-net cages placed directly in the ocean, where farm waste, chemicals, disease, and parasites are released into the surrounding waters, harming other marine life.

> *Sea lice are small ectoparasitic copepods [small crustaceans] that attach onto the scales of fish, feeding on tissue, mucus, and sometimes blood. In naturally occurring systems, lice infestation usually occurs in adults whilst they are at sea. Since sea lice cannot survive in fresh water, they fall off the adult salmon or die when they return to freshwater spawning streams. The problem for farmed salmon is*

> *that they are confined to a limited area. If a louse originating from a wild salmon infects a farmed salmon, the farmed salmon never migrates to freshwater and so can't shed the lice. In farmed salmon, typically kept in high densities, lice can easily increase to levels not normally experienced in the natural environment.*[278]

Since the start of large-scale salmon farming in the 1970s, sea lice have primarily been managed using chemotherapy. This has been effective and simple to use but also creates unwanted environmental effects, occupational hazards, and drug resistance problems.[279]

> *Fish excreta [waste] and uneaten feeds from fed aquaculture diminish water quality. Increased production has combined with greater use of antibiotics, fungicides, and anti-fouling agents, which in turn may contribute to pollute downstream eco-systems.*[280]

When over half a million or more farmed salmon are crowded in a small area, an enormous amount of fish feces and waste feed is produced. For example, in Scotland, the discharge of untreated organic waste from salmon production equals 75% of Scotland's human population's pollution.[281] This can significantly impact the ocean bottom and surrounding ecosystems and make the farmed salmon more vulnerable to sea lice. Sea lice proliferate on salmon farms and spread to surrounding waters, attacking younger vulnerable wild salmon as they swim in their natural environment, causing disease in the wild salmon populations.

> *The evidence that salmon farms are the most significant source of... sea lice on juvenile wild salmonids [a fish of the salmon family] in Europe and North America is now convincing. Farms may contain millions of fishes almost year round in coastal waters and, unless lice control is effective, may provide a continuous source of sea lice, although the amount of infestation pressure will vary over time owing to seasonal and farm management practices.*[282]

Another inescapable consequence of open-pen salmon farming is that predators are naturally attracted to the vast quantities of fish confined in artificial environments. Just as ranchers shoot predators, such as wolves and coyotes, salmon farmers shoot predators such as seals and sea lions that endanger their salmon stocks.

> *Department of Fisheries and Oceans (DFO) statistics show that since 1990, the B.C. [British Columbia] industry has shot and killed more than 7,000 of our marine mammals: almost 6,000 harbour seals, 1,200 California sea lions, and 363 endangered Steller sea lions.*[283]

The United States imports about 86% of its seafood, about half of which is grown in aquaculture. Almost 90% of farmed fish and shellfish come from Asia.[284] China produces about 70% of the farmed fish in the world, employing 4.5 million fish farmers. Fish farms can be found in lakes, ponds, rivers, reservoirs, or sizeable rectangular fish ponds dug into the earth.

Fish and shellfish in Asian countries are often raised in filthy, overcrowded conditions. In China, water supplies are contaminated by sewage, industrial waste, and agricultural runoff, including pesticides. Half of the rivers in China are too polluted to even serve as a source of drinking water. Nearby coastal waters, which are also heavily fish farmed, are polluted with oil, lead, mercury, and copper.[285] Farmers have been known to feed tilapia fish with the feces of pigs and geese to lower production costs.[286] The heavily polluted Yangtze River, contaminated with heavy metals, fertilizers, pesticides, and weed killers, is bordered by 20,000 chemical plants[287] and is also lined with fish farms. In 2007, one fish farm along that river sent about 2.7 million catfish fillets each year to the United States through an importer in Virginia.[288]

In China and other countries with lax environmental standards, to keep the animals alive in these low-quality water conditions and overflowing farms, they are liberally dosed with powerful and often illegal antibiotics and pesticides. This keeps the animals living yet leaves poisonous and carcinogenic residues in seafood.[289] The contaminated output from all of these farms also has a tremendous impact on the surrounding environment. Reservoirs used by fish farmers become toxic waste dumps because of the liberal use of animal manure, fertilizers, and antibiotics.[290]

> *Industrial fish farming has destroyed mangrove forests in Thailand, Vietnam, and China, heavily polluted waterways, and radically altered the ecological balance of coastal areas, mostly through the discharge of wastewater. Aquaculture waste contains fish feces, rotting fish feed, and residues of pesticides and veterinary drugs as well as other pollutants that were already mixed into the poor quality water supplied to farmers.*[291]

In the United States, most shrimp is imported, grown in industrial-sized, man-made ponds along the coasts of Southeast Asia and South and Central America. Thailand is the leading shrimp exporter to the United States, followed by Ecuador, Indonesia, China, Mexico, and Vietnam.[292] Coastal mangroves, which provide a thriving habitat for many species, are frequently demolished to make way for shrimp ponds. The construction of shrimp ponds, considered the foremost cause of coastal mangrove destruction, has led to the destruction of nearly 10% of the world's mangrove forests.[293][294] In addition, these shrimp farms produce a tremendous amount of waste that pollutes the surrounding land and water, depleting the freshwater supply. In Bangladesh, shrimp aquaculture generates 600 metric tons (660 tons) of waste per day.[295] After an average of seven years, the ponds, crammed with millions of shrimp, become so polluted with shrimp waste and chemicals that shrimp farmers move on to build new ponds, leaving behind abandoned wastelands.

The explosive growth of shrimp farms in Ecuador, Honduras, and Mexico has negatively impacted water quality and destroyed large areas of mangroves. Nearly 75% of the fish caught at sea hatch or breed in mangroves, or relied on the intact mangrove system. The destruction of these mangroves has caused a decline in wild fish catch, with studies indicating that for every hectare of mangrove forest destroyed, an estimated 757 kilograms (1,669 pounds) of commercial fish are lost.[296]

In addition to their own environmental and health hazards, farmed fish are often fed large quantities of wild-caught fish. Fishmeal made from this wild-caught sea life utilizing "annihilation trawling" is used to feed farmed tiger prawns or salmon that are then sold to stores and restaurants.[297]
According to a short film about fishing and fish meal:

> *By dragging small mesh nets along the bottom of the sea, the trawler pulls up virtually every living thing in its way. Everything from endangered species, such as sharks, to seahorses and sea snakes. Most of the catch consists of small fish. These are not only vital to larger fish, but if left to grow could be used as food by people living there. This is how ecosystems are destroyed, and a large number of animals, like dolphins and seabirds, can no longer find enough food.*[298]

Many intensive aquaculture farms use between two to five times more fish feed to fatten their farmed fish than is actually produced. So, for example, it can take 2.8 pounds of wild fish to produce a pound of shrimp, 3.2 pounds to produce a pound of salmon, and 4.7 pounds to produce a pound of eel.[299] So in many instances, aquaculture, instead of reducing stress on our oceans, may actually be increasing it.

Ocean life collapse

The world's oceans are under increasing stress, with nearly 6 billion metric tons (6.6 billion tons) — equal to more than the weight of 18,000 Empire State Buildings — of fish and invertebrates that have already been taken from the oceans since 1950.[300] Dr. Boris Worm, a Marine Research Ecologist and Associate Professor at Dalhousie University, Canada, led an international team of researchers who found that fishery decline is closely tied to a broader marine biodiversity loss. These fisheries experts and ecologists predict that if fishing around the world continues at its current pace, species will vanish, marine ecosystems will unravel, and there will be a global collapse of all species currently fished, possibly as soon as mid-century.

> *Our data highlight the societal consequences of an ongoing erosion of diversity that appears to be accelerating on a global scale. This trend is of serious concern because it projects the global collapse of all taxa [group of organisms] currently fished by the*

mid–21st century (based on the extrapolation of regression to 100% in the year 2048).[301]

The decrease of biodiversity, or variety of life, in the oceans tends to reduce local fish stocks' size and robustness. It is this loss of biodiversity that is driving the declines in fish populations that are seen in large-scale studies.

> *"The image I use to explain why biodiversity is so important is that marine life is a bit like a house of cards,"* said Dr. Worm. *"All parts of it are integral to the structure; if you remove parts, particularly at the bottom, it's detrimental to everything on top and threatens the whole structure."* [302]

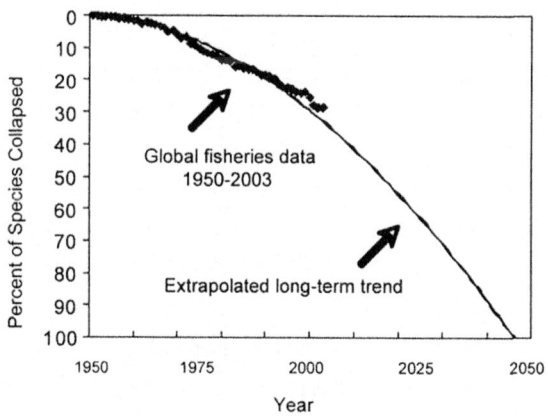

Projection to 2050 of ocean species collapse.

The global collapse of species being fished may come even earlier. According to Cyrill Gutsch, the founder of Parley for the Oceans, fish stocks' viability may face a point of no return by 2030 if humanity keeps overfishing and does not step up efforts to reduce water pollution.

> *I would say the deadline of 2048 was a very optimistic one, we are already in the sixth mass extinction and we are losing 400 species every day. At Parley, we believe that if we do not turn things around in the next 10 years, we are reaching the point of no return. We are*

really thinking of 2030 deadline. 10 years is nothing and it means we have to organize ourselves really well. We have to innovate and prepare ourselves that we are not meeting this deadline.[303]

In 1883 Professor T. H. Huxley, a nineteenth-century English biologist, and anthropologist, came to the following conclusion, based on what he saw as the insignificant human impact on the inconceivably enormous numbers of fish in the ocean:

I believe, then, that the cod fishery, the herring fishery, the pilchard fishery, the mackerel fishery, and probably all the great sea-fisheries, are inexhaustible; that is to say that nothing we do seriously affects the number of the fish.[304]

Ultimately time would prove Professor Huxley wrong. The unrelenting expansion of fishing operations due to the market-driven competition for resources means that the risks of overfishing and unsustainable natural resource use are tending to increase, despite efforts to promote sustainable fishing and fish farming worldwide. Fueled by increased public demand as part of a desire for a "healthy and diversified" diet, the market for fish continues to soar, rising from 9 kilograms (19.8 pounds) per year in 1961 to over 20 kilograms (44 pounds) per year in 2015.[305] Market demand for tuna is still high, and the significant overcapacity of tuna fishing fleets remains. Growing seafood demand, combined with fewer fish, inadequate traceability systems, and a vast ocean virtually impossible to patrol, provide significant incentives for those willing to catch fish illegally and funnel them into the legitimate supply chain.

Pollution, environmental degradation, climate change, diseases, and natural and human-induced disasters threaten livelihoods for those that rely on fishing. Shrinking catches and declining fish stocks, combined with pressure from growing coastal populations, are particularly affecting smaller fishing communities in many developing countries, where social protection and other employment opportunities are often lacking.

Ever-expanding global aquaculture is fraught with its own environmental impacts that are also negatively impacting wild fish populations. With a blind emphasis on economic growth, fish farms have destroyed mangroves, polluted waterways, and created seafood contaminated with antibiotics, pesticides, heavy metals, and potentially other dangerous chemicals, as well as encouraging overfishing/trawling of wild-caught fish used as fishmeal.

As human populations continue to grow, so does the insatiable appetite for seafood. The fish supply has been slowly collapsing for decades as bigger and better technologies are utilized to harvest every bit of the ocean. The once "inexhaustible manna" has proven to be finite and buckling under the pressure of human activities. *Without severe and worldwide transformations as to how we manage our marine resources, this strain may reach a point in the near future where we may find a mostly dead ocean devoid of fish.*

What you can do!

Life in oceans across the entire planet is under severe pressure from human activity and faces collapse. Here are some simple steps that you can take at a personal level to make a difference.

Move away from wild-caught seafood – For decades, vast numbers of profit-seeking commercial fishing fleets, armed with ever-improving technologies, have expanded to virtually every part of the ocean. Removal of fish from the world's oceans peaked in the mid-1990s and has been declining since, with only 6% of the world's fish stocks not overfished or at maximum. Simultaneously, vast numbers of mammals, sharks, seabirds, turtles, and more have been decimated either directly or because these vessels leave precious little for populations to eat, causing many to starve to death. Each year lost or discarded fishing equipment only magnifies the destruction of "ghost fishing," devastating life across the oceans.

Largely ignored, many tens of thousands of poor people work incredibly long hours as literal slaves to supply more affluent countries with cheap seafood. Slavery, trafficking, physical, mental and sexual violence are all entangled with decimating life in our oceans.

With the focus on "human nutrition" and "productive jobs," the ever-increasing demand for fish[306] continues to decimate life in the oceans. The human notion that there is no ceiling in the endless "sustainability" of the market ignores all other life on the planet. It could result in a mostly fishless ocean by mid-century or even as early as within 10 to 20 years.

Switch from fish to a primarily plant-based diet – It can supply a person with plenty of protein and nutrients. Eat foods such as quinoa, lentils, chickpeas, kidney beans, tempeh, broccoli, nuts, and nut butters, to get proper, healthy amounts of protein.[307] Eat foods such as chia seeds, Brussels sprouts, hemp seeds, walnuts, and flax seeds, to get healthy amounts of omega-3 fatty acids.[308] Omega-3 fatty acids (DHA and EPA) are also available from sustainably grown algae and seaweed.

Move away from farmed seafood – Aquaculture continues to grow across the globe with devastating environmental impacts such as destroying mangroves and local aquatic environments. The use of "annihilation trawling" to scoop up everything in the oceans to make fishmeal to feed farmed fish is an enormously destructive practice wiping out entire ecosystems. Because it requires three times or more of the weight of wild fish to produce farmed fish for people to eat, it makes this an environmentally unsustainable practice. As recommended, when moving away from eating wild-caught fish, change to a primarily plant-based diet.

Choose sustainable seafood – If you decide to eat seafood, choose fish that are the most sustainable. Use the Good Fish Guide from the Marine Conservation Society or other resources to determine which fish are the most sustainable.

Avoid industrial-raised foods – Indiscriminately caught fish is turned into fishmeal to feed fish, chicken, or other livestock. When consuming animal products, purchase those that use organic and sustainable methods. Switching to a primarily plant-based diet reduces the probability of inadvertently using products that are destroying our oceans.

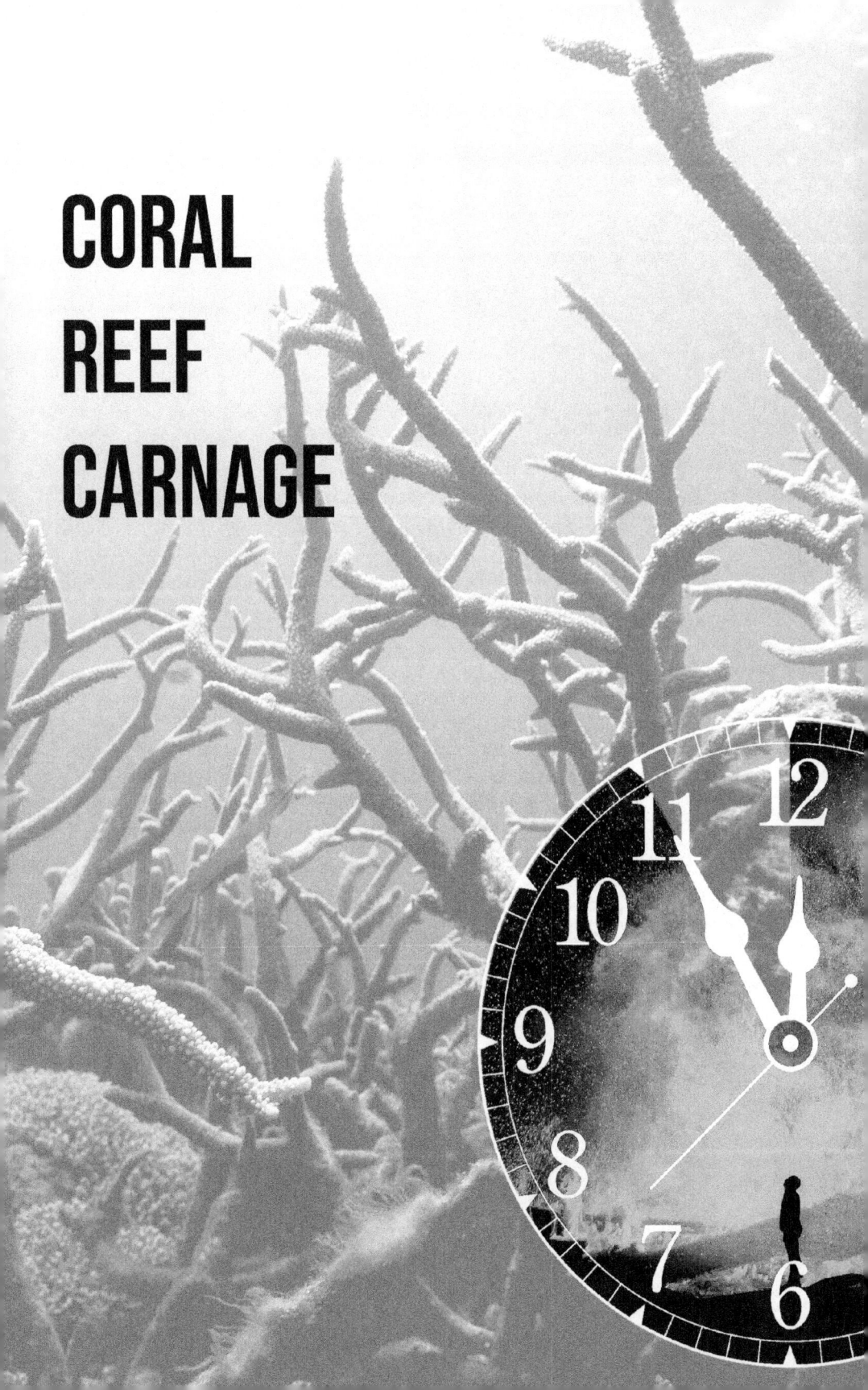

CORAL REEF CARNAGE

Beauty, beauty, beauty everywhere, on the sea, on land, in the sky, and down beneath the ocean's breast as well; for spread out on these coral reefs and sandy shoals are the most wonderful submarine gardens that eye ever beheld.
– Gordon Stables, Jack Locke: A Tale of the War and the Wave

If we don't manage this resource, we will be left with a diet of jellyfish and plankton stew.
– Daniel Pauly

The canary, a member of the finch family, is very sensitive to carbon monoxide, which is a colorless, odorless, and tasteless gas. This deadly gas was frequently found in coal mines after fires and explosions. At one time, miners and rescue workers used the canary to test the mine atmosphere, and when the bird showed signs of distress, the workers knew there was a problem with the air.[309]

> *A canary in a small portable cage is an almost indispensable adjunct to rescue work after a mine explosion or during a mine fire... The symptoms manifested underground by a canary in the presence of carbon monoxide are an increased rate of breathing, often accompanied by opening of the mouth, fluttering and unsteadiness on perch, and last – sometimes almost instantly – unconsciousness and death.[310]*

Rainforests under the waves

Coral reefs are like desert oases in the vast, mostly nutrient-devoid tropical oceans. They provide the shelter and nourishing framework that allow reefs to teem with an extraordinary variety of life. Coral reefs have one of the most incredibly diverse yet vulnerable ecosystems on our planet, second only to tropical rainforests in biodiversity.

As many as 100,000 species have been identified inhabiting coral reefs. However, the total number may actually be anywhere between 500,000 to 2,000,000 or more.[311] The world's reefs measure a total of 284,300 square kilometers (109,768 square miles), slightly less than the size of Italy. Coral reefs account for less than 1.2% of the world's continental shelf area and less than 0.1% of the world's oceans,[312] yet they account for 25% of all marine life. For this reason, coral reefs are often referred to as the "rainforests of the sea."[313]

> *The beauty of color and form and the overwhelming variety of life on coral reefs are both legendary and real. Nowhere else in the seas is there such a bewildering range of living things, and perhaps*

nowhere else is the physical and biological pattern so uniform, characteristic, and widespread as in the coral reef.[314]

Coral reefs are built by and made up of thousands of tiny animals called polyps, which live in large colonies. The polyp uses calcium and carbonate ions from seawater to build itself a hard, cup-shaped skeleton made of calcium carbonate (also known as limestone), protecting the polyp's soft, delicate body. Most coral polyps have transparent bodies, and the skeletons they build are bone white. The coral's brilliant colors come from tiny plants (microscopic marine algae called zooxanthellae) living inside their tissues. The animal and plant components live together, where the animal provides protection and raw materials for the plant to produce food through photosynthesis. The symbiosis (a mutually beneficial relationship) between the polyps and zooxanthellae provides an abundant nutrient-dense habitat that supports the remarkable assortment of life on the reef.

The coral reef ecosystem has existed for at least 200 million years.[315] Corals have built the primary structure of entire reefs, islands, and such massive oceanic structures that they can even be seen from outer space. One such formation that can be seen and monitored from space is off Australia's coast – the Great Barrier Reef (GBR). It is the most massive living structure on the planet.[316]

Coral reefs develop very slowly. It may take up to a hundred years for a coral reef to grow about one meter (three feet).[317] Reefs thrive in highly stable environments, with approximately 500 million people depending on coral reefs for food, coastal protection, building materials, and tourism income.[318]

Vanishing corals

Research by Associate Professor John Pandolfi, a Chief Investigator in the ARC Centre of Excellence for Coral Reef Studies (CoECRS) and The University of Queensland, warns that the world's reefs are now experiencing transformations that they have never experienced before. Recent human activity has dramatically degraded the Caribbean coral reefs that were remarkably stable over many tens of thousands of years, measuring up to 220,000 years ago.[319]

> *In the past, you would have seen an overwhelming dominance of Elkhorn coral. It was one of the most beautiful and striking features of the Caribbean reefs. Now, [that species] has virtually disappeared, and the same reefs are dominated by algae and seaweed. There are precious few large fish, turtles, dugongs, or sharks. It is totally different to the past.*[320]

Human impacts have been the most influential driving force in coral reef decline over the last several hundred years.[321] Large branching coral species that once were common have been decimated all over the Caribbean. In 1918 there were reports of expansive tracts of Elkhorn coral near the island of Barbados. By 1960 this species was only a minor part of the reef. By 1987 this coral comprised less than 1% of coral cover. The substantial degradation in many coral reefs actually began before the year 1900. This pre-1900 decline was probably mainly due to overfishing, pollution, and other human activities.[322]

> *The greatest impacts on Barbados are probably related to increased turbidity resulting from land clearing and the development of sugar cane agriculture in the mid-17th century.*[323]

Comparing early British nautical charts to modern coral habitat maps demonstrates that entire sections of the Florida reef that were present before European settlement are now mostly gone. An estimated 88% of Florida Bay corals have been lost, a massive cumulative loss of reef coral in the Florida Keys over the last 240 years.[324]

Coral reefs are highly susceptible to environmental disturbances. Reefs are greatly influenced by increased seawater temperature, overfishing, alterations in reef fish populations, eutrophication (excessive nutrient enrichment),[325] disease, and heavy metal pollution.[326]

Overfishing has become so widespread that there are few if any, reefs in the world which are not threatened. This, combined with such destructive practices as blast fishing, is shifting the patterns and balances of life in many reef ecosystems. From onshore, a much greater suite of damaging activities is taking place. Often remote from reefs, deforestation, urban development, and intensive agriculture are now producing vast quantities of sediments and pollutants which are pouring into the sea and rapidly degrading coral reefs in close proximity to many shores.[327]

When corals experience stressful conditions, such as high temperatures, it causes them to expel their symbiotic algae, which are the source of their brilliant colors. Once the algae are gone, only the polyps' bone-white skeletons remain, giving the coral a bleached look. Prolonged bleaching events lead to coral deaths because the plant part of the coral is no longer present to provide food for the coral's animal component. The coral can die from starvation or become so weakened by a lack of food that it succumbs to harmful bacteria and seaweeds. Coral can also be more susceptible to bleaching by excessive sewage and fertilizers associated with increasing coastal human populations.[328,329]

Coral deaths resulting from bleaching encourage seaweed (macroalgae) to quickly inhabit the area and take over even the few remaining patches of live coral reefs. This radically transforms the space from a coral to a seaweed-dominant environment. Major coral bleaching events occur during high sea surface temperatures associated with El Niño, an extremely warm weather pattern. But other factors such as cold conditions, elevated solar radiation, and pollution have been found to cause minor local coral bleaching. Not all bleaching events result in coral deaths. If the bleaching is mild, coral reef systems can recover by surviving long enough to reacquire new algal partners.

Since the 1970s, coral reef degradation has accelerated. Many reefs have suffered an enormous, long-term decline in abundance, diversity, and habitat structure due to many negative human-generated factors. These losses were more recently compounded by substantial mortality due to disease and coral bleaching. By 2003 the reefs in the Caribbean had already experienced a considerable decline, which continues today.

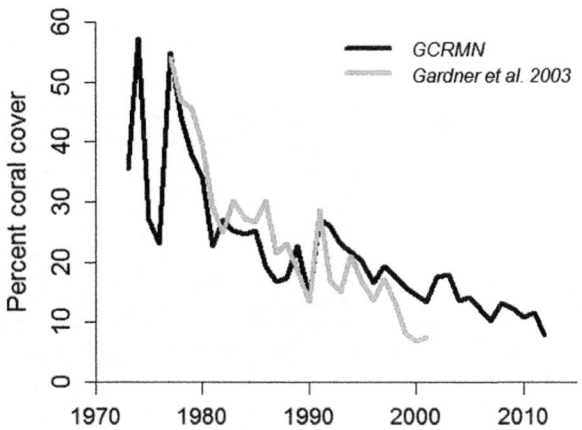

Decline in percent coral cover on Caribbean coral reefs from 1963 to 2012. Data is from a 2003 study (Gardner) and from the Global Coral Reef Monitoring Network (GCRMN) in 2012.

> *We report a massive region-wide decline of corals across the entire Caribbean basin, with the average hard coral cover on reefs being reduced by 80%, from about 50% to 10% cover, in three decades... Although the rate of coral loss has slowed in the past decade compared to the 1980s, significant declines are persisting. The ability of Caribbean coral reefs to cope with future local and global environmental change may be irretrievably compromised.[330]*

In the Florida Keys, coral reefs were already severely degraded in the early 1970s.[331] Total coral cover in the Florida Keys, US Virgin Islands, and Puerto Rico has progressively declined from 25% to 35% coverage in 1970 to less than 15% today. There was a gradual decline in coral in the United States Virgin Islands and Belize in the 1970s and 1980s. This was followed by a catastrophic collapse in the 1980s and 1990s.[332]

Carysfort Reef is located east of Key Largo, Florida. Analysis has shown that between 1974 and 1999, living coral cover on the Carysfort Reef declined by 92%.[333] Images of the reef from 1975 to 2014 show the devastating breakdown in this reef. A once thriving reef now looks like little more than ocean rubble. Florida's reefs have declined by at least 70% since the 1970s. Staghorn corals, once common in shallow water, have been reduced by an estimated 98% and are now listed as a threatened species under the Endangered Species Act.[334]

Carysfort Reef, Florida Keys 1975 (top) to 2014 (bottom).

Off the coast of Galveston, Texas, sits the Flower Garden Banks National Marine Sanctuary, home to spectacular coral formations. In 2016 divers saw cloudy water, lifeless sea urchins with their spines falling out, and other dead crustaceans. In the sanctuary's most severely impacted area, as much as 70% of the corals were killed. It was termed a "hot zone" because it was almost as if the corals had been incinerated, exposing the coral's skeletons.[335] Emma Hickerson, the sanctuary research coordinator, had for years felt confident that Flower Garden Banks was relatively safe.

> *What she saw stunned her: dead and dying corals, many of them tens of centuries old, some the size of SUVs. And it was more than corals that were affected. Sponges and clams were dead and dying too. And there were no fish to be seen, which is unusual around a coral reef. "It was just shock, disbelief," Hickerson says.[336]*

Destructive fishing

One of the significant reasons for reef deterioration is the decline of grazing animals such as parrotfish and sea urchins. Grazers are essential in the marine ecosystem as they eat the algae that can smother corals. Disease led to the mass death of the sea urchin in the 1980s. Overfishing throughout the twentieth century brought the parrotfish population to the brink of extinction in some regions.[337] A 2014 report by Global Coral Reef Monitoring Network (GCRMN) indicated that most Caribbean coral reefs will disappear unless this problem is addressed within the next 20 years. Jeremy Jackson, a lead author of the report, noted that,

> *Even if we could somehow make climate change disappear tomorrow, these reefs would continue their decline. We must immediately address the grazing problem for the reefs to stand any chance of surviving future climate shifts.*[338]

Blast fishing is the most damaging method of fishing on the reefs. Explosives are usually homemade from fertilizers, although dynamite is occasionally employed. The explosive is often thrown near the reef, which detonates on the surface of the water. The shock wave from the blast kills or stuns most fish, with most of the fish sinking to the bottom. The fisherman can effortlessly gather the small number of dead fish that float to the surface. This type of fishing is non-selective killing and causes massive damage to the reef itself.[339] The Mergui archipelago on the Thai-Myanmar border has been damaged by this destructive fishing.

> *The region was once famous for its coral reefs; these have been ravaged by dynamite-fishing and climate-change induced bleaching.*[340]

A large-scale survey in 2015 of East Africa's coral reefs showed places where destructive fishing practices have wreaked havoc in the area.

> *On more than one occasion, we dived up against massive monofilament gill nets, un-fondly referred to as 'hanging walls of death.' In addition, the destructive effects of dynamite fishing have devastated*

the corals in many areas, an illegal practice entered into more boldly than one might expect... The destruction is total and devastating, leaving behind eerie uninhabited craters of coral rubble.[341]

Coral-eating starfish

Large outbreaks of coral-eating crown-of-thorns starfish (CoTS) also severely threaten some reefs. The coral-eating starfish plays an important role, feeding on the fastest-growing corals, allowing slower growing coral species to form colonies, increasing coral diversity. However, due to several factors, including terrestrial runoff, which increases the phytoplankton that the larvae of CoTS feed on, and the loss of CoTS predators, major outbreaks of CoTS have dramatically increased.

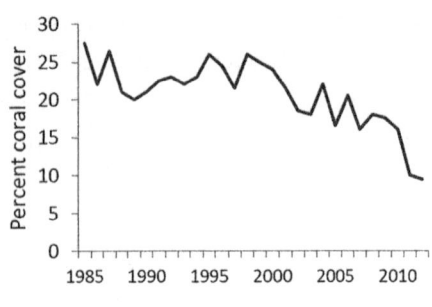

Decline in percentage of reef surface covered by live coral across the Great Barrier Reef, Australia, 1985-2012.

> *"Coral-eating starfish threaten Great Barrier Reef,"* explains marine biologist Bernard Degnan, at the University of Queensland. *And although these pests are native to the reef, scientists believe they have prospered in recent years because overfishing has left few starfish predators and starfish larvae may now gorge on huge supplies of plankton supported by agricultural run off.*[342]

Like the Caribbean reefs, other of the world's reefs show a decline. The Great Barrier Reef has been classified as the world's least threatened coral reef due to its distance from the relatively small human population centers and strong legal protection. However, the GBR has been subject to severe disturbances, including CoTS outbreaks, mass coral bleaching, and declining growth rates of coral due to increasing seawater temperatures, terrestrial runoff, tropical cyclones, and coral diseases. The runoff of soils, fertilizers, and pesticides from agricultural and coastal development have significantly affected inshore coral reefs.[343] From the 1980s to 2012, live coral cover on the Great Barrier Reef has decreased by approximately 50%, with CoTS outbreaks causing up 42% of this decline.[344]

> ...coral cover has declined considerably, seagrass health in the central GBR is in poor shape, dugong numbers have declined precipitously, shark populations are in serious decline (although perhaps recent management has reduced the rate of decline), many other large fish on the GBR have had large population declines (although data on many are incomplete) and the fourth wave of CoTS outbreaks has commenced. Most notably, coral bleaching has become more frequent, widespread, and damaging, and coral calcification has started to decline due to ocean acidification.[345]

Unbridled human development

Coral reefs are widespread in the South China Sea (SCS) but have also declined dramatically over the last 50 years. According to a 2013 study, China's unrestrained economic development has caused a devastating decline of 80% in China's coral reefs over the last 30 years.[346] Coastal development, pollution, overfishing, and destructive fishing practices have been the primary factors causing this decline. Professor Terry Hughes, who conducted this research, noted that:

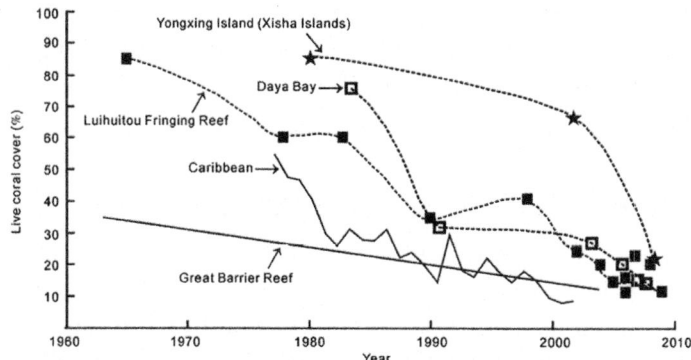

Live coral cover declines in the South China Sea in comparison to long-term trends in the Great Barrier Reef and the Caribbean.

Typically, when a coral reef degrades, it is taken over by seaweeds – and from there, experience has shown, it is very hard to return it to its natural coral cover. The window of opportunity to recover the reefs of the South China Sea is closing rapidly, given the state of degradation revealed in this study.[347]

Long-term monitoring suggests that the coral reefs of the SCS [South China Sea] have suffered a dramatic decline over the past 50 years, reflected in the decrease of live coral cover. Live coral cover in Daya Bay (northern SCS) declined from 76.6% to only 15.3% between 1983/1984 and 2008; Luhuitou fringing reef (Hainan Island, northern SCS) coral cover decreased from ~80%–90% to ~12% between 1960 and 2009; and that in Yongxing Island (the Xisha Islands, central SCS) decreased from 90% to ~10% between 1980 and 2008.[348]

The year 2016 marked the opening of an expanded Panama Canal, which now allowed a new class of ships to transport twice as much cargo as before.[349] These larger cargo ships called for deepening and widening projects for ports around the world. Coastal dredging and construction are known causes of significant destruction to coral reef ecosystems as they expose the corals to vast amounts of dredging sediments. One such project in the Port of Miami created sediment plumes that extended 228 square kilometers (88 square miles). An estimated 4.2 million cubic meters (5.5 million cubic yards) of material was dredged up [350] - enough to fill over 300,000 tandem dump trucks.[351] This caused 50-90% of the reef to be covered in dredging sediment, causing at least half a million corals to die within 500 meters (1,600 feet) of the dredging activities. It is likely multiple millions of more corals further away also died from coastal dredging and construction near them.

In 2017 a study also implicated ocean dead zones (low or reduced oxygen concentration areas) in coral die-offs. Land-based sewage and agricultural runoff spur the growth of algae. When the algae die, they decay, consuming oxygen faster than it can be brought down from the surface. As a result, sea life (including corals in the area) that can't move away can suffocate and die.

> *"Based on our analyses, we think dead zones may be underreported by an order of magnitude,"* said Nancy Knowlton of the Smithsonian's National Museum of Natural History and study co-author. *"For every one dead zone in the tropics, there are probably 10 — nine of which have yet to be identified,"* she said.[352]

In hot water

Although many factors can result in coral reef bleaching, higher Sea Surface Temperature (SST) and stronger El Niño have been accepted as the primary causes of large-scale coral reef bleaching events. SST increased dramatically in recent times, with the 1990s being 2.2°C (4°F) warmer than 1,500 years ago. Global ocean temperatures have risen by 0.7°C (1.3°F) since the late 19th century and are climbing.[353] This is significant because it indicates that recent human-caused global warming is a substantial factor in reef bleaching. The extremely high SST in the 1998 El Niño year resulted in 16% of global coral reef degradation. It impacted almost all coral reef sites worldwide, especially on the GBR, where several well-established and mature coral colonies, some up to 1,000 years old, were killed.[354]

> *...the 1997-1998 global coral bleaching event caused mortalities of up to 80% in some of Tanzania's reefs. This worldwide bleaching event was caused by elevated sea surface temperature due to El Nino. Sea surface temperatures were 2°C higher than average (over 30°C). The Misali and Tutia reefs in Pemba and Mafia Islands were the most affected, with about 90% of these reefs suffering coral mortality.*[355]

Mass coral reef bleaching episodes that resulted in large-scale coral mortality were first recorded in the early 1980s. Before that, reports of coral bleaching were scattered or almost nonexistent in the scientific literature. However, analysis of dated corals from the SCS has shown that coral deaths within the last 200 years correlated with past El Niño events. These coral deaths were probably caused by temperature-induced bleaching, indicating that this is not a new phenomenon produced by recent global warming.[356] However, the analysis also has shown that there has been a drastic change in coral community composition from long-term stability in recent declines. This indicates that the extent of mass bleaching events is much worse today than in the past.[357]

Scientists noted another devastating loss of coral due to spikes in sea temperatures in 2015 and 2016. A recent Japanese Environment Ministry survey of the Sekiseishoko coral reef, just off the coast of Okinawa, Japan, found 70% of the coral dead due to an inflow of red soil into the ocean, seawater contamination, as well as higher water temperatures. The water temperature was 1-2 °C (1.8-3.6°F), higher than usual.[358]

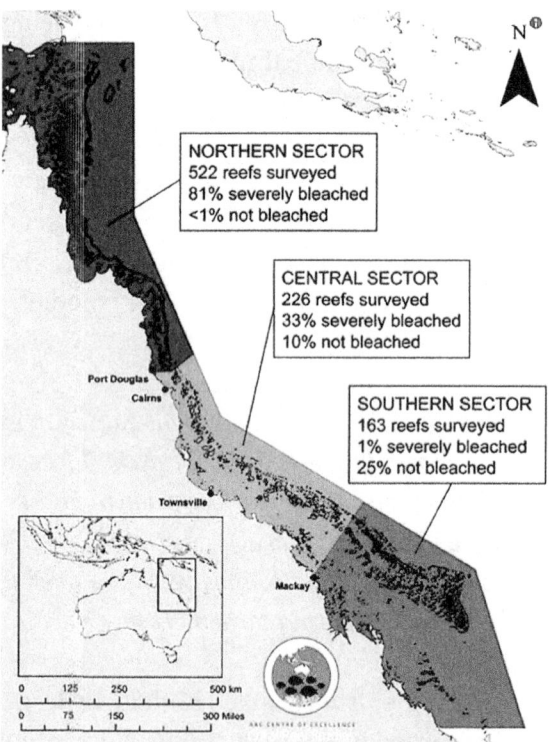

Results of aerial surveys of bleaching of the Great Barrier Reef.

A 2018 government survey of Sekisei Lagoon found that only 1.4% of the coral reef was healthy.[359] At the Dongsha Atoll in the South China Sea, 40% of coral has died. According to experts, nothing this severe has happened on Dongsha Atoll for at least 40 years. Anne Cohen of Woods Hole Oceanographic Institution in Massachusetts said,

> *The 2015/2016 El Nino was devastating for reef systems in other parts of the world as well, including Dongsha Atoll and reefs in the central Pacific, where some of the most pristine coral reefs are located and of course, the US Pacific Remote Marine National Monument. We observed devastating bleaching in that area as well.*[360]

In 2016, two-thirds of corals in the northern sector of the GBR died after severe bleaching from unusually warm waters. This was followed by another major bleaching event in 2017. With these back to back bleaching events, nine hundred miles of the GBR experienced severe bleaching at some point during those two years.[361] Terry Hughes, director at ARC Centre of Excellence for Coral Reef Studies at James Cook University in Queensland, noted that,

> *Last year [2016] we lost 67 percent on average of the corals in the northern 700 kilometers (430 miles) of the barrier reef, between March and October. That's obviously an enormous loss over two-thirds of the Great Barrier Reef. I wouldn't say the barrier reef is dying. But clearly, we're measuring serious losses here. And the reason it's happening is global warming.*[362]

The 2016 bleaching was tied with El Niño. However, the lack of El Niño in 2017 indicates that warming ocean temperatures from global warming is likely the primary reason for the bleaching. The GBR has experienced bleaching events in 1998, 2002, 2016, and in 2017. Typically, these events have been localized in scale, and if bleaching is mild, the coral can survive long enough to reacquire new algal companions. Bleaching in itself is not something new, but the observation of mass coral bleaching on a vast scale certainly appears to be, and it represents a whole new level of coral reef decline. An analysis published in the journal Nature noted reefs that suffered this severe heat exposure had undergone a "catastrophic die-off" and an "unprecedented ecological collapse" in the upper third of the GBR.[363]

The die-off of corals drove a radical shift in the composition and functional traits of coral assemblages on hundreds of individual reefs, transforming large swaths of the Great Barrier Reef from mature and diverse assemblages to a highly altered, degraded system.[364]

Unfortunately, it takes over a decade for reefs to recover from a bleaching event. With two consecutive years of bleaching, the chances of significant recovery become severely limited. Kim Cobb, a coral reef expert at Georgia Tech, stated that this event "may very well mark the beginning of the end for many iconic reefs as we know them today, at least for the foreseeable future."[365] Some ecologists and marine biologists fear that the GBR is in terminal decline.[366]

"I don't think the Great Barrier Reef will ever again be as great as it used to be — at least not in our lifetimes," said C. Mark Eakin with the National Oceanic and Atmospheric Administration.[367]

In 2014 the GBR long-term outlook was rated as "poor" by Australia's Great Barrier Reef Marine Park Authority. In 2019, that rating was downgraded to "very poor."[368]

Plastic curse

The massive amount of plastics polluting the ocean is another severe threat that corals are facing. With their irregular and pitted surface, plastic particles absorb pollutants such as PCBs (polychlorinated biphenyls) to levels up to 1,000,000 times the surrounding seawater. They can also be home to bacteria and other microbes. Scientists examined 125,000 corals across the Asia-Pacific region and found 89% of those contaminated by plastic were suffering from a disease. On plastic-free reefs, only 4% of the corals were diseased.

"Corals are animals just like me and you – they become wounded and then infected," said Joleah Lamb, at Cornell University in the US, *"Plastics are ideal vessels for microorganisms, with pits and pores,*

so it's like cutting yourself with a really dirty knife. It's like getting gangrene on your toe and watching it eat your body. There's not much you can do to stop it. If a piece of plastic happens to entangle on a coral, it [the coral] has a pretty bad chance of survival."[369]

Toxic sunscreen

A toxic ingredient commonly found in sunscreen, oxybenzone, is also killing corals around the world. Oxybenzone has been found in bird eggs, fish, coral, humans, and other marine mammals. It is a widespread environmental contaminant found in streams, rivers, lakes, and marine environments from the Arctic Circle to beaches and equatorial coral reefs.[370] Oxybenzone breaks coral down by leaching it of nutrients and eventually turning it white. Not only does it harm coral, but it has also been shown to disrupt the development of fish and other marine life.[371] Each year between 6,000 and 14,000 tons of mostly oxybenzone-laden sunscreen lotion ends up in coral reefs worldwide.[372]

Coral reef contamination of oxybenzone in the U.S. Virgin Islands ranged from 75 μg/L [micrograms/liter] to 1.4 mg/L [milligrams/liter], whereas Hawaiian sites were contaminated between 0.8 and 19.2 μg/L.[373]

Sunscreens are also often made up of nanoparticles that allow the products to appear transparent and smooth when applied on the skin[374] but can disrupt coral reproduction and growth cycles, ultimately leading to bleaching.[375]

Oxybenzone, for example, is toxic in four different ways: it causes damage to the DNA that may lead to cancer and developmental abnormalities, it is an endocrine disruptor, it causes deformations in juvenile corals, and, lastly, it leads to bleaching.[376]

Oxybenzone is found in more than 3,500 sunscreen products worldwide, as well as lipstick, mascara, shampoo, and many other products.[377]

Mystery disease

Since 2014, the Florida coast reefs have been suffering from an unknown condition that appears as white patches that spread until they consume and kill the entire coral.[378] The disease seems to have primarily affected brain and star coral species. The cause of the condition unknown to scientists has been described as "the largest and longest reef infection on record." As a result, almost all brain and star corals are dead in some areas, which scientist William Precht noted is "essentially equivalent to a local extinction."[379]

Coral reef demise

The majority of coral reefs worldwide are threatened by human activities, and many show signs of degradation. However, while some coral reefs have not been harmed, less than half of the world's reefs are regarded as being relatively healthy and not under any immediate threat of destruction.[380]

There are certain well-established no-take zones where some reefs are in near pristine condition. These have high fish diversity and well-developed coral communities with large and old corals. Yet, the ever-increasing perils from pollution, disease, and coral bleaching indicate that coral reef ecosystems will not survive for more than a few decades unless they are promptly and massively protected from human activity fallout. Although there has been some global attention to the ongoing degradation of coral reefs due to overfishing, pollution, and climate change, the reefs' deterioration continues.

> *Without significant changes to the rates of disturbance and coral growth, coral cover in the central and southern regions of the GBR is likely to decline to 5–10% by 2022.*[381]

Coral reefs are deteriorating not from a single cause but from multiple factors. With the focus almost singularly on global warming as the cause of the decline of coral reefs worldwide, it obscures the other numerous manmade factors that have caused their immense devastation over many decades. Reefs are exposed to a combination of stresses, including destructive fishing practices; overfishing or loss of herbivorous fish and other grazing organisms; increased discharge from the land of sediment, nutrients, and pesticides; coral predator outbreaks; increased bleaching associated with global climate change; chemical assaults; and increased incidence of and severity of coral diseases magnified by oceanic plastic pollution.

The world has lost roughly half its coral reefs in the last 30 years, and even if global warming stopped now, scientists still expect that more than 90% of corals will die by 2050. Biologist Ove Hoegh-Guldberg, director of the Global Change Institute at Australia's University of Queensland, noted,

> *Whether you're living in North America or Europe or Australia, you should be concerned. This is not just some distant dive destination, a holiday destination. This is the fabric of the ecosystem that supports us. You couldn't be more dumb... to erode the very thing that life depends on – the ecosystem – and hope that you'll get away with it.*[382]

The continued loss of coral reefs is much more than losing a beautiful work of nature. Coral reefs provide food and resources for over 500 million people in 94 countries and territories. The decline of the structurally complex corals means the reef will be much flatter, with a decreased habitat for the hundreds of thousands of species that live in the coral reefs.[383] Reefs also create a shoreline barrier to guard against incoming storms and mitigate the damage done by surging seas.[384] The coral reefs are experiencing nothing less than a full-blown ecological catastrophe. Over the centuries, reefs have suffered massive deterioration due to human activity and have, in some cases, been entirely eradicated.

As reefs, which are a storehouse of biodiversity, continue to degrade, they will be supplanted with a seaweed-dominated ocean bottom. There will be fewer and fewer fish but more jellyfish.[385] Almost no large animals will survive in the low water quality that ensues as the dead or dying large corals are replaced by soft corals and seaweed. The oceans will likely eventually turn into little more than rubble, seaweed, and slime.[386]

Like the canary in the coal mine, the environmentally sensitive reefs are a warning for the health of the rest of the oceans and the planet. We ignore this aquatic canary at our own peril.

What you can do!

Coral reefs are in danger of vanishing due to a variety of different human activities. Here are some simple steps that you can take at a personal level to make a difference.

Buy local – According to the World Trade Organization, global trade has exploded over the last 30 years, with world merchandise trade quadrupling between 1980 and 2011.[387] Increased global trade provides economic benefits and allows consumers to obtain products from all over the world, but it also creates substantial environmental harm, including the destruction of coral reefs. Read the chapter "Consumerism Gone Mad" to find out more.

Move away from wild-caught seafood – Overfishing, destructive fishing practices, and plastic fishing gear are all endangering coral reefs. Read the chapter "Fished Out" to find out more.

Move away from using plastics – Plastics can harbor microorganisms that harm and kill corals. Read the chapter "Plastic Oceans" to find out more.

Move away from industrial agriculture – Fertilizers and pesticides from agriculture have greatly affected coral reefs. Read the chapter "Dead Zones" to find out more.

Decrease your carbon footprint – According to the Florida Keys National Marine Sanctuary, "One of the most important threats facing coral reefs on a global scale is a big one: climate change. Scientists agree that climate change is real, and this spells real trouble for the world's coral reefs." Read the chapter "Acid Seas" to find out more.

Rethink your use of sunscreen – Even if you don't go in the water after applying sunscreen, it can go down the drain when you shower. Aerosol sunscreens can spray large amounts of sunscreen into the environment, where it eventually gets washed into the oceans, potentially harming corals.[388] Not only are they harming corals, oxybenzone "has been found in 96% of urine samples in the US and that several UV-filters have been found in 85% of Swiss breast milk samples."[389]

Oxybenzone acts as a barrier to UV light and is entirely replaceable by other, less marine-toxic ingredients. Use "reef-safe" sunscreen free of oxybenzone and formulas that use "non-nano" particles. Also, consider using hats, shirts, and other clothing to reduce your need for sunscreens.

Note that vitamin D, an important hormone, is produced in the skin from solar radiation.[390] That's why vitamin D is often referred to as the sunshine vitamin. Vitamin D deficiency is attributed to low sunlight exposure (specifically Ultraviolet-B or UVB radiation), which is required to induce vitamin D synthesis in the skin. Vitamin D deficiency is very prevalent in the United States and worldwide. Approximately 30% of apparently healthy middle-aged and elderly adults are deficient.[391] A study in 2018 showed that most patients lacked the metabolized endproduct of vitamin D, which is abbreviated as 25(OH)D, and over 40% were severely or very severely deficient.[392] Your body's vitamin D production from healthy amounts of sunlight exposure is something essential to consider.

Coral Reef Carnage

DEAD ZONES

All of us have in our veins the exact same percentage of salt in our blood that exists in the ocean, and, therefore, we have salt in our blood, in our sweat, in our tears. We are tied to the ocean. And when we go back to the sea - whether it is to sail or to watch it - we are going back from whence we came.
– John F. Kennedy

It is a curious situation that the sea, from which life first arose should now be threatened by the activities of one form of that life. But the sea, though changed in a sinister way, will continue to exist; the threat is rather to life itself.
– Rachel Carson, The Sea Around Us

If you can't breathe, nothing else matters.
– American Lung Association

Suffocating discovery

In the fall of 2006, a low-oxygen ocean zone appeared off the coast of Oregon. As the oxygen levels dropped, the fish that could leave the area escaped. Some fish weren't as lucky and slowly suffocated and died. Less mobile sea creatures such as crabs, sea stars, and sea worms had no chance at all and died in huge numbers. In partnership with others, Oregon State University (OSU) deployed an underwater vehicle to investigate the extent of what was going on under the Pacific Ocean's surface just off of Cape Perpetua.[393]

Ordinarily, the area would be teeming with rockfish, lingcod, and kelp greenling. The seafloor also typically crawls with large populations of Dungeness crabs, sea stars, sea anemones, and other marine life. Instead, there were no fish, and the bottom was covered with massive amounts of dead ocean creatures' remains.

> "We saw a crab graveyard and no fish the entire day," said Jane Lubchenco, the Valley Professor of Marine Biology at OSU. "Thousands and thousands of dead crab and molts were littering the ocean floor, many sea stars were dead, and the fish have either left the area or have died and been washed away. Seeing so much carnage on the video screens was shocking and depressing."[394]

The university survey found that the water's oxygen level was 10 to 30 times lower than usual. This was the fifth year in a row for this reduced oxygen water area off the Oregon coast, but this was the most significant drop to date, and it caused the suffocation of marine life on a massive scale. The immense growth of microscopic plants, called phytoplankton, contributed to this low oxygen region and turned parts of the ocean into a dirty chocolate brown.

As they pulled up their crab traps, fishermen found silver dollar-sized octopuses inching their way up the lines toward the buoys floating on the surface. Dennis Krulich, a longtime fisherman in Newport, Oregon, later realized that these babies were coming up from oxygen-depleted waters that hover near the seafloor to survive. Krulich noted that in 30 years of crabbing, he had never seen anything like it before.[395]

During the same year, severely low oxygen levels were also detected across the shallow waters off the coast of California. Surveys of those waters found the complete absence of all fish that would normally inhabit rocky reefs in that area and a "near-complete mortality" of bottom invertebrates such as crabs.[396] Five decades of available records on oxygen levels show little evidence of this low oxygen level phenomenon before the year 2000. In 2017 the worst low oxygen zone in a decade was found off the Oregon coast.

> *When the oxygen levels were high, the crabs were happy, and then the oxygen started to decline, and then the crabs started to slow down and not move so much and over time they died. They suffocated on the sea floor, said OSU marine ecologist Francis Chan.*[397]

Industrial Agriculture

Every summer, enormous amounts of industrial fertilizer and pesticides run off Midwest land, ending up in the Mississippi River. These chemicals flow down the river and into the Gulf of Mexico. The discharging freshwater from the river flows over the top of the denser and heavier salty gulf waters. This freshwater on the surface forms a barrier, preventing oxygen from penetrating the water's bottom layers. The nitrogen and phosphorus in the fertilizer stimulate the rapid growth of phytoplankton that naturally grows in the sea's top layer. When the phytoplanktons consume all the nutrients, they die and sink to the bottom, where they are decomposed by bacteria. As the bacteria do their work, they use up much of the remaining oxygen in that bottom layer.

Dissolved oxygen of 4-6 parts per million (PPM) in water is considered normal. Where oxygen is below 4 PPM, fish try to relocate to waters where they can breathe more easily. Below 2 PPM is called hypoxia, which is a deficiency in the amount of oxygen reaching most aquatic animals' tissues. Anoxia means there is no oxygen at all. The excessive nutrients that cause these large phytoplankton blooms and the resulting loss of oxygen from the water is a process known as eutrophication.

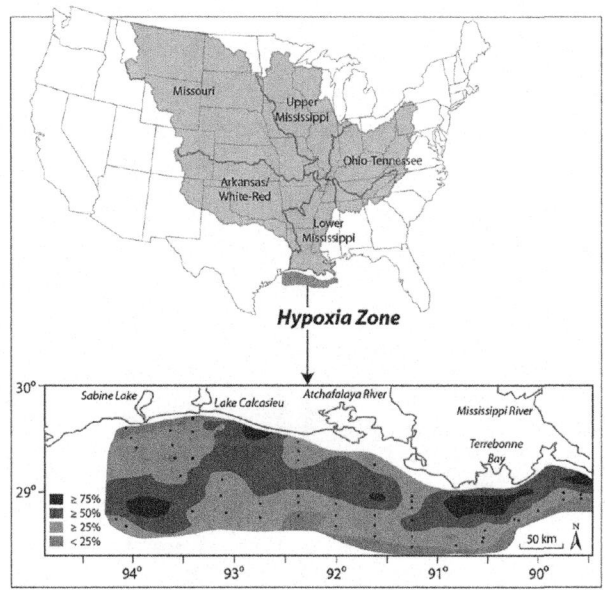

Gulf of Mexico drainage basin. Inset: Frequency of midsummer bottom-water hypoxia (2 mg/L O_2) off the coast of Louisiana and Texas for 60 to 80 stations (small dots) sampled during the summer from 1985 to 2008.

In 2002, the Gulf of Mexico's hypoxic zone reached 21,973 kilometers (8,484 square miles), bigger than Massachusetts.[398] Each year, the region's size fluctuates based on several different factors, including weather and rainfall, which determine how much water pollutants make it to the Gulf. In 2015 the gulf hypoxia zone was 16,767 square kilometers (6,474 square miles), or roughly the size of Connecticut and Rhode Island combined.[399] In 2017 the area expanded to at least 22,730 square kilometers (8,776 square miles), approximately the size of New Jersey. It was the largest recorded since tracking began in 1985.[400]

Low oxygen levels radically alter the ecology of coastal systems. Fish and mobile invertebrates (like shrimp and crabs) can migrate out of hypoxic areas. Animals that are slow-moving or attached to the bottom (like clams, worms, and starfish) cannot escape from the dangers of hypoxic water and die with extended exposure. These extremely-low oxygen ocean areas have been given the ominous, although the appropriate name of "dead zones."

> *In the northern Gulf of Mexico, the occurrence and extent of the dead zone are tightly coupled with freshwater discharge from the Mississippi River, which delivers large quantities of nutrients from U.S. agricultural activities. During years with low river flow, the area of hypoxia shrinks to < 5000 km2 [1,930 mi2], only to increase to > 15,000 km2 [> 5,790 mi2] when river flow is high.*[401]

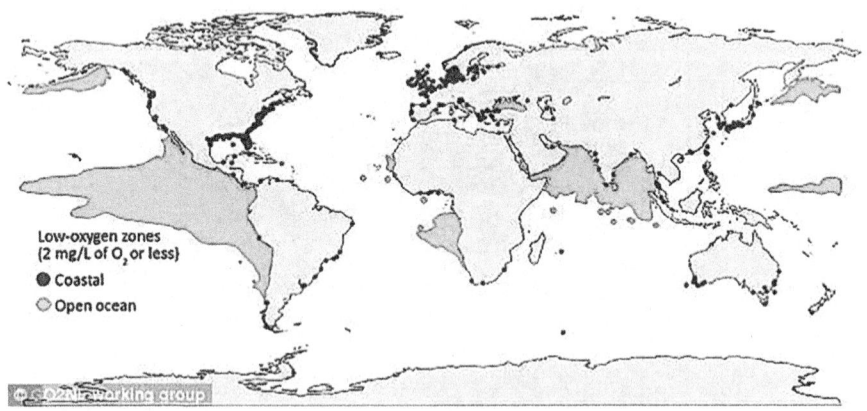

Dead zones are on the rise globally, with 95,000 square miles affected across the world's oceans. Pictured is a map showing coastal (black) and open ocean (grey) dead zones worldwide.

What is extremely worrisome is that dead zones have been appearing worldwide at an increasing rate. Nutrient over-enrichment is the leading cause of these dead zones, and nutrient-fed hypoxia is now widely considered a significant threat to the health of aquatic ecosystems. Before 1970, there were only scattered reports of coastal hypoxic zones in Europe and North America, with only 49 dead zones identified in the 1960s.[402] However, by 1995, there were over 195 cases reported worldwide. This number doubled to just over 400 zones by 2008. In 2011 an additional 115 sites in the Baltic Sea were added to the list.[403] At least 500 dead zones have now been reported near coasts. However, a lack of monitoring in many regions means the actual number may be much higher.[404]

These worldwide dead zones cover an area four times bigger than they were in 1950. The largest coastal dead zone is in the partially enclosed Baltic Sea in Europe, which often covers more than 51,800 square kilometers (20,000 square miles).[405]

The observed increase in these dead zones has lagged about 10 years behind the increased use of industrially manufactured nitrogen-based fertilizers that began in the late 1940s, with explosive growth in the 1960s and 1970s. As it is known, the Green revolution, which started between 1950 and the late 1960s, significantly increased agricultural production with the use of high-yielding varieties of grains, especially wheat and rice. However, these new grains require large amounts of chemical fertilizers and pesticides to produce their high yields.

Since the start of the Green Revolution, the number of dead zones has approximately doubled each decade.[406] The leading causes of these dead zones are linked to nutrient use (particularly nitrogen and phosphorus in fertilizers) in agricultural production, wastewater from human populations and industrial sources, and fossil fuel burning.

Agricultural and urbanization activities are the primary drivers of nitrogen pollution in coastal waters. The green revolution has led to the use of synthetic nitrogen fertilizers, creating reactive nitrogen at a rate four times greater than fossil fuel combustion.[407] An estimated 118 teragrams (118 million metric tons or 130 million tons) of reactive nitrogen are deposited globally from fertilizer, the leading source of this increase.

> Reactive nitrogen is a term used for a variety of nitrogen compounds that support growth directly or indirectly.

In the United States, fossil fuel combustion emits approximately 6.9 million metric tons of nitrogen per year to the environment, which is about 60% of the country's rate of nitrogen fertilizer use. Most of the nitrogen emitted from fossil fuel is deposited back to the ground through rain,

> A teragram is equal to 1 million metric tons.

with a significant contribution to nutrient pollution in coastal waters. Approximately half of the fossil fuel nitrogen emission comes from mobile sources, including automobiles, buses, trucks, and off-road vehicles. Electric power generation produces about 42% of the nitrogen.

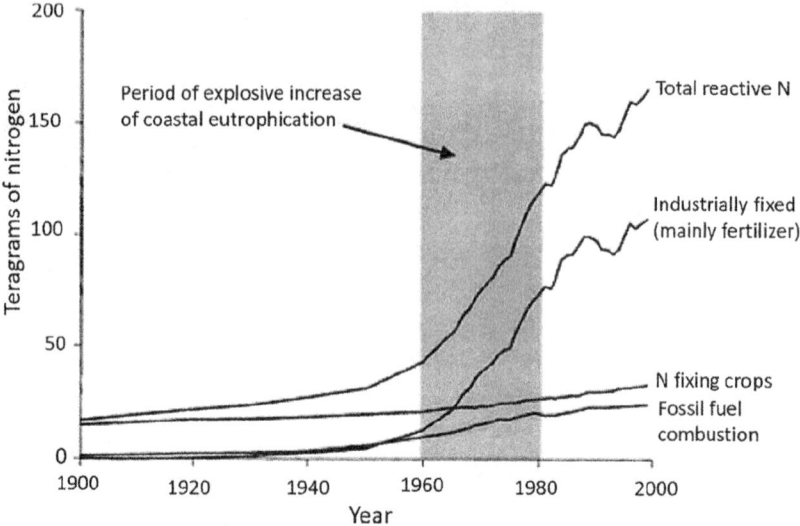

Period of the explosive increase in coastal eutrophication in relation to global additions of nitrogen.

Triple the amount of nitrogen and phosphorus is being deposited into the world's oceans today than was in pre-industrial times.[408] Each year, more than one million tons (the weight of over 600,000 cars) of nitrogen flows through the Mississippi River system and into the Gulf of Mexico.

> *Forty-one percent of the continental United States (1.2 million square miles) drains into the Mississippi River and then out to the Gulf of Mexico. The majority of the land in Mississippi's watershed is farm land. Seventy percent of nutrient loads that cause hypoxia are a result of agricultural runoff caused by rain washing fertilizer off of the land and into streams and rivers. Additionally, 12 million people live in urban areas that border the Mississippi, and these areas constantly discharge treated sewage into rivers.[409]*

In pre-industrial times, most coastal and offshore ecosystems rarely became hypoxic. A combination of natural processes and possible human activities such as land clearing may have resulted in the development of algal blooms and associated low-oxygen bottom water on the Louisiana shelf before 1950. However, analysis indicates that low-oxygen events have been more extreme in the last few decades than in the previous 150 years.[410]

Meat and biofuels

According to a 2017 report by the environmental group Mighty, a significant factor in the generation of the Gulf of Mexico dead zone is American's voracious appetite for meat.[411] People eat roughly 87 kilograms (193 pounds) of beef, pork, and chicken a year in the United States.[412] By 2025, total meat consumption is projected to increase to nearly 99 kilograms (219 pounds) per person per year.[413]

To satisfy this ever-increasing demand for meat, a highly industrialized and centralized factory farm system has been developed to grow the soy and corn needed to feed millions of farm animals. Vast tracts of native grassland in the Midwest have been converted into these crops predominantly used for livestock. Toxins from manure and fertilizers that wash into waterways exacerbate the harmful algal blooms that create oxygen-deprived stretches down in the Gulf. Tyson Foods is identified by the report as a "dominant" influence in pollution due to its market strength in chicken, beef, and pork.

> *To keep up with orders from companies like McDonald's and Walmart, Tyson slaughters 125,000 head of cattle, 35 million chickens, and 415,000 hogs every week—nearly equal to the human population of California. To raise all of this meat, Tyson requires an estimated five million acres of corn—greater than the size of New Jersey—each year, not to mention other feed like soybeans, which it buys from the major feed suppliers.*[414]

One-third of the planet's arable land is occupied by livestock-feed crop cultivation.[415] In the United States, more than 90 million acres (over 140,000 square miles or about the size of the state of Montana) are used to grow corn.[416] About the same amount of land is used to grow soy.[417] Just over 70% of the soybeans[418] and nearly half of the corn[419] grown are used as animal feed.

In addition to the fertilizers used to raise the feed, the waste generated by the massive number of animals is contributing to the downstream problem. Tyson alone produced 55 million tons (the weight of over 33,000,000 cars) of manure in 2016. According to the Environmental Protection Agency (EPA), Tyson also dumped 104 million pounds (the weight of over 31,000 cars) of pollutants directly into waterways from 2010 to 2014, making it the second biggest polluter in the United States just after AK Steel Holding Corporation.[420]

The raw materials of the global food system are controlled by a tiny number of large multinational corporations that don't have much of a public reputation. Known as the ABCD of food, ADM, Bunge, Cargill, and (Louis) Dreyfus account for between 75% and 90% of the global grain trade.[421] These companies don't sell directly to individual consumers but supply the feed for the animals that end up in restaurants and grocery stores. Yet, these companies bear no responsibility for the soil erosion and run-off from enormous portions of America's crop fields washing into the waterways, causing environmental havoc, including dead zones.[422]

The irony is that the human food supplied by the oceans is, in part, jeopardized by modern industrial land-based agriculture. Around 75,000 metric tons (83,000 tons) of sea life are lost in the Chesapeake Bay dead zone each year. The Gulf of Mexico's dead zone causes the annual loss of 212,000 metric tons (235,000 tons) of food.[423] A recent study shows that one of the most significant contributors to the Gulf of Mexico's dead zone, nitrogen, which flows downriver from Midwest farms, has been responsible for up to $2.4 billion in damages to Gulf fish stocks and their habitat every year for more than 30 years.[424]

Increasingly, parts of the United States are being plowed up to raise corn and soy, primarily to feed livestock and create biofuels. China has become the world's biggest importer of soybeans, with 1.171 billion bushels exported from the United States in 2014-2015.[425] China's soybean imports are driven by increased demand for meat and expanding livestock production, resulting from China's rapid economic growth and rising incomes since the early 2000s.[426]

The biofuel industry is also decreasing prairies and replacing them with fertilized fields of corn and soy. A 2015 study showed that just over four years, 7 million acres (10,900 square miles or larger than the state of Massachusetts) of new land were converted in the United States for crops to create biofuels.[427] Nearly 30% of the corn grown in the United States is used to produce ethanol.[428]

Heavy rainfall, soil erosion, and the destruction of wetlands and grasslands that absorb runoff, are all factors that allow chemical fertilizers and manure applied to these fields to eventually end up polluting surrounding waterways. Because crops only take up on average half the nitrogen applied to the fields, the remainder washes into streams and finally into the Mississippi River.[429] The amount of nitrogen that flows down the Mississippi River and into the Gulf of Mexico has averaged about 1.26 million metric tons per year (1.39 million tons or the weight of over 842,000 cars). The vast dead zone in the Gulf of Mexico and the 500 or so other dead zones in the world are mainly due to the animal agriculture and biofuel industries.

According to a team of scientists, industrial agriculture has destroyed a staggering amount of America's Midwestern prairie's rich soil.[430] Their research shows that the most fertile topsoil has entirely disappeared from a third of all the land devoted to growing crops across the upper Midwest.

> Some of their colleagues, however, remain skeptical about the methods that produced this result. Even the study's critics, though, agree that topsoil is endangered. "To me, it's not important whether it's exactly a third," says Anna Cates, Minnesota's state soil health spe-

cialist. "Maybe it's twenty percent, maybe it's forty percent. There's a lot of topsoil gone from the hills." Cates says that farmers already know that these eroded hilltops are less productive, and many of them are looking for solutions. "We're essentially trying to make up for many years of fairly thoughtless practices," she says.[431]

Warming waters

While many dead zones, such as the one in the Gulf of Mexico, are primarily caused by agricultural runoff and other modern human-related activities, the dead zone off Oregon's coast was different. There wasn't enough farming and associated fertilizer runoff in the area to explain it. There had to be other factors in the case of Oregon's massive sea life die-off in 2006 and 2017.

Scientists hypothesize that climate change may be a contributing factor in the formation of dead zones. Increasing land-sea temperature differences can drive an increase in coastal winds.[432] The stronger winds produce a longer coastal upwelling, which causes an increase in the number of phytoplankton. The excess of phytoplankton, which isn't eaten, dies and drifts down to the seafloor, rotting and using up oxygen in the process.[433] A 2014 analysis in the journal Science indicates a trend of windier conditions over the last 60 years off the west coast of North America, the coast of Peru and Chile, and the west coast of southern Africa.[434] Researchers can't definitively say that climate change is to blame. Yet, they noted finding a consistent pattern across several parts of the planet provides a strong indicator that it is a factor. There are not only shifts in coastal wind intensity but there are also other disturbing changes taking place.

Surface ocean waters down to about 100 meters (325 feet) generally have oxygen concentrations close to equilibrium with the Earth's atmosphere. Oxygen enters the ocean in the surface water through contact with the atmosphere and aquatic plants' photosynthesis. The mixing of surface waters by wind and waves increases the rate of oxygen absorbed from the air into the water. Farther out to sea, beyond the continental shelf, water at a depth of roughly 600–1,200 meters (1,970-3,940 feet) is permanently oxygen-deprived. Called an oxygen minimum zone (OMZ), this layer is a regular feature in many parts of the ocean, which are too deep to mix with the oxygen-rich surface waters. Oxygen mainly enters these deeper parts of the ocean by the slow motion of water currents.

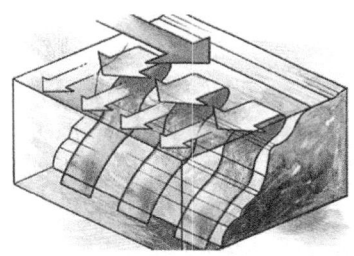

Usually, in the spring, occasional periods of northerly wind blow surface waters offshore, allowing cool waters, rich in nutrients but low in oxygen, to upwell from deeper, offshore layers. That nutrient-rich upwelling is what makes Oregon's fisheries so productive. Researchers have found that over the last 30 years, the oxygen level of the OMZ off the Oregon coast is steadily losing oxygen. These are the waters that move up towards the shallower waters in the spring and summer. Because of this decreasing oxygen content, the chances of seeing a hypoxic event off the Oregon coast increased from 10% to 60%.[435] This, in part, helped explain the increasing low oxygen water die-offs that had been occurring where they hadn't before. But why was the oxygen level dropping in this naturally occurring OMZ?

Illustration of the coastal upwelling process, in which winds blowing along the shore causes nutrient-poor surface waters to be replaced with nutrient-rich, cold water from deep in the ocean.

Because the oceans are vast and complex, no one is certain why OMZs have increased in size. However, climate models have suggested that ocean oxygen concentrations have and will continue to decline in the future because of the warming of ocean waters.

> ...the ocean likely will lose a substantial amount of oxygen in the coming decades and centuries in response to global warming, a process termed "ocean deoxygenation." Global models suggest a loss of between 1 to 7% for this century for a business as usual scenario.[436]

If ocean circulation slows over the coming decades, as it is theorized to do, there will be less ocean oxygen mixing, causing many OMZs to continue increasing in size. This raises the risk that upwelling currents, which carry low oxygen waters to shore areas, will increase dead zones like those that occur off the Oregon coast. While parts of the ocean have shown some rise in oxygen levels, other areas have decreased particularly in the tropics.

> ...results show expanding low-oxygen-minimum zones in all three tropical oceans between the time periods 1960–1974 and 1990–2008. The low oxygen zones expand both horizontally and vertically.[437]

Although not definitively proven, it appears that warming oceans are at least in part causing the increase in these low oxygen zones. If true, this can expand naturally occurring OMZs into shallower coastal waters, further damaging fisheries.

> ...our analysis strongly supports the notion that if anthropogenic [human caused] climate change continues to evolve unabated, the ocean is bound to deoxygenate with poorly understood consequences for marine life. This is a source of concern, especially when considering that ocean deoxygenation is not occurring in isolation, but together with ocean acidification and ocean warming.[438]

> Our investigation of the OMZ in the tropical eastern North Atlantic reveals significant deoxygenation in the core of the OMZ... if continued, the OMZ would go anoxic [no oxygen] in less than 100 years... links to global warming and possible changes in the hydrologic cycle as the causes for the long-term observed temperature and salinity changes in the Atlantic.[439]

Large-scale warming causes the oceans to absorb more heat, resulting in less dissolved oxygen at the surface since a smaller amount of oxygen dissolves in warmer water. Increased layering between the upper oxygen-rich layers of the ocean and the lower oxygen-poor layers also occurs, which reduces the mixing and transfer of oxygen from the atmosphere into the deeper waters.[440] The decrease in oxygen movement to lower layers happens because as the water heats up, it expands, becoming lighter than the water below it and less likely to sink.

> "The natural thing to expect is that as the ocean gets warmer, circulation will slow down and get more sluggish, and the waters going into the deep ocean will hang around longer," says Curtis Deutsch, a chemical oceanography professor at the University of Washington, in Seattle. "And indeed, oxygen seems to be declining."[441]

Oxygen must be supplied through downward transport from the oxygen-rich surface waters to the deeper ocean to make them habitable. A reduction in oxygen levels at the surface and reduced downward transport can change oxygen levels over time.[442] This appears to have, at least in part, caused the observed expansion of OMZ areas.

Because there is a lot of variability in the natural warming and cooling of the ocean, oxygen concentrations are continually changing, making it difficult to detect a warming climate's effects on ocean oxygen levels. However, a 2016 study indicates that forced ocean deoxygenation due to climate change should already be evident in the southern Indian Ocean and parts of the eastern tropical Pacific and Atlantic oceans.[443] This decline in oxygen due to climate warming appears to be manifesting in the Indian Ocean, corroborating this study. At the end of 2016, a multinational team of scientists reported that a large dead zone had appeared in the Bay of Bengal. This zone, which seems to be increasing in size, already spans some 60,000 square kilometers (23,160 square miles), or roughly the combined areas of New Jersey, Maryland, and Delaware.

The bay has also been under assault, with several large rivers emptying their contents of untreated sewage, plastic, and aquaculture, as well as industrial and agricultural waste into it for years.[444]

> *The impact of this pollution could be catastrophic. The high load of organic pollutants, coupled with the diminution of the fish that keep them in control, could lead to massive plankton blooms, further reducing the water's oxygen content.*[445]

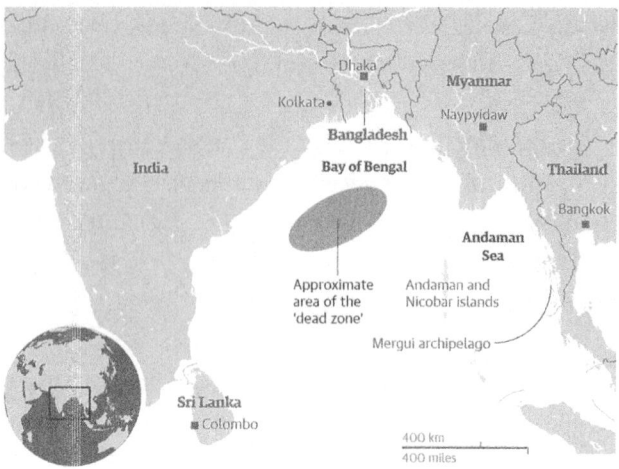

The dead zone of the Bay of Bengal is now at a point where a further reduction in its oxygen content could strip the water of nitrogen, a key nutrient.

Approximately 200 million people living along the Bay of Bengal are wholly or partly reliant on fishing. If the Bay's fisheries collapse, it would be nothing short of catastrophic. The scientists who identified the Bay's dead zone warn that this stretch of ocean is approaching a tipping point. If the last of the oxygen is removed due to continued nutrient input from rivers and the effects of climate change, it could have severe consequences for the planet's oceans and the global nitrogen cycle.[446] A 2017 report presented to the United Nations by marine scientists at the University of Oxford highlighted the Bay's highly precarious situation.

> *Nutrients like nitrate act as a fertiliser to algae, stimulating bacteria growth, which competes with fish and marine organisms for oxygen. If oxygen levels in the Bay of Bengal decrease any further, the area is at risk of flipping to a 'no oxygen' status. This would result in the formation of new bacteria that then remove nitrates from the water, destabilising the bay's ocean ecosystem. The denitrified water could then be carried away by ocean currents and reduce productivity elsewhere.*[447]

Expanding dead zones

Another massive dead zone exists in the Gulf of Oman, just off the coasts of Iran, Pakistan, Oman, and the United Arab Emirates. Agricultural runoff and sewage have created a dire environmental problem, with increasing ocean temperatures making the situation even worse. Research led by Doctor Bastien Queste from UEA's School of Environmental Sciences in the United Kingdom found a dead zone almost entirely devoid of oxygen greater than the size of Scotland.[448] At 102,515 square kilometers (39,500 square miles),[449] it is about 4.5 times larger than the Gulf of Mexico dead zone's maximum recorded size. Dr. Queste commented on what he and his team found,

> *Our research shows that the situation is actually worse than feared -- and that the area of dead zone is vast and growing. The ocean is suffocating.*[450]

Inexpensive fertilizers, used on crops to feed rising populations in a hungry world, have partly led countries to apply fertilizer with reckless abandon. In China, nitrogen-use efficiency has dropped from more than 60% in 1961 to just 25%. The result is that the Chinese ecosystems are under siege.[451]

> *Nitrogen kills fish in huge numbers from the Yellow River in the north to the Pearl River in the south. Algal blooms are reported in a third of the country's lakes. Massive "red tides" of toxic algae spread from river estuaries across the East China Sea.*[452]

Over the last 100 years, Florida has been significantly altered by substantial population growth, urbanization, and agriculture, becoming the largest sugarcane producer in the United States. Massive changes in land use and run-off from fertilizers contribute to algae blooms along the coasts.[453] In 2018 a dead zone along about 150 miles of Florida's Gulf Coast resulted in over 260 tons of marine life washing up onto Florida's white, sandy beaches.[454] While this "toxic algae bloom" is not entirely new, scientists believe it may be getting worse after decades of unchecked development, water mismanagement, and a changing climate.[455]

> *Generations of sugar cane farming has altered the chemistry of Florida's biggest lake [Lake Okeechobee] and a vast system of dikes and dams built to "drain the swamp" and create a retirement wonderland has killed half of the Everglades and put the rest of this vital wetland on life support. In the wet season, Florida dumps massive amounts of Okeechobee's nutrient-rich water into the most delicate ecosystems, while in the dry season, that water is diverted to farms and cities.*[456]

Currently, hypoxia and anoxia are among the most widespread harmful human-caused influences on estuarine and marine environments. They now rank with overfishing, habitat loss, and harmful algal blooms as major global environmental problems. The expansion of OMZs into shallow waters, as seen off Oregon's coast, may interact with nutrient-induced areas just off the coast to intensify and increase the total number of dead zones.[457] Keryn Gedan, a marine ecologist at the University of Maryland, cautioned,

> *If an area has low oxygen to begin with, then any change is going to have fairly significant ecological repercussions. We know that the shallow, coastal ocean is warming faster than the open ocean, especially in estuaries that are fairly sheltered. We're seeing numerous dead zones pop up all around the world, and that's going to become more common.*[458]

Researchers warn that it is likely that the open-ocean dead zone just off the West African coast may at some point flood the Cape Verde archipelago with low-oxygen water. If this happens, it would put severe stress on the coastal ecosystems and cause marine life die-offs.[459]

> *'Climate change will drive expansion of dead zones, and has likely contributed to the observed spread of dead zones over recent decades,' Altieri and Gedan write in a new paper that appears today in Global Change Biology... As temperatures increase, animals such as fish and crabs require more oxygen to survive. But with less oxygen available, 'that could quickly cause stress and mortality and, at*

larger scales, drive an ecosystem to collapse,' Altieri and Gedan warn.[460]

Around the globe, many OMZs are losing oxygen and expanding horizontally and vertically. According to Lisa Levin, a marine biologist at the Scripps Institution of Oceanography, deeper waters off the continental shelf have experienced a 20-30% decline in oxygen.

> *"Right now, it's still out of sight, out of mind," says Levin of the low-oxygen water. But the hypoxic conditions are creeping up the water column, rising by as much as 90 meters [300 feet] between 1984 and 2006 off the coast of Santa Barbara, California.*[461]

The dead zone off the coast of West Africa is the size of the continental United States. It has grown by 15% since 1960 and by 10% just since 1995. A recent study that measured this dead zone's oxygen level found the lowest oxygen levels ever recorded in the open Atlantic.

> *...the minimum levels of oxygen now measured are some 20 times lower than the previous minimum, making the dead zones nearly void of all oxygen and unsuitable for most marine animals.*[462]

Globally, low-oxygen areas have expanded by more than 4.5 million square kilometers (1.7 million square miles) in the past 50 years,[463] with the open oceans losing 77 billion metric tons (84 billion tons) or about 2% of oxygen,[464] equal to the weight of over 230,000 Empire State Buildings or around 50 billion cars.[465] As these areas have expanded vertically, they have pushed diving marine creatures such as sailfish, sharks, tuna, swordfish, and Pacific cod, as well as the smaller sardines, herring, shad, and mackerel that they eat, into ever narrower bands of oxygen-rich water closer to the surface.

Smothering the oceans

One key to reducing dead zones will be to keep fertilizers on the land and out of the sea. For agricultural systems, methods need to be developed that close the nutrient cycle from soil to crop and back to the ground. Cover crops can reduce nitrogen runoff by 30%, yet in 2016 Iowa farmers planted cover crops on less than 3% of the state's cropped land.[466] The resultant dead zones can be reversed when these fertilizers and other pollution sources are kept away from the oceans.

> *From 1973 to 1990, the hypoxic zone on the northwestern continental shelf of the Black Sea had expanded to 40,000 km2; however, since 1989, the loss of fertilizer subsidies from the former Soviet Union reduced nutrient loading by a factor of 2 to 4, with the result that, by 1995, the hypoxic zone had gone.[467]*

Since 1995, over $30 billion in federal conservation funding and voluntary limits on fertilizer use have failed to reduce the size of the Gulf of Mexico's dead zone, with it reaching a record size in 2017.[468] Demands on industrial fertilizer-intensive farming are only increasing and will only raise the amount of nitrogen and other chemicals dumped into waterways. The human production of nitrogen is already five times higher than it was 60 years ago.[469]

> *Farmers rely so heavily on fertilizer to boost yield and profits. It's cheap insurance for expensive seed, and billion-dollar industries have formed around its unbridled use—from fertilizer manufacturers to equipment makers and ag [agriculture] consultants.[470]*

The Energy Independence and Security Act (EISA) passed by the U.S. Congress set a required renewable fuel standard, requiring that at least 136 billion liters of biofuels be used by 2022. That mandate will increase the total nitrogen flux by 21% to more than 100%.[471] This governmental mandate for biofuels alone may raise nitrogen spilling into the Gulf of Mexico to a whopping 3 million metric tons (the weight of over 2,000,000 cars) per year. Total world fertilizer consumption by major crops is projected to increase from 166 million metric tons in 2007 to 263 million metric tons (the weight of over 175,000,000 cars) by 2050 – a 58% increase.[472] Less and less nitrogen poured onto fields is being incorporated into crops, with more than half washing from fields into rivers. Worldwide, farmers' nitrogen-use efficiency has decreased from more than 50% in 1961 to about 42% today.[473]

On top of already enormous agricultural-related nitrogen increases expected to be drained into the Gulf, a 2017 study indicates climate change may still raise this amount further. Climate change models project that both total and extreme precipitation in the Northeast and the United States' corn belt will substantially increase. This will raise the amount of fertilizer washed off fields, causing total nitrogen loading of the Mississippi River Basin to increase by 18% by the end of the century.[474]

In early 2018 nitrogen experts worldwide met to discuss what a nitrogen-soaked planet might look like. Many concluded that the amount of nitrogen being dumped into the environment should be reduced by 50% by 2050, or ecosystems will face epidemics of toxic tides, lifeless rivers, and dead oceans.[475] Unfortunately, rather than decreasing nitrogen use, all indicators show fertilizer use is on an ever-growing trajectory.

This increasing fertilizer overload, along with the growing use of fossil fuels and wastewater from human and industrial sources, will no doubt increase the number and size of dead zones. If warming-induced changes advance, observed decreasing ocean oxygen trends may very well continue, causing a further expansion in OMZs.

> *The multiplicative effects of oxygen stress on shelf systems are predicted to yield ecosystem-level changes. Increases in jellyfish blooms are likely be part of this response. Long-term consequences may include impacts on ocean CO_2 uptake and commercial fisheries.*[476]

Climate models have replicated some of the oxygen changes that have already occurred. They predict that the oxygen in the world's oceans will drop between 1 and 7% by the next century. According to Daniel Pauly, a fisheries biologist at the University of British Columbia, that could be enough to profoundly affect sea life.[477] Jellyfish, which can tolerate lower oxygen levels than fish, may thrive in the new conditions. Pauly and his colleagues predict that the drop in the ocean's oxygen and pH levels will together decrease the world's fish catch by 20 to 30% by 2050.

> *The links between human activity and local jellyfish blooms are strong. In the Black Sea, invasive comb jellies dumped from the ballast of tankers have spawned deliriously and destroyed the region's fishing industry. In the Sea of Japan, fertiliser run-off has left an oxygen-depleted sea where little other than jellies can thrive.*[478]

The present deoxygenation of the ocean is similar to an event that occurred 93-94 million years ago, known as the Oceanic Anoxic Event-2 (OEA-2).[479] OAE-2 developed over about 50,000 years and was believed to be caused by undersea volcanic activity triggering an extinction event that suffocated about 27% of marine invertebrates in Earth's oceans.[480] Dr. Sune Nielsen, of Woods Hole Oceanographic Institution, commented,

> *Our results show that marine deoxygenation rates prior to the ancient event were likely occurring over tens of thousands of years, and surprisingly similar to the two percent oxygen depletion trend we're seeing induced by anthropogenic activity over the last 50 years. We don't know if the ocean is headed toward another global anoxic event, but the trend is, of course, worrying.*[481]

Increasing fertilizer-intensive industrial agriculture, primarily driven by consumer appetite for animal protein and desire for biofuels, combined with a warming climate, will continue to be an ever-increasing oxygen-depleting assault on the ocean. These onslaughts will have dire consequences to life in the oceans, with growing threats to the world's food stocks. Professor Robert Diaz at the Virginia Institute of Marine Science noted that the speed of ocean suffocation already seen was breathtaking.[482]

No other variable of such ecological importance to coastal ecosystems has changed so drastically in such a short period of time from human activities as dissolved oxygen.[483]

For centuries, people have been exploiting the seas with minimal restriction, with proof of extensive deterioration of the marine environment hidden beneath the waves. Unfortunately, our past and current actions result in an overall decline in the ocean's health and resilience. *Now, humankind faces an immediate choice between exerting ecological restraint and proceeding towards increasing global sea life-smothering catastrophes.*

What you can do!

Maintaining the health of our oceans needs to be a worldwide effort. Here are some simple steps that you can take at a personal level to make a difference.

Eat a primarily plant-based diet – Huge amounts of pollutants enter our waterways each year, significantly affecting our oceans and other bodies of water. One of the primary drivers of the problem is industrial fertilizers used to grow massive amounts of crops to feed animals. Switching to a primarily plant-based diet reduces the demand for industrially raised animals and the crops used to feed them.

Switch to organic farming – In the United States, only 0.17% of soy[484] and 0.26% of corn[485] crops are organic. Various organic agricultural techniques have been used for approximately 6,000 years. They have been shown to conserve soil, water, energy and eliminate nitrogen fertilizers, herbicides, insecticides, and fungicides to improve public health and the environment.[486] Fossil fuel energy inputs for organic production are 30% lower than in conventionally produced corn, helping to decrease food's global carbon footprint.

Support sustainable local agriculture – According to a study by the Worldwatch Institute in the United States, food now travels between

1,500 to 2,500 miles from farm to table. Americans are spending far more energy transporting the food than we get from eating the food.[487] "A head of lettuce grown in Salinas Valley of California and shipped nearly 3,000 miles to Washington DC requires about 36 times as much fossil fuel energy in transport as it provides in food energy when it arrives."

Buying locally at farmer's markets reduces energy impact. It is also generally fresher as much food shipped from hundreds of miles away needs to be picked while unripe and can be gassed to "ripen" it after being transported. It can be factory-processed with preservatives and irradiated.[488]

Grow an organic garden – Buying locally grown food is a great way to minimize your environmental impact, but growing your own food takes it to the next level. By raising your own food, you reduce your impact on the environment and reap the benefits of connecting with nature.

Plant cover crops – When possible, plant cover crops and native plants. These will help retain nutrients that might be lost by being leached out of the soil into groundwater and lost through surface runoff.

Decrease your carbon footprint – Decreasing our global environmental impact will help mitigate the ever-accelerating harms happening to our oceans. Read the chapter "Acid Seas" to find out more.

Reduce usage of plastics – Decreasing the use of plastics in everyday life (plastic bags, bottled water, and drinks, straws, take-out containers, utensils, etc.) will reduce the amount of plastics ending up in the ocean. Read the chapter "Plastic Oceans" to find out more.

ACID SEAS

*Way down in the mine, your tears turn to mud
And you can't catch your breath for the dust in your lungs
Loading hillbilly gold where the sun never shines
Twelve hours a day, diggin' your grave
Way down in the mine.
– Dierks Bentley, Down in the Mine*

*We are blessed with a magnificent and miraculous world ocean on this planet. But we are also stressing it in ways that we are not even close to bringing under control.
– Carl Safina*

Coal's deadly and toxic history

During the 1800s, the Western world was in the midst of the Industrial Revolution. What propelled this revolution was energy in the form of fossil fuels and, in particular, coal. Countless souls were employed in coal mines to extract this vital material used to heat buildings, power industries, and fuel transportation. Poor men, as well as women, prisoners, slaves, and ex-slaves, labored exceedingly long hours in horrifying conditions to extract this vital sustenance, feeding the maw of the newly unleashed and growing energy colossus. Even young children were subject to the grueling and endless drudgery in the mines.

> William Slater: "Is six years old; draws the empty corves [small wagons for carrying coal] with a hook." Adam Widowson: "Is seven years old; has worked in a pit one year." Aram Richardson: "Is seven years old; works in the soft coal-pit; has done so for nearly a year." [489]

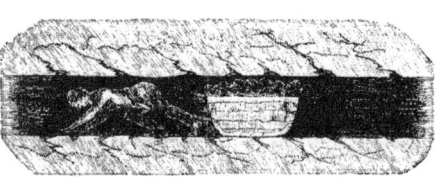

Boys called carters are employed in narrow veins of coal in parts of Monmouthshire [England]; their occupation is to drag the carts or skips of coal from the working to the main roads. In this mode of labour the leather girdle passes round the body, and the chain is, between the legs, attached to the cart, and the lads drag on all-fours.

As young as four years old, boys worked naked, often in mud and water, dragging sleds filled with coal for twelve to fifteen hours a day.[490] Women and girls were harnessed to coal-carts, creeping on all fours through coal mines' cramped spaces.[491]

> Women and girls have been known to wear men's clothes and to take their place side by side with men in the coal and iron mines or in ditches of any kind. A day's work is often from sixteen to twenty hours in duration, rain, or shine. Food is very poor, and clothes are scant.[492]

Cities in Britain, Europe, and America consumed vast amounts of coal. By the end of the 19th century, the United States and England were mining 450 million metric tons (496 million tons) of coal yearly.[493]

Coal mining was not only arduous, but it was also health-damaging and very often deadly. In the United States, from 1900 to 2016, there were 104,851 recorded coal mining-related fatalities.[494] Throughout the decades and worldwide, large numbers of the destitute perished from falling, explosions, being crushed, drowning, suffocating, electrocution, and other horrific manners of death.[495,496,497]

...number of persons employed in coal-mining operations the world over to be 2,500,000, we have it that on the average, almost 5,000 persons are annually killed in the production of the world's coal supply.[498]

With these numbers, it can be reasonably estimated that roughly half a million people died over more than a century of extracting coal from the ground. It's also estimated that there were one hundred or so injuries for every death, two of them permanently incapacitated.[499]

Breaker boys, Woodward Coal Mines, Kingston, Pennsylvania, ca. 1900.

These sobering figures show that forgotten millions forfeited their health and lives to push the world ahead into our present fossil fuel-dominated world.

Not only were millions of people subject to appalling and dangerous working conditions extracting coal from the Earth, but the burning of coal also had severe impacts on health and the environment. Cities and towns became notoriously polluted as endless smokestacks belched out dark clouds of soot and ash, making this bleak and dreary cityscape a symbol of the new modern industrial metropolis.[500] In 1880 meteorologist Rollo Russell wrote of the pollution in London,

In winter, more than a million chimneys breathe forth simultaneously smoke, soot, sulphurous acid, vapour of water, and carbonic acid gas, and the whole town fumes like a vast crater, at the bottom

of which its unhappy citizens must creep and live as best they can.[501]

The famous London fog was not a low-lying cloud of water vapor. In fact, the London fog was made up of soot, and smoke spewed into the air by the massive amount of coal being burned. In 1902, the daily smoke that went up household chimneys and was emitted by the 14,500 factories in London was estimated to total 7 million tons.[502] The sunshine that reached London's streets was a fraction of what it was in the countryside, often keeping the city dark and miserable.

October 1919: A man braves the blinding fog to deliver ice around London. Thick smog regularly fell upon the city from the onset of winter in October until the beginning of spring.

In London, the great fog of 1880 increased the number of deaths by 2,994 over three weeks, and in 1892 caused an excess of 1,484 deaths in one week.[503] Between 1800 and 1900, air pollution may have killed people in Great Britain four to seven times the rate it killed people worldwide.[504] A visitor from India wrote of her London experience in 1882,

> *A London fog is a thick mist -- people in our country cannot imagine what a typical foggy day is really like. Other parts of England also experience the fog, but it is not as dense and dirty as a London fog... London has so many mechanised wagons and factories, and in winter, every home spews smoke out of its chimneys so that on particular days the smoke becomes heavier than the air, cannot rise up and therefore settles over the city and sometimes engulfs large areas and darkens almost everything. On particular days this sort of fog persists through the day and assumes different hues - sometimes ashen - sometimes black - sometimes yellow... One walks in the streets, vis-*

ibility so poor that one moves almost by instinct. Darkness more horrible than that at night has descended at noon, and no artificial light can really illuminate the blackness created by a fog. It is difficult to breathe; one is suffocated by tiny black, oily particles that clog the nose.[505]

The deadly coal-burning vapors continued throughout the 1800s and into the 1900s. In October 1948, a killer smog containing airborne pollutants emitted from nearby zinc smelting plants and steel mills hovered over Donora, Pennsylvania, killing 20 people and making thousands ill.[506] In December 1952, a mass of dry, cold air settled over London, trapping smog and almost entirely immobilizing its nine million residents. Known as the Great Smog of 1952, 4,000 people died from the blinding and suffocating toxic gases over four days, and many thousands more became seriously ill.[507] In New York City, an estimated 220–240 deaths were caused by the six-day 1953 smog, and an estimated 300–405 deaths were caused by the two-week 1963 smog.[508]

1952 Smog along the Strand, London, which almost completely obscures the midday sun.

During this metamorphosis of Western societies from agrarian to industrial, mankind attained the ability to create wondrous machines that altered the way people would live and utilize energy. The introduction of the steam engine and a series of technological advances shifted the production of goods from homes and small operations to large industrial factories, often augmenting or largely replacing manual labor. The application of primarily coal power to the industrial processes and the railroads helped accelerate this historic societal transformation.

In the pre-modern era, communities were often unhygienic and dirty. With the advent of new technologies, life changed and improved for many as piped water, sewer systems, and electricity was introduced. Yet, with these technological and societal shifts, humans also attained the ability to subject the environment to large-scale contamination and destruction, bringing forth a new modern phenomenon of widespread and persistent environmental pollution.

Carbon let loose

Starting in those early years of the Industrial Revolution, carbon output from coal and new fossil fuels increased. From 1750 to 2010, approximately 356 gigatons (1 gigaton = 1 billion tons) of carbon (1,305 gigatons of carbon dioxide or CO_2) have been released into the atmosphere, primarily from consuming fossil fuels and cement production throughout the world.[509] Cement is the source of about 8% of the world's CO_2 emissions. It contributes more CO_2 than aviation fuel (2.5%) and is not far behind the global agriculture business (12%).[510]

> Science refers to the carbon cycle and often put emissions in the amount of the element carbon. However, the carbon entering the atmosphere is in the form of carbon dioxide, which is one carbon atom combined with two oxygen atoms. This means carbon dioxide (CO_2) is about 3.67 times heavier than carbon alone.
>
> "Why do carbon dioxide emissions weigh more than the original fuel?" U.S. Energy Information Administration, https://www.eia.gov

This massive amount of CO_2 that has entered the environment is equivalent to the weight of nearly 4,000,000 Empire State Buildings,[511] with more than half of industrial CO_2 pollution having been emitted since 1988.[512] Approximately 9.86 gigatons of carbon (36.2 gigatons of CO_2 equal to nearly 110,000 Empire State Buildings or over 64,500,000 Airbus A380 Passenger Jets[513]) are released by human activity into the atmosphere each year.[514] With expanding world populations and economies, the production of the significant sources of CO_2 global emissions (coal, oil, gas, and cement) continues to increase with no realistic end in sight.

Flaring occurs when unwanted natural gas released in oil extraction is burned. While flaring only accounts for 1% of CO_2 emissions, it still translates to a global burn of approximately 150 billion cubic meters of natural gas annually, causing more than 400 million tons of CO_2 to be emitted into the atmosphere.

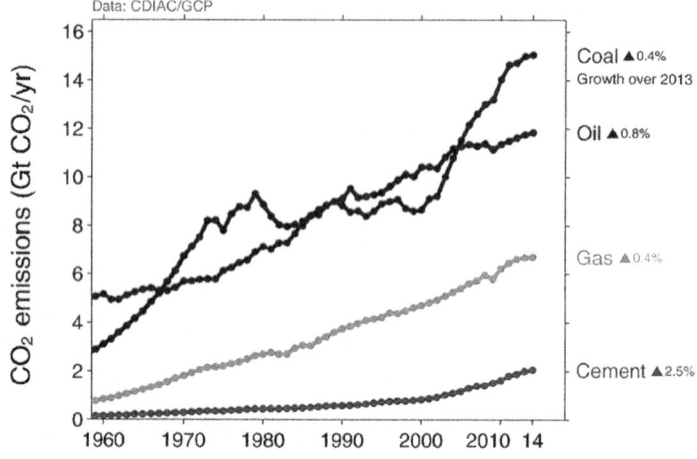

Share of global emissions of CO_2 in 2014: coal (42%), oil (33%), gas (19%), cement (6%), flaring (1%, not shown).

> *Estimates from satellite data show global gas flaring increased to levels not seen in more than a decade, to 150 billion cubic meters (bcm), equivalent to the total annual gas consumption of Sub-Saharan Africa.* [515]

The use of fossil fuels and other human activities have resulted in the global CO_2 atmospheric concentration increasing from approximately 277 parts per million (ppm) in 1750 to 397 ppm in 2014, hitting 400 ppm in March 2015[516] and 410 ppm in May 2018, which is the highest level seen in 800,000 years.[517]

If humans continue to emit greenhouse gases at current rates, scientists estimate that atmospheric carbon dioxide could reach 550 ppm to 800 ppm by 2100.[518] Over the past century, the rate of change in global temperature and atmospheric CO_2 is 100 to 1,000 times higher than in the past 420,000 years.[519]

Turning acidic

Not all of the carbon dioxide that has entered the atmosphere has remained there. Oceans have absorbed up to 30% of human-made carbon dioxide or about 476 billion metric tons (525 billion tons)[520], which has increased ocean water acidity worldwide. This acidification of the oceans occurs as the carbon dioxide from fossil fuels dissolves in seawater, producing carbonic acid. Carbonic acid breaks down and increases hydrogen ions (H+), which lower the water's pH.

What is referred to as the acidity of a liquid solution is the concentration of hydrogen ions (H+). This acidity is measured using the pH scale, ranging from 0 (strong acid) to 14 (strong base) and where a value of 7 is considered neutral. This scale is logarithmic, so that a slight change in pH is actually a significant change in hydrogen ion concentration. These seemingly small pH changes can substantially impact many organisms that are very sensitive to pH, requiring them to be within a narrow range to thrive or even survive.[521] For instance, human blood pH typically falls within the range of 7.35–7.45.

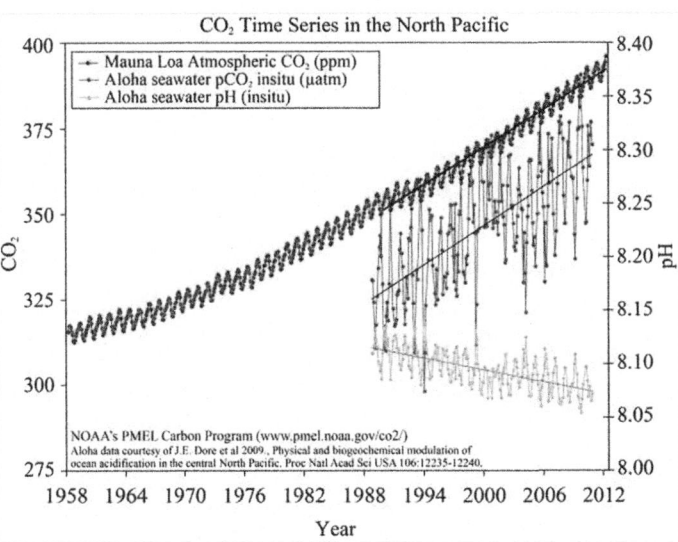

CO_2 Time Series in the North Pacific from the Mauna Loa observatory in Hawaii showing that as atmospheric CO_2 levels have increased, ocean pH has decreased.

A slight variation outside of this range can result in rather serious health consequences. For example, if a human's pH drops to 6.9, a person will be in a coma, and at 6.8, a person will die.[522]

Since preindustrial times, the oceans' pH has dropped from an average of 8.2 to 8.1 today, meaning the oceans are 30% more acidic.[523] Climate change projections estimate that by the year 2100, this number will drop further to around 7.8, or 170% more acidic, since the start of the Industrial Revolution.[524]

> *Ocean acidification (OA)—a result of too much carbon dioxide reacting with seawater to form carbonic acid—has been dubbed "the other CO_2 problem." As the water becomes more acidic, corals and animals such as clams and mussels have trouble building their skeletons and shells. But even more sinister, the acidity can interfere with basic bodily functions for all marine animals, shelled or not. By disrupting processes as fundamental as growth and reproduction, ocean acidification threatens the animals' health and even the survival of species.*[525]

Although the amount of CO_2 that has been emitted is large, it may seem small compared to the enormous volume of water that oceans contain, which is at 1.3 billion cubic kilometers (0.3 billion cubic miles).[526] However, the CO_2 in the atmosphere does not directly interact with all this ocean water.

The ocean's surface layer, which is in direct contact with the atmosphere, is mixed by wave action to a depth of typically about 100 meters (300 feet). Exposed to the sun, this top layer is warmer and less dense than the water beneath it, making it resistant to mixing with the bulk of deeper water. With an average ocean depth of about 3700 meters (2.3 miles), this top ocean layer can be compared to the less than ½ inch thickness of the icing on top of an 18-inch tall wedding cake. It's at this top thin layer where the exchange of atmospheric CO_2 occurs, with the movement of surface layer CO_2 to the deep ocean taking centuries. In this upper layer of the ocean, pH has decreased, with the deeper oceans remaining virtually unchanged.[527]

When combined with CO_2, the ocean surface's pH drops to levels that will potentially compromise or prevent calcium carbonate ($CaCO_3$) accretion (gradual growth) by a wide range of organisms, including reef corals and calcifying algae. Many organisms depend on the relatively stable ocean pH, which has endured for millions of years until the Industrial Revolution.[528]

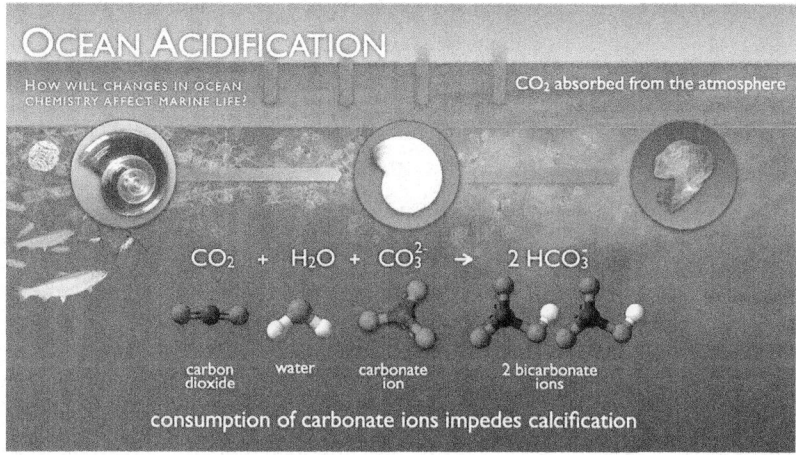

How carbon dioxide changes ocean chemistry, impacting the ability of animals to build skeletons.

The California Current Large Marine Ecosystem (CCLME) is an oceanic ecosystem in the eastern North Pacific Ocean spanning nearly 3,000 kilometers (1,864 miles) from southern British Columbia and Canada to Baja California and Mexico.[529] This area is among the most productive globally, supporting a large percentage of global fishery yield.[530]

Across the CCLME, scientists have observed near-shore pH that fell well below the current global mean surface ocean pH of 8.1. The pH reached as low as 7.43 at the most acidified site, and up to 18% of the values recorded by scientists fell below 7.8 during the upwelling season. These levels are among the lowest reported to date for the surface ocean and matching levels not projected for the global surface ocean until atmospheric CO_2 exceeds 850 ppm, double present levels of over 400 ppm. Ocean deoxygenation and increasing upwelling from growing offshore winds may be accelerating the rising acidity levels near-shore. In the most severely acidic spots, suboptimal conditions for calcifying organisms encompassed up to 56% of the summer season.[531] The ever-increasing acidic and hypoxic coastal waters are an escalating threat to the CCLME and other coastal waters, with severe consequences for marine ecosystems and the fisheries they support.

Coral, plankton, and other life

Many scientific studies have shown that a wide range of marine organisms are sensitive to pH changes. Their physiology, fitness, and survival are almost always affected negatively.[532] Laboratory and field experiments, along with observations of naturally high CO_2 marine environments, have shown lower rates of growth, survival, or other performance measures for many organisms in acidified waters, although with considerable variation between species.[533]

The declining pH of upper seawater layers, due to absorption of increasing atmospheric CO_2, has been added to the list of coral reef threats that include land-based sediment discharges, coral predators, sea surface temperatures, and overfishing. In general, corals do not appear to have the capacity to adapt fast enough to these relatively sudden environmental changes. Coral bleaching is often associated with increased ocean temperatures. However, studies show that ocean acidification alone can also create bleaching in certain reef organisms.

> *Our results indicated that prolonged CO_2 dosing causes bleaching (loss of symbiotic algae) in two key groups of reef-building organisms. The bleaching results indicate that future predictions of bleaching in response to global warming must also take account of the additional effect of acidification.*[534]

Corals require an environment where they can form their skeletons. Studies show that coral calcification decreases with declining pH, which has hindered reefs' ability to build their structures. Research suggests that this reduction in growth is a response to ocean acidification.[535]

> *Seawater acidification partially results from an increase of atmospheric CO_2 and is thought to reduce a reef coral's calcification ability. For example, coral calcification rates in the Great Barrier Reef have reduced by 14%–21% since the 1990s, which is unprecedented in at least the last 400 years.*[536]

A 2009 study showed that the growth and calcification of massive porites, a genus of stone coral, is declining in the Great Barrier Reef (GBR). If the declining calcification of this type of coral is similar to other reef-building corals, then the maintenance of the calcium carbonate structure that is the foundation of the GBR will be severely compromised, causing widespread ecosystem degradation.[537] As atmospheric CO_2 levels rise, the resulting decline in pH in the oceans may make coral reefs unsustainable in the not too distant future.[538]

> *These organisms [corals] are central to the formation and function of ecosystems and food webs, and precipitous changes in the biodiversity and productivity of the world's oceans may be imminent.*[539]

A 2020 study showed that from 2009 to 2019, seawater CO_2 has risen 6% and matches the rate of CO_2 increases in the atmosphere.[540] The study reveals that ocean acidification is no longer a somber future prediction for the GBR but a present-day reality. The research shows that acidification is rapidly changing the conditions that support the growth of coral on the Reef. Dr. Katharina Fabricius, lead author and Senior Principal Research Scientist at the Australian Institute of Marine Science (AIMS), commented,

> *People talk about ocean acidification in terms of 50 years' time, but for the first time our study shows how fast ocean acidification is already happening on the Reef.*[541]

Phytoplanktons are microscopic marine single-cell plants. They form the base of several aquatic food webs by directly providing food for a wide range of herbivorous marine creatures. In turn, other animals eat these herbivores, from small predators like sardines and up the food chain to top predators like sharks and humans. Phytoplankton use energy from the sun to convert carbon dioxide and nutrients into complex organic compounds through a process known as photosynthesis. As these plants die and sink to the ocean floor, a small portion of their organic carbon is buried. The carbon remains there for millions of years in the form of substances like oil, coal, and shale until it is converted into energy through human activity and released back into the atmosphere.

A 2015 study published in the journal Nature Climate Change, reported that by the year 2100, increased ocean acidification will cause changes in phytoplankton. Some species will die out, and some will flourish, although it is hypothesized that an increase in CO_2 could be an overall benefit to phytoplankton.[542] However, this will still alter the balance of phytoplankton around the entire world. Stephanie Dutkiewicz, a principal research scientist at MIT's Center for Global Change Science, noted that,

> *The fact that there are so many different possible changes, that different phytoplankton respond differently, means there might be some quite traumatic changes in the communities over the course of the 21st century. A whole rearrangement of the communities means something to both the food web further up, but also for things like cycling of carbon.*[543]

The study also predicted that as the oceans warm many phytoplankton species will move toward the poles, creating an ocean environment that may look quite different than today. These significant changes at the bottom of the food web may have considerable ramifications further up the food chain.

> *Generally, a polar bear eats things that start feeding on a diatom [a common type of phytoplankton]... Dutkiewicz says. The whole food chain is going to be different.*[544]

The studies investigating the effects of high CO_2 on phytoplankton growth, which have, in some cases, shown that certain phytoplankton seemed to benefit from increased CO_2 concentrations, had been conducted under high-iron conditions. However, a study published in 2018 found that the rising concentrations of atmospheric CO_2, which acidifies the ocean and decreases carbonate, affected phytoplankton's ability to obtain enough of the vital nutrient iron needed for growth.[545] The drop in carbonate concentrations made it harder for the phytoplankton to utilize iron and grow. Consequently, these high concentrations of atmospheric CO_2 could have more of a negative effect on phytoplankton growth than was initially thought.

> *Ultimately, our study reveals the possibility of a 'feedback mechanism' operating in parts of the ocean where iron already constrains the growth of phytoplankton," said Jeff McQuaid, lead author of the study who made the discoveries as a Ph.D. student at Scripps Oceanography. "In these regions, high concentrations of atmospheric CO_2 could decrease phytoplankton growth, restricting the ability of the ocean to absorb CO_2 and thus leading to ever higher concentrations of CO_2 accumulating in the atmosphere.[546]*

Bryozoans are a family of small filter-feeding invertebrates that live as colonies, superficially similar but not related to corals. They are abundant in California kelp forests, and they build their honeycomb-shaped skeletons from calcium carbonate. A 2017 study showed that when these animals were exposed to warmer water and increased acidity, they dissolved in as little as two months. Lead author Dan Swezey was surprised by these results.

> *We thought there would be some thinning or reduced mass, but whole features just dissolved practically before our eyes.[547]*

As of 2021, there might not be much of the iconic Northern California coast ecosystem remaining to be studied. As little as a decade ago, hundreds of miles of forests of amber-green bull kelp swayed beneath the waves home to many fish species. Satellite images show that those forests have collapsed, decreasing by 95% since 2013.[548] The ecosystem destruction began with the mysterious death of sunflower sea stars, which eat the purple sea urchins. Unchecked, this allowed the purple sea urchin numbers to explode, letting them eat their favorite food, kelp. The relentless attack of urchins combined with successive recent marine heatwaves made it harder for the cold water-loving kelp to survive. As a result, the once-massive kelp forests have almost entirely been eradicated, replaced with barren stretches of seafloor covered with spiky purple sea urchins.

A three-year study published in 2017 by Oregon State University found that the California and Oregon coasts' pH levels were among the lowest ever recorded. Team member and marine ecologist Francis Chan found the results concerning because acidified ocean water impacts coastal species.[549]

> The oyster industry is who really sounded the alarm, he said. About 10 years ago, they stopped being able to successfully grow the seed oysters they need for their industry. It turns out the water had absorbed so much carbon dioxide.[550]

For several years the Pacific Northwest oyster industry has struggled with significant losses. Oyster larvae encountered higher mortality rates sufficient to make production economically unworkable. Researchers at Oregon State University have documented why oysters appear so sensitive to increasing ocean acidification.[551] The acidity level isn't high enough to dissolve adult shells. Instead, it is a case of water high in CO_2 altering shell formation rates, energy usage, and, ultimately, the young oysters' growth and survival. With exposure to increasingly acidified water, it becomes more energetically expensive for organisms to build shells. Adult oysters and other bivalves may grow slower when exposed to rising CO_2 levels, but larvae in the first two days of life do not have the luxury of delayed growth.

> They must build their first shell quickly on a limited amount of energy – and along with the shell comes the organ to capture external food more effectively," said Waldbusser, who is in OSU's College of Earth, Ocean, and Atmospheric Sciences. "It becomes a death race of sorts. Can the oyster build its shell quickly enough to allow its feeding mechanisms to develop before it runs out of energy from the egg?[552]

Oyster hatcheries have altered their working practices to avoid using very low pH seawater, either by recirculating their seawater or treating their water during upwelling events. With these new practices, the northwest coast oyster hatcheries are producing near to full capacity again.[553] However, in 2018 academics in England found oysters in New South Wales have become smaller and fewer in number because of coastal acidification.[554] As CO_2 levels in the atmosphere continue to rise, ocean water will become even more acidic, creating more shellfish problems. By the year 2100, mussels are expected to calcify their shells 25% slower than they currently do, and oysters 10% slower.[555]

An eight-year study by more than 250 international researchers found that infant sea creatures will be especially harmed by changes in ocean pH. They determined that the number of baby cod maturing to adulthood could fall by the year 2100.[556] The number of baby cod making it to adulthood could fall 25% to even 9% of today's figures. The study's lead author, Professor Ulf Riebesell from the GEOMAR Helmholtz Centre for Ocean Research in Kiel, Germany, is a world authority on ocean acidification. Riebesell noted,

> *Acidification affects marine life across all groups, although to different degrees. Warm-water corals are generally more sensitive than cold-water corals. Clams and snails are more sensitive than crustaceans. And we found that early life stages are generally more affected than adult organisms. But even if an organism isn't directly harmed by acidification, it may be affected indirectly through changes in its habitat or changes in the food web. At the end of the day, these changes will affect the many services the ocean provides to us.*[557]

The great dying

The Earth has weathered five mass extinction events. The most famous extinction event is the one that annihilated the dinosaurs some 66 million years ago. However, the worst mass extinction event called the Permo-Triassic Boundary (PTB), also known as "The Great Dying," happened 251 million years ago over the span of 60,000 years. During this mass extinction, 96% of all species were lost. Today's life descended from just the remaining 4% of the surviving species.

This cataclysm was caused by enormous volcanic eruptions that filled the air with carbon dioxide, driving warming and causing the oceans to become more acidic.[558] As a result, the oceans lost all their oxygen and effectively became one massive dead zone.[559]

> *Bacteria ate the oversupply of dead bodies, producing hydrogen sulfide gas, creating a toxic atmosphere. The hydrogen sulfide oxidized in the atmosphere to form sulfur dioxide, creating acid rain, which killed much of the plant life on Earth.*

While the amount of carbon released into the atmosphere today does not reach the Great Dying level, the rate that it is being injected into the atmosphere is similar to what it was then. Matthew Clarkson, while at the University of Edinburgh, noted,

> *Scientists have long suspected that an ocean acidification event occurred during the greatest mass extinction of all time, but direct evidence has been lacking until now. This is a worrying finding, considering that we can already see an increase in ocean acidity today that is the result of human carbon emissions.*[560]

Marine life today has not experienced such a rapid shift in ocean pH in millions of years.[561] While some species will not be directly affected by increasing acidity, some will be severely impacted with insufficient time to adapt.[562] Human-generated increases in atmospheric CO_2 and ocean chemistry alterations will take tens to hundreds of thousands of years to return to preindustrial values.[563]

> *Eventually, the sediments in the oceans will buffer these chemical changes, but chemical recovery from such events may take tens of thousands of years while a return to the biological status quo, even if possible, could take millions of years.*[564]

Increasingly, scientists recommend limiting atmospheric CO_2 to prevent dangerous levels of global temperature increases. However, restricting the output of CO_2 should be set with the effects on ocean acidification in mind as well. We've emitted so much carbon dioxide that it is being absorbed by the ocean, changing the seawater's very chemistry. Persistent and increasing acidification could completely restructure marine ecosystems with domino effects across the entire food chain.

Ocean acidification is occurring in concert with other climate-related stressors, such as ocean warming and sea-level rise. In conjunction with other non-climate-related impacts, including overfishing and pollution, acidification adds pressure to already strained marine ecosystems, which provide food for human consumption. Fish stocks, which are already declining in many areas due to overfishing and habitat destruction, are now faced with the additional threat posed by ocean acidification.

Coral reefs that create the habitat for 25% of all marine life on the planet are already at severe risk and are incredibly delicate and prone to the effects of ocean acidification. If CO_2 levels keep increasing, coral reef erosion will outpace building even if other coral reef damaging issues are addressed. Since over 400 million people worldwide live within 100 kilometers of coral reefs, with many reliant on them for their livelihoods and food security, coral reefs' health is of paramount concern.

The impacts of ocean acidification are beginning to be felt in some places, but future forecasts indicate even more widespread deleterious consequences if action is not taken. The apparent solution to ocean acidification's potential threats is to make rapid and substantial cuts to anthropogenic CO_2 atmospheric emissions and, hence, oceanic CO_2 concentrations. Unfortunately, ocean acidification is not a temporary problem and could take many thousands of years to return to preindustrial levels even if carbon emissions are curbed. *Unfortunately, it is now nearly a certainty that within the next 50 to 100 years, continued anthropogenic carbon dioxide emissions will further increase ocean acidity to levels that will have mostly detrimental widespread impacts on marine ecosystems.*

What you can do!

Reducing the amount of CO_2 emitted into the atmosphere will reduce ocean acidification. Here are some simple steps that you can take at a personal level to make a difference.

Buy local – Buying products produced locally, in your own city or the surrounding area, helps reduce the carbon footprint created by shipping them long distances.

Reduce food waste – A great deal of food is bought is wasted, as well as all the energy that went to produce, package, and ship that food. As much as possible, purchase only the food you and your family can actually eat. Read the chapter "Consumerism Gone Mad" to find out more.

Buy less – Carefully consider before purchasing clothes, electronics, and other products you could do without. All these products need to be manufactured and often times shipped long distances. Buy what you genuinely need and when you are done with an item, donate it where possible. Read the chapter "Consumerism Gone Mad" to find out more.

Eat a primarily plant-based diet – Livestock production accounts for 70% of all agricultural land and 30% of the land surface of the planet. Transitioning towards a more plant-based diet could reduce food-related greenhouse gas emissions by up to 70%.[565]

Drive less – When possible, choose to walk, bike, skateboard, or use other human-powered machines to avoid producing any carbon at all. As an additional benefit, you will be getting exercise that helps control your weight, combats health conditions such as high blood pressure, improves mood, boost your energy, and promotes better sleep.[566] Carpooling and public transportation also drastically reduces your carbon footprint by spreading fuel usage across multiple riders.

Wash in colder water, and line-dry your clothes – Depending on how you clean your clothes and how many loads you get through each week, laundry can contribute a surprising amount to your carbon footprint. A load of laundry washed at 60°C (140°F) along with using a drier produces over 5 times more CO_2 equivalent than washed at 30°C (86°F) and then line dried.[567] We also sometimes throw things into the laundry out of habit. Use a towel multiple times, letting it dry between uses, and only wash other items when they need it.

Home heating and cooling – Poorly insulated housing requires more energy to keep warm and cool. Do what you can to improve the energy efficiency of where you live.

Innovate! – There are many ways to save energy and use fewer of the world's resources. Use your own critical thinking and imagination to think of ways to positively impact the planet.

Acid Seas

A man who keeps company with glaciers comes to feel tolerably insignificant by and by. The Alps and the glaciers together are able to take every bit of conceit out of a man and reduce his self-importance to zero if he will only remain within the influence of their sublime presence long enough to give it a fair and reasonable chance to do its work.
– Mark Twain, A Tramp Abroad

There is a glacier in Iceland, Solheimar, which has retreated a great deal, and every time I go back there and see what's not there any more, it does something to the heart. It makes you realize it's possible for a gigantic natural element to just disappear.
– James Balog

The Iceman Cometh

In 1991, while traversing an Alpine glacier on the Austrian-Italian frontier, hikers discovered a man's remains in the melting ice.[568] At first, it was believed this human body might be the victim of a recent mountaineering accident. Later, Swiss researchers determined that this was actually a naturally mummified man that had been covered by snow shortly after his death roughly 5,300 years ago.

This 5300-year-old man, nicknamed "The Iceman" and later named Ötzi, is perhaps the most celebrated and scientifically interesting mummy that has recently been found. Other bodies are now also emerging from icy graves for the first time since they were lost long ago. Artifacts, too, are being exposed, such as a centuries-old moss, Roman coins, an Iron Age horse,[569] and even a well-preserved leather shoe last worn 3,400 years ago during the Bronze Age.[570] Bodies and items left on the surface had long since been buried in snow, eventually compacted to ice. This protected them from decay and disturbance for hundreds or thousands of years. The ice acts as a type of time machine, providing archeologists an opportunity to learn about the past.

Glaciers

The reason for all of these recent discoveries is that all around the world, glaciers are melting and retreating. While providing a treasure trove of archaeological discovery, the melting glaciers are a disturbingly visual indication of a warming planet. While these discoveries are exciting, the boon is likely short-lived as the ice thaws and vanishes from places where it has been for thousands of years.

> *Glaciers that exist today are remnants of the last ice age. Thick sheets of ice advanced and retreated across most continents several times before withdrawing to the polar regions about 10,000 years ago; continental ice sheets still cover Greenland and Antarctica.*[571]

Glaciers are slow-moving rivers of ice that are pushed along by accumulating snow and ice at higher altitudes. They flow downhill at a speed of a hundred to thousands of meters per year.[572] Glaciers grow through snowfall and shrink through melting and when water evaporates directly from solid ice - which is a process called sublimation. Glaciers that end in a lake or the ocean also shrink when ice chunks from their edges break off into the body of water—a process known as iceberg calving.

Despite abundant water quantities on Earth, most of it is not directly usable by many living things, including humans, making freshwater a relatively scarce resource. While almost three-quarters of the Earth's surface is covered with water, over 96.5% is in the salty oceans and seas, with the remaining 3.5% freshwater. Over 68% of that freshwater is found in the frozen Arctic, Antarctic icecaps, and mountain glaciers, with just over 30% found in groundwater. Only about 0.3% is located in the surface water of lakes, rivers, and swamps.[573] The 171,000 glaciers worldwide, excluding the Greenland and Antarctic ice sheets, contain nearly 170,000 cubic kilometers (41,000 cubic miles)[574] of ice, equal in volume to over 65 million Great Pyramids of Giza.[575] This makes mountain glaciers one of the most critical reservoirs that provide reliable freshwater to natural ecosystems and billions of people.

In Asia, the Tibetan Plateau is the largest high-altitude landmass on Earth, covering 2.5 million square kilometers [965,000 square miles].[576] The size of the Tibetan Plateau is equal to about the area of Spain, France, Germany, Italy, the Netherlands, Belgium, Switzerland, Austria, Poland, the Czech Republic, Slovakia, Hungary, and Slovenia combined.[577]

The Tibetan Plateau is referred to by scientists as the "Third Pole" because the ice covering it holds the largest reserve of freshwater outside of the Antarctic at the South Pole and the Arctic at the North Pole. Its immense glacial ice sheets are made up of more than 45,000 glaciers[578] covering an area of over 100,000 square kilometers (38,600 square miles),[579] roughly the size of the state of Kentucky. The Himalayas that separate the plains of the Indian subcontinent from the Tibetan Plateau contain 100 times as much ice as the Alps, providing for more than half of the drinking water for 40% of the world's population.[580]

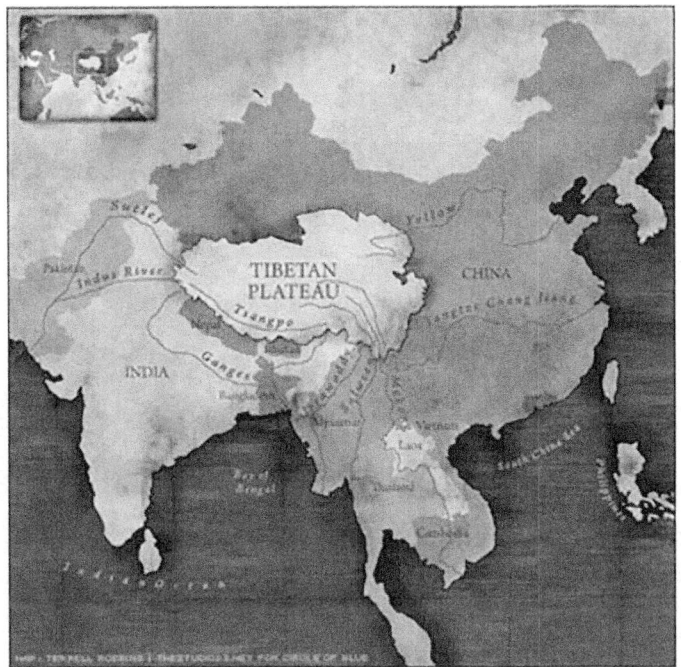

The Tibetan Plateau

This landmass is literally the roof of the world with a mean altitude of 4,000 meters [13,000 feet]. It is dotted with the world's highest mountains, glaciers, rivers, grassland, forests, and lakes to form one of the outstanding zones for biological diversity on earth.[581]

Considered "water towers of the world," the meltwater from the Third Pole's glaciers feeds nine of Asia's biggest rivers, including the Ganges, Yangtze, Yellow, Brahmaputra, and the Mekong. These rivers supply a water lifeline to twelve of Asia's most densely populated nations,[582] including China and India. Societies along these rivers have grown, expanded, and been nurtured by the supply of abundant water that supports daily life, agriculture, livestock, and industry. Millions of people of Asia are dependent on water flowing off the Tibetan Plateau for their very existence.

The glacial vanishing act

Temperatures on the plateau are rising twice as fast as in other parts of the world. Over western China, the atmosphere has warmed 0.7°C [1.26°F] over the past 50 years.[583] Temperatures measured in 2006 on the Tibetan Plateau were the warmest since records for the region began to be kept in 1951, while analysis of ice cores from glaciers indicates that the 1990s was the warmest decade of the previous 1,000 years.[584]

Heat in the Himalayas has increased by 1°C [1.8°F] since 1950, causing thousands of glaciers to retreat by an average of 30 meters [98 feet] each year.[585] From 1930 to 1998, the Western Sichuan's Hailuogou Glacier retreated 1,700 meters [1 mile]. The Kunlun Mountains form the northern edge of the Tibetan Plateau. Glaciers in the eastern part of those mountains have receded by 17% over the past 30 years.[586]

Dramatic retreat of Ata Glacier, southeast Tibetan Plateau, from 1933 (upper) to 2006 (lower) due to global climate changes.

Comparisons of photographs taken in the 1930s, 1970s, 1980s, and 2006 show that both ice volume and glacial surface conditions of the Ata Glacier have dramatically changed.[587] Measurements from the early 1960s to 1999 show the glacier has lost almost 8% of its original area and has retreated by nearly 10%.[588] In the 1990s, the glaciers shortened 1.9 times faster than in the approximately prior 30 years.

The Rongbuk Glacier on Mount Everest's north face has retreated 170 to 270 meters [558 to 886 feet] from 1966 to 1997. Since Sir Edmund Hillary and Tenzing Norgay were the first to reach Everest's summit in 1953, much of that glacier has turned to meltwater, retreating about 5 kilometers [3 miles] up the mountain.

> *Back in 1953, when Hillary and Tenzing set off to climb Everest, they stepped out of their base camp and straight on to the ice. You would now have to walk for over two hours [from the same site] to get on to the ice.*[589]

The new snow that falls each year cannot make up for the ice loss and, as a result, glaciers retreat upslope and shorten. Some smaller glaciers are shrinking so fast they are predicted to entirely disappear within the next few decades. Since the 1980s, the total amount of glacial ice in the Tibetan and the Himalayan regions has decreased by about 6%.[590] According to Jeff Kargel of the U.S. Geological Survey,

> *"Glaciers in the Himalayas are wasting at alarming and accelerating rates, as indicated by comparisons of satellite and historic data, and as shown by widespread, rapid growth of lakes on the glacier surfaces."*[591]

Many Himalayan glacial lakes are forming and are quickly filling up because of this accelerated melting. The melting glaciers cause floods and landslides by increasing river runoff and the volume of lakes, which then grow and collapse. Glacial-lake outburst floods, a type of megaflood, and glacial debris-flow hazards have become more severe since the 1980s.[592]

> *...glacier ice mass loss since the 1980s has increased the magnitude and frequency of flash floods in the Gezhe river because ablation [the loss of snow and ice by melting or evaporation] of the glacier surface accelerated as a consequence of atmospheric warming.*[593]

Glacial lakes have broken their banks in Nepal more than 20 times since the early 1960s. Tsho Rolpa Lake in Nepal has grown sixfold since the 1950s. A flood from this lake could cause severe damage as far as 100 kilometers [62 miles] downstream in the village of Tribeni, threatening 10,000 lives.[594] The Imja Tsho glacial lake in Nepal was drained in 2016 to safe levels[595] because if it burst its banks, it would have sent a 100 meter [328 foot] wall of water down the valley.

Data from the Himalayan mountains shows that the region's annual temperature has increased, with rising temperatures contributing to glacial shrinkage and large ice mass deficits.[596] Several human actions have helped to accelerate the melting of glaciers and exacerbate the resulting problems. "Black carbon" is the soot-like byproduct of the use of coal, burning of agricultural waste, or forest fires. It amplifies the adverse effects of climate change as it deposits on the glaciers, causing them to absorb more solar heat, accelerating melting.[597] According to James Hansen, who is an American adjunct professor directing the Program on Climate Science, Awareness and Solutions of the Earth Institute at Columbia University:

> *Tibet's glaciers are retreating at an alarming rate. Black soot is probably responsible for as much as half of the glacial melt, and greenhouse gases are responsible for the rest.*[598]

Tibet has lost nearly 50% of its forests, decreasing from 25.2 million hectares in 1959 down to 13.57 by 1985.[599] This large-scale deforestation has caused Bangladesh and China to face unprecedented floods.[600] Deforestation adds to global warming significantly because once trees are felled, the sequestered carbon is released, and the trees are no longer absorbing carbon dioxide from the atmosphere.

> *One of the driving factors for the Chinese invasion in 1949 [of Tibet] was the quest for rare and precious minerals to fuel industrial growth. According to Chinese statistics, there are over 126 different minerals in Tibet, including copper, gold, uranium, borax, and iron, which Tibet has in abundance and which are rapidly being depleted in China itself. Tibet's landscape has been transformed with roads, railways, mines, housing, and hydropower plants to support the harvesting of these materials, and, as a result, rich agricultural land has been lost, water sources polluted, and soil contaminated. According to Chinese estimates, at least 30 percent of Tibet's arable land is now considered degraded.*[601]

If this melting continues at the current rate, two-thirds of the existing Himalayan glaciers will be gone by 2050,[602] with a projected 43% average decrease in the glacial area by 2070 and a 75% decrease by 2100.[603] In India, glaciers in the Garhwal Himalaya are retreating so rapidly that researchers believe that most central and eastern Himalayan glaciers could virtually disappear by 2035.[604]

To the north of the Tibetan Plateau, across the Tarim Basin and Taklimakan Desert, lies the Tian Shan, also called the Celestial Mountains. The Tian Shan extends 2,500 km (1,550 miles) east-to-west across Central Asia. The Tian Shan glaciers are melting at an alarming rate. They have lost 27% of their mass since 1961, losing an average of 5.4 gigatons of ice per year.[605] A male African elephant might weigh at most 6.8 metric tons,[606] meaning that the Tian Shan is losing the equivalent of the weight of nearly 800 million African elephants or over 3.6 billion cars worth of ice annually.

GFZ German Research Centre for Geosciences at Rennes University noted that "currently, the Tien Shan is losing ice at a pace that is roughly twice the annual water consumption of entire Germany."[607] Snow and glacier melt from the Tien Shan are essential for the water supply of Kazakhstan, Kyrgyzstan, Uzbekistan, and parts of China. Because meltwater from the glaciers supplies the Fergana Valley, one of the largest irrigated areas on earth, the impact on farmers could be enormous. The melt there is four times the global average, and because of rising summer temperatures, the remaining ice could fall by another 50% by 2050.[608]

According to a 2006 United Nations report, Asia already has less freshwater per person than any other continent outside Antarctica.[609] In Africa, the average household water usage is 47 liters per person per day. In Asia, the average is 95 liters per person per day. In the United Kingdom, the average is 334 liters per person per day. In the United States, the average is 578 liters per person per day.[610]

As temperatures increase and the ice melts, this essential water supply is threatened with massive impacts on ecology and human livelihoods. The nearly 2 billion people in China, India, Pakistan, Bangladesh, and Bhutan will undoubtedly be at serious risk of acute water shortages as demand rapidly increases and the glacial water that supplies vital rivers disappears.

There are somewhere between 150,000 and 200,000 mountain glaciers worldwide, and nearly all of them are shrinking. Mount Kilimanjaro and Glacier National Park in Montana are well-publicized examples of the glacial vanishing act. The fresh snow that falls annually isn't enough to compensate for the melt causing the world's glaciers to retreat and grow smaller at, on average, about 10 meters (33 feet) each year.[611]

Shepard Glacier, Glacier National Park, MT, 1913 (top) and 2005 (bottom).

The famed snows of Kilimanjaro have melted more than 80% since 1912. When President Taft signed the bill establishing Glacier National Park in 1910, there were an estimated 150 glaciers. Today that number has decreased to fewer than 30, most of which have shrunk in area by two-thirds. Daniel B. Farge of the U.S. Geological Survey predicts that most, if not all, of the park's glaciers will disappear by mid-century.[612]

> *"Things that normally happen in geologic time are happening during the span of a human lifetime," says Fagre. "It's like watching the Statue of Liberty melt."*[613]

According to scientists, it is inevitable that the contiguous United States will lose all of its glaciers within a few decades.[614] In western North America, excluding Alaska, glaciers are melting four times faster than in the previous decade. In Denali National Park, Alaska, Mount Hunter is producing 60 times more snowmelt than 150 years ago.[615] The more than 1,000-year-old Arikaree glacier, located in Roosevelt National

Forest in Colorado, has been thinning by about 1 meter (3.2 feet) annually over the past decade and is expected to vanish within 25 years. Dr. Farge stated that glaciers started to shrink from around 1910 and then entered "rapid and continual" melting from the 1970s onwards:

> *This is the first time in 7,000 years they've experienced this temperature and precipitation. There's no hope for them to survive. We'd need a major reversal where it would get cooler, not just stop getting warmer. There's nothing to suggest that will happen.*[616]

In Canada, the Peyto Glacier in the Canadian Rockies in Banff National Park is losing as much as 3.5 million cubic meters of water each year.[617] The diminishing snowpack threatens the health of Western watersheds and increases the risk of forest fires. Approximately 70% of the glacier is already gone, and its retreat will severely impact the Mistaya and North Saskatchewan rivers that it feeds. David Hik, an ecology professor at Simon Fraser University, noted that "Probably 80 per cent of the mountain glaciers in Alberta and B.C. will disappear in the next 50 years."[618]

> *British Columbia's 17,000 glaciers — both in the Rockies and along the Pacific coast — are losing 22 billion cubic meters of water annually. That's equivalent to refilling a 60,000-seat football stadium 8,300 times.*[619]

In Colombia, mountaintop glaciers have lost 18% of their area in just seven years, with six glaciers shrinking from 45 square kilometers (17.4 square miles) in 2010 to 37 kilometers (14.3 square miles) in 2017.[620] To the south in Peru, over the past 40 years, the glaciers have shrunk by 40%. The meltwater from these retreating glaciers has formed nearly 1,000 high altitude lakes, which can cause severe destruction to towns and infrastructure.[621] In 1970 Peru, an earthquake triggered a glacial avalanche of ice, rock, and mud that buried the village of Yungay, killing 20,000 people.[622] In 2010 a massive portion of a glacier broke off and plunged into a lake in Peru. The resulting 23-meter [75-foot] tsunami wave killed three people and destroyed a local water processing plant.[623]

> It was one of the most concrete signs yet that glaciers are disappearing in Peru, home to 70 percent of the world's tropical icefields. Scientists say warmer temperatures will cause them to melt away altogether within 20 years.[624]

Peru's 2,679 glaciers are the source of the vast majority of the country's drinking and agricultural water.[625] The government has used this water source to irrigate the desert, turning it into more than 100,000 acres of farmland. Items produced here, such as asparagus and blueberries, are shipped to far-off places like Denmark and Delaware.[626] Those glaciers continue to retreat and vanish, with scientists projecting that many of them will be gone by 2050. Once the freshwater supply is gone, the bountiful farmland will revert back to an empty desert.

> "We're talking about the disappearance of frozen water towers that have supported vast populations," said Jeffrey Bury, a professor at the University of California at Santa Cruz who has spent years studying the effects of glacier melt on Peruvian agriculture. "That is the big picture question related to climate change right now."[627]

Not only is the water supply threatened, but as Peruvian glaciers retreat, metal-rich rocks are being exposed for the first time in thousands, maybe millions, of years. Meltwater carries these metals, such as lead, arsenic, cadmium, manganese, and iron downstream, contaminating the water and soil.[628] The metals are leaking into the groundwater, turning entire streams red, killing livestock and crops, and making the water undrinkable.[629]

The analysis of 45,000 observations recorded since 1894 of 2,000 glaciers showed that from 2001 to 2010, glaciers lost, on average, 75 centimeters [2½ feet] of their thickness each year. This recorded rate from 2001 to 2010 was twice as fast as it was in the 1990s and three times as fast as it was in the 1980s, showing a rapid acceleration of glacial ice loss.[630] Lead author Michael Zemp of the World Glacier Monitoring Service at the University of Zurich, Switzerland, highlighted the seriousness of the study:

The first decade of the 21st century, from 2000 to 2010, saw the greatest decadal loss of glacier ice ever measured. It's without precedent... globally, we're now losing treble the total ice volume of the European Alps each year. We were shocked.[631]

According to NASA, the world has lost approximately 400 billion tons of ice from mountain glaciers since 1994,[632] or over the weight of 1,200,000 Empire State Buildings or 265 billion cars.[633] In the 2017 State of the Climate report published by the American Meteorological Society, the cumulative mass balance loss from 1980 to 2016 was 19.9 meters [65 feet].[634] By 2018 that number had increased to 21.7 meters [71 feet.][635]

In 2021, a team from the universities of Edinburgh, Leeds, and University College London said the rate at which ice is melting across the world's polar regions and mountains match "worst-case climate warming scenarios."[636] The rate of loss has risen from 0.8 trillion metric tons per year in the 1990s to 1.3 trillion metric tons per year by 2017. Using satellite data, the experts found the Earth lost 28 trillion metric tons of ice between 1994 and 2017, or the weight of over 84 million Empire State Buildings or over 18.7 trillion cars. Climate scientists estimate that if the world continues warming at the same rate, 90% of its glaciers will be gone by 2100.[637] Archeologist Lars Pilö notes with a sense of certainty, "they've been here for 6,000 years, so it's incredible to think they'll be gone in 100."

Glaciers are somewhat slow to react to a warming climate and aren't directly responding to recent warming. Instead, current ice loss results from the warming of the planet's climate system over the last century.

Today, many glaciologists are more concerned with predicting when various glaciers will disappear. In many parts of the world—including the western United States, South America, China, and India.[638]

In the Hindu Kush-Himalaya (HKH) region, the glaciers that feed the rivers that 1.65 billion people rely on in India, Pakistan, China, and other countries, will melt by 36% by 2100 if global warming is limited to

1.5°C. If emissions are not cut and global temperatures increase further, 66% could melt and vanish.[639] Daniel Farinotti, a researcher at the Swiss Federal Institute for Forest, Snow, and Landscape Research in Birmensdorf, Switzerland, noted that even if we halted carbon dioxide emissions tomorrow, the glaciers would continue to melt for several decades because of the carbon dioxide already in the atmosphere.[640] According to Ben Marzeion of the Institute of Geography at the University of Bremen,

> *Around 36 percent of the ice still stored in glaciers today would melt even without further emissions of greenhouse gases. That means: more than a third of the glacier ice that still exists today in mountain glaciers can no longer be saved even with the most ambitious measures.*[641]

A certain amount of melting has already occurred and will continue to happen no matter what action is taken. As glaciers continue to disintegrate, the icy white surfaces that once reflected sunlight back up into space are replaced with exposed darker surfaces that absorb that heat and raise temperatures. This positive feedback loop accelerates warming as these darker surfaces are exposed.

While there is evidence that some of the glacial melt is independent of human-induced factors, scientists have found "unambiguous evidence of anthropogenic [human-caused] glacier mass loss in recent decades."[642] Yet, altering human actions that impact glaciers can reduce the amount of damage to make it less catastrophic. Georg Kaser, a climate scientist at the University of Innsbruck in Austria, noted that "our current behavior has an impact on the long-term evolution of the glaciers -- we should be aware of this." Scientists calculated that every kilogram (2.2 pounds) of CO_2 emitted causes 15 kilograms (33 pounds) of glacial melt over the long term. This means that one kilogram (2.2 pounds) of glacier ice is lost for every five hundred meters (1/3 of a mile) driven by a conventional car.[643]

Sea-level rise

As temperatures rise, ocean water warms and expands in volume, and more water flows to the seas from melting glaciers and ice caps. This combination of effects has played a significant role in raising the average global sea level. Although glaciers store less than 1% of global ice mass, they are responsible for most of the sea-level rise.[644]

Since 1900 the global sea level has risen by approximately 20 centimeters (7.9 inches). A 2012 study found that between 1902 and 2009, melting glaciers contributed 11 centimeters (4.3 inches) to a little more than half of the total sea-level rise.[645] The rest was from warming, expanding seawater, melting Greenland and Antarctic ice sheets, and changing terrestrial water storage in dammed lakes and groundwater reservoirs. Future glacial melting is projected to raise sea level by an additional 15 to 22 centimeters (5.9 to 8.7 inches) by 2100. If all the glaciers were to melt, global sea levels would rise about 43 centimeters (17 inches).[646] Compared with the potential sea-level rise from the Greenland and Antarctic ice sheets, the volume of land-based glaciers is slight. However, according to Matthias Huss, a glaciologist at the University of Fribourg in Switzerland,

> ...*mountain glaciers are still a concern because they react very fast to higher temperatures, and a considerable retreat is very likely in the next decades."*[647]

Water is life

Time and water are running out for the world, and the picture is very bleak. If there are vast water shortages, which are almost certainly going to happen in the foreseeable future, we are talking about a colossal human disaster. The United Nations estimates that two-thirds of the world's population will live in water-stress areas within the next 20 years, with much of that population living in Asia.[648]

Rapid population and economic growth are putting a strain on the two linked resources of energy and water. Energy production produces greenhouse gases that heat up the atmosphere while simultaneously diminishing and contaminating water sources vital to life. China's unquenchable thirst for water and energy has caused extensive deterioration of its own major rivers. This may bring "catastrophic consequences for future generations" because some of the harm already done is irreversible. [649]

The state of glaciers is closely linked to sea-level rise and questions of water availability and, thus, food security. Without this mass storehouse of frozen freshwater, our way of life, based on climate as we have known it for thousands of years, will be radically altered. The mindset of "economic growth at any cost" is ultimately going to be immensely expensive as the ice that has existed for centuries vanishes in just a relatively few years. Glaciers have been retreating for decades, and it is no longer an issue of if they are going to melt, but how fast, and how humans and life will be able to adapt to these severe and unyielding changes.

Mountain glaciers provide a powerful demonstration of how climate change has and still continues to reshape the world. *If society cannot adapt to glacial retreat in the coming decades, it raises the grim specter of water wars in Asia and other parts of the world. Addressing glacial retreat is a question of providing water stability for future generations, ensuring the survival of billions of people.*

What you can do!

Due to the alterations to the atmosphere that have already occurred, a large amount of glacier loss is inevitable. However, over the long term, our actions will make a difference. By decreasing our carbon footprint, we can reduce the stress on our planet. Read the chapter "Acid Seas" to find out more.

Conserve water – Water wastage is something that may occur in the household without being noticed. However, Water-saving is something that every person can do by taking responsibility for the amount of water you use. Taking responsibility and performing a home audit of your home or workplace can significantly impact the amount of water-saving that can be achieved. Here are some tips.

• Turn off the tap when you are not physically using the water. Turn off the tap when brushing teeth, shaving, doing dishes, etc.

• Check for potential water leaks such as running toilets, dripping faucets, etc.

• Run washing machines and dishwashers only when they are full. When possible, wash clothes on the cold cycle, which saves energy.

• Replace old shower heads with low-flow showerheads.

• If you wash dishes by hand, place a plug in the sink and use the collected water. Try not to wash dishes and vegetables by leaving the tap to run water straight down the sink. Vegetables can be washed in a large bowl, and then that water can be used to water the garden outside.

• Catch and use rainwater to water indoor and outside plants.

• When staying in hotels, you can opt to have your sheets and towels washed every couple of days instead of daily. This saves water and energy.

• Teach our children about the importance of water and how we cannot live without this valuable resource. Let them learn by your example how to live with water-saving methods.

Glaciers Going, Going, Gone...

FLATTENED FORESTS

Climb the mountains and get their good tidings. Nature's peace will flow into you as sunshine flows into trees. The winds will blow their own freshness into you, and the storms their energy, while cares will drop away from you like the leaves of Autumn.
– John Muir, The Mountains of California

It broke one's heart to think of man, the civiliser, wasting treasures in a few years to which savages and animals had done no harm for centuries.
– Marianne North, 1875

What we are doing to the forests of the world is but a mirror reflection of what we are doing to ourselves and to one another.
– Chris Maser, Forest Primeval: The Natural History of an Ancient Forest

The world's largest rainforest

The Amazon River Basin is an immense region made up of a mosaic of ecosystems ranging from savannas to swamps and is home to the largest contiguous rainforest on Earth. With its intricacy and vastness, the Amazon contains a magnificent diversity of life, providing a home to one out of every five mammal, fish, bird, and tree species in the world.[650] While the Amazon rainforest covers only 4% of the Earth's surface[651], the more than 400 mammal, 40,000 plant, and 2,500,000 identified insect species and a myriad of other invertebrates, microbial and fungal life forms, make it the most abundant habitat on the planet.[652]

The Amazon Rainforest is defined by the vast Amazon River. It is the second-largest river in the world, flowing for more than 6,600 kilometers (4,100 miles.)[653] It is the world's most voluminous river delivering more water into the oceans than the next eight biggest rivers in the world combined,[654] accounting for 15-20% of the global freshwater input into the oceans.[655] The Amazon River Basins' hundreds of tributaries and streams contain the largest number of freshwater fish species in the world.[656] The more than 2,000 identified species of fish are greater than that of the entire Atlantic Ocean.[657] While 60% of the Amazon lies in Brazil, the rainforest spans nine countries covering about 40% of South America. It is equal in size to approximately two-thirds of the continental United States.[658]

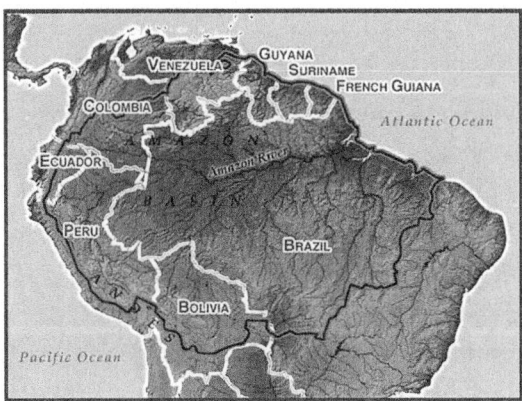

The Amazon

Known as the "Lungs of the World," the Amazon is one of Earth's most critical environmental filters, absorbing carbon dioxide (CO_2) from the atmosphere to use in a biochemical process called photosynthesis.[659] Each leaf in the forest uses the sun's energy and chlorophyll to convert water (H_2O) and CO_2 into sugars used by the plants for energy through photosynthesis. During this process, twelve molecules of H_2O and six molecules of CO_2 react to form one glucose molecule ($C_6H_{12}O_6$), six molecules of oxygen (O_2), and six molecules of H_2O.[660]

> Photosynthesis can be represented by the following chemical equation:
>
> $12H_2O + 6CO_2 +$ light energy $= C_6H_{12}O_6 + 6H_2O + 6O_2$.

Through the action of each leaf of almost 400 billion trees, made up of 16,000 tree species,[661] the Amazon produces 20% of the world's oxygen. [662] Also, by absorbing CO_2, the Amazon has effectively negated the fossil fuel emissions and emissions from forest loss and degradation from the nine Amazon nations of Brazil, Bolivia, Colombia, Ecuador, French Guiana, Guyana, Peru, Suriname, and Venezuela since the 1980s.[663] The Amazon is a vast carbon repository. Between 81 to 127 gigatons (90 to 140 billion tons) of carbon are stored in the Amazon Basin,[664] equal to the total carbon emissions discharged into the atmosphere by all human activity over an 8 to 13 year period.[665]

Flying rivers

While creating energy to support its life, each tropical tree pumps a geyser of water into the atmosphere. About 760 liters (200 gallons) of water evaporates each year from all the leaves of each tree that make up the Amazon canopy, or roughly 76,000 liters (20,000 gallons) for each acre of canopy trees.[666] This amounts to about 18.1 billion metric tons (20 billion tons) of water emitted by all of the Amazon's trees every day,[667] which is more water than the Amazon River's daily discharge into the Atlantic Ocean and is equal in weight to over 32 million Airbus A380 Passenger Jets.[668] This tree-generated moisture helps create the thick cloud cover that hangs over the forest. It is through this process that the Amazon generates as much as 50% of its own rainfall. The remaining moisture comes from winds that bring it from the ocean.[669]

Rainfall in much of South America depends on the moisture coming from the tropical Atlantic Ocean. Westward-blowing trade winds carry water transporting it across the Amazon Basin. The clouds eventually collide with the Andes Mountains 3,200 kilometers (2,000 miles) to the west, where they are then forced to move south and east to the rest of the continent.[670] As the air stream travels west from the coast, it picks up water vapor that has evaporated from the forest and deposits it as rain further on.

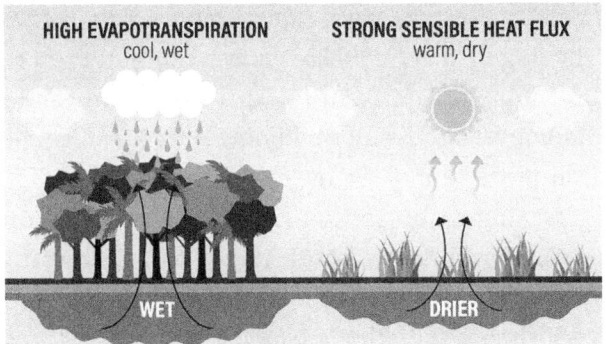

Trees pull water from the ground and release water vapor through their leaves, generating atmospheric rivers of moisture.

It may be more than just the tropical trade winds that simply carry moisture from the east coast across the continent and then south still further. According to the "biotic pump" theory, the forests, through the enormous amount of water that evaporates from the trees, create the mechanism in which the rainforest "sucks in" moist air from the ocean.[671] The combined force of the billions of trees in the rainforest irrigates the atmosphere, generating massive, invisible, flying rivers and creating ocean-to-land winds that pump the moisture-laden currents over the continent.[672] This theory explains how deep interiors of forested continents get as much rain as the coast.[673] According to Doug Sheil of the Norwegian University of Life Sciences, near Oslo,

> Traditionally, people have said areas like the Congo and the Amazon have high rainfall because they are located in parts of the world that experience high precipitation. But the forests cause the rainfall, and if they weren't there the interior of these continental areas would be deserts.[674]

Deforestation

The entire Amazon rainforest is estimated to have initially covered a staggering 5.7 million square kilometers (2.2 million square miles), with approximately 12% of that area lost since the dawn of industrialization.[675] By 1970, about 2.4% of the rainforest had been cleared. Since that time, it has continuously been stripped of forest at varying rates.[676] In 1995 loggers, cattle ranchers, and farmers set a new deforestation record, having cleared 29,000 square kilometers (11,200 square miles.) In 2005, annual deforestation turned a corner, and rates began to decrease, falling to 4,575 square kilometers (1,766 square miles) by 2012 – a fall of more than 83% over eight years.[677] Yet, after about a decade of decreasing deforestation rates, they reversed direction and are now growing. From August 2015 to July 2016, an estimated 7,989 square kilometers (3,085 square miles) were deforested, equal to 3 times the size of the state of Rhode Island.[678]

> The policy director of Greenpeace, Marcio Astrini, says among the causes of the increased deforestation were actions taken by the federal [Brazilian] government between 2012 and 2015, such as the waiving of fines for illegal deforestation, the abandonment of protected areas — that is, 'conservation units' and indigenous lands — and the announcement, which he calls 'shameful,' that the government doesn't plan to completely stop illegal deforestation until the year 2030.

In the Brazilian Amazon, ranching has been the most common land use for over four decades and is the leading source of deforestation.[679] Brazil has the world's largest commercial cattle herd and is the largest exporter of beef and leather.[680] Cattle ranching accounts for about 70%

of land use requiring two acres of cleared rainforest to raise one steer.[681] Other land uses such as soybeans, sugar cane, and palm oil for biofuels, cotton, and rice are all expanding. Brazil is second only to the United States as a global producer of soybeans.[682] Only about 6% of soybeans grown worldwide are turned directly into food products for human consumption. The rest is used as animal feed, vegetable oil, or non-food products such as biodiesel. About 75% of the world's soy ends up as feed for chickens, pigs, cows, and farmed fish.[683]

Rainforest land is cleared for agricultural purposes. While tropical rainforest trees are well-adapted to their environment, the soils they grow in are actually very thin and lacking in nutrients.[684] Farmers use slash-and-burn agriculture burning the trees and vegetation in a deforested area to create a fertilizing layer of ash. However, with tropical high temperatures and heavy rains over time, most minerals are washed from the soil. The land overtaken by weeds often becomes unable to support crops in just a few years.[685] The deforestation caused by cattle ranching is responsible for releasing 308 metric tons (340 million tons) of carbon into the atmosphere every year, making up about 3.4% of current global emissions.[686]

Once trees have been removed and replaced with grass cattle pastures, it rapidly dries out during periods of limited rains, significantly increasing fire risk. Frequent fires expose the land to heavy rains resulting in high levels of soil erosion.[687] Forest fires alone destroyed more than 85,500 square kilometers (33,000 square miles) of the Amazon rainforest between 1999 and 2010, an area larger than the state of South Carolina.[688] Since 2000, the region has been hit by three unprecedented droughts, which have led to substantially worse fires. Anja Rammig, of the Potsdam Institute for Climate Impact Research, Germany, said,

> *Today, the wet season is getting wetter and the dry season drier in southern and eastern Amazonia due to changing sea-surface temperatures that influence moisture transport across the tropics. It is unclear whether this will continue, but recent projections constrained with observations indicate that widespread drying during the dry season is possible in the region.*[689]

Once a tract of land can no longer support agriculture, farmers clear adjoining forest sections, leaving swaths of unproductive deforested land in their wake, leading to a vicious cycle of deforestation. Untouched rainforest land has little direct monetary value, but cleared pastureland can produce cattle or be sold to large-scale farmers. The cattle supply domestic and international markets with beef, leather products, and items such as rawhide dog treats. Many of these products are shipped to Europe, the Middle East, and Russia.[690] China is the leader in growing affluent classes, which consume increasing amounts of beef and other types of meat. In 2018, Brazil exported an estimated 1.6 million metric tons (1.7 million tons) of beef, driven by a strong demand from Hong Kong and China that imported 44% of that amount. In 2019, Brazil was expected to increase its beef export to 1.8 million metric tons (2 million tons).[691]

> *Meat from the cattle is canned, packaged and processed into convenience foods. Hides become leather for shoes and trainers. Fat stripped from the carcasses is rendered and used to make toothpaste, face creams and soap. Gelatin squeezed from bones, intestines and ligaments thickens yoghurt and makes chewy sweets.*[692]

In the Amazon, land grabbing and illegal encroachment are rampant. A preferred method to lay claim to an area is to deforest it, making that parcel of land 5 to 10 times more valuable. After that piece of land has become unproductive, it can be sold to larger landholders who then use massive amounts of industrial fertilizers and pesticides to make it profitable. The original settlers then move on to grab another part of the rainforest.[693] In addition to cattle farming, soy plantations, logging, biofuel crops; mining for diamonds, bauxite, manganese, iron, tin, copper, lead, and gold; and construction of dams and roads all contribute to deforestation.[694] In Peru, between 2005 and 2011, some 7,000 hectares (27 square miles) of Amazon forest was leveled to create three palm oil plantations by just one Palm oil company.[695] In 2017 alone, over 155,000 hectares (600 square miles) of Peruvian forests were cut down.[696]

Poverty plays a role in deforestation as some of the most impoverished communities are located near rainforests. To the rural poor, the rainforest becomes a source of survival where few other options exist.[697] Because they lack the resources to improve the existing cleared land, they are forced to move and deforest another piece of land, continuing the destruction cycle. These poor farm households have little incentive to care about the environmental effects of their actions.

> *Deforestation is affected mainly by the uneven distribution of wealth. Shifting cultivators at the forest frontier are among the poorest and most marginalized sections of the population. They usually own no land and have little capital. Consequently they have no option but to clear the virgin forest. Deforestation including clearing for agricultural activities is often the only option available for the livelihoods of farmers living in forested areas.*[698]

Gold mining has rapidly increased in parts of the Amazon, with small clandestine operations making up more than 50% of all gold mining activities. In the Madre de Dios region of the Peruvian Amazon from 1999 to 2012, the area of gold mining increased by 400%, and following the 2008 global economic recession, gold mining tripled. From 2008 to 2012, mines expand by deforesting approximately 61 square kilometers (23 square miles) per year in just this area.[699] The 500 square kilometers (193 square miles) of cleared rainforest is equal to double the size of the city of Boston. Illegal gold mining is rampant and is either simply informal or run by criminal mining organizations driving other criminal activities such as human trafficking, forced labor, prostitution, and money laundering.[700] In Peru, illegal mining is an enormous business estimated in 2012 to be worth more financially than the cocaine trade.[701]

With this Peruvian gold rush, as many as 30,000 miners[702] swarmed along the Madre de Dios River, not only clearing forests but use mercury as the easiest, yet highly toxic, way to extract the gold.[703] Artisanal, or small-scale, miners release an estimated 1.32 kilograms (2.9 pounds) of mercury into waterways for every 1 kilogram (2.2 pounds) of gold extracted.[704] Worldwide, artisanal gold mining releases 900 metric tons (1,000 tons) of mercury into the environment every year.[705] This accounts for one-third of all environmental mercury contamination and is the second-worst source of mercury pollution globally after the burning of fossil fuels.[706] Mercury enters the air when it is burned off during the gold extraction process or enters waterways, where it is eventually taken up by plants, fish, and humans. The mercury damages the nervous system of miners and their families. It also travels thousands of miles in the atmosphere, settling in oceans and river beds worldwide, and then it biomagnifies up the food chain.

> *"The continued use of mercury in gold mining threatens millions of people all over the world, since mercury is a global air pollutant," said Michael Bender, a coordinator for the Zero Mercury Working Group, a coalition of 40 groups worldwide that campaigns to reduce mercury use. "We're talking about a neurotoxin that science clearly shows threatens pregnant women, their fetus and those who eat large amounts of fish."[707]*

In 1965, only one highway existed in the Brazilian Amazon. Now numerous roads penetrate into the rainforest.[708] As of 2012, there were 96,500 kilometers (60,000 miles) of roads in the Amazon Basin,[709] with 51,000 kilometers (over 31,600 miles) built just in Brazil between 2004 and 2007.[710] Oil and gas access roads are also creeping deeper into the western Amazon. Oil and gas blocks are now covering 730,000 square kilometers (281,800 square miles), which is larger than the state of Texas.[711]

Roads are the primary drivers of deforestation and environmental degradation. They are often built by industries that directly cause deforestation, such as mining, energy exploitation, and commercial logging. Roads also allow access to settlers, land speculators, ranchers, farmers, and illegal miners and loggers. The hot winds that dry out forests along the edges of roadways can result in wildfires and forest degradation that weaken ecosystems.[712] According to Dr. William Laurance of the Centre for Tropical Environmental & Sustainability Science at James Cook University in Australia,

> *The incredible expansion of roads into the last remaining tropical wildernesses—this is a true environmental crisis, because such roads open up a Pandora's Box of problems, often leading to large-scale forest destruction and degradation.[713]*

The completion of more paved roads allowed settlers to make their way north and west into the forest. Along these highways in the southern part of the Amazon basin is where most of the Amazon deforestation has occurred. It is known as the "Arc of Deforestation."

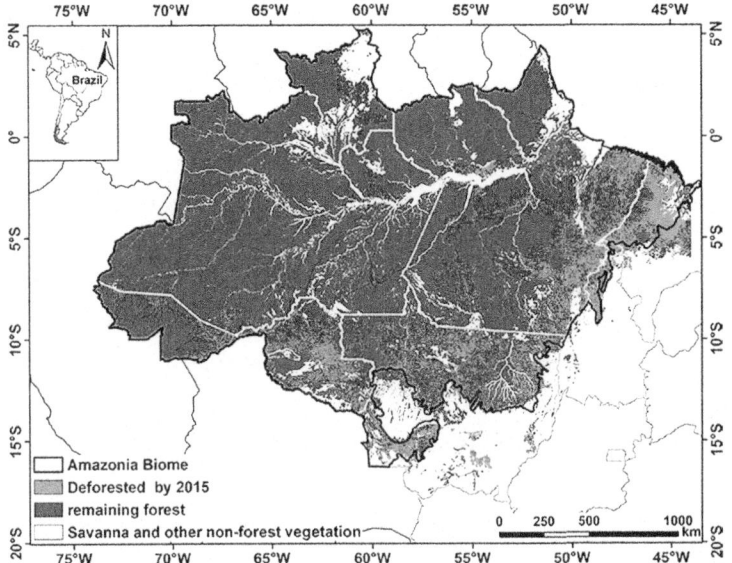

Amazon Deforestation through 2015 in Legal Amazonia and the Amazonia biome. The "arc of deforestation" is the heavily impacted crescent-shaped area along the eastern and southern edges of the forest.

Opening roads inevitably sets in motion a chain of land invasion, land speculation, and deforestation that quickly escapes government control. An urgent example of this is the planned reopening of the abandoned Manaus-Porto Velho highway, which, along with existing and planned roads linking to this highway, would open about half of what is left of Brazil's Amazon rainforest to the soy growers, ranchers, loggers, and others from the notorious "arc of deforestation" that stretches along the southern edge of the region.[714]

End of the Amazon

From 1970 to 2017, nearly 20% of the Brazilian Amazon rainforest has been lost, with deforestation still continuing.[715] By 2016, Brazil's Amazon forest, which was initially the size of Western Europe, had been deforested by 784,666 square kilometers (over 300,000 square miles) or the size of France and the United Kingdom combined.[716] The ever-growing deforestation of the Amazon will reach a point that scientists warn will have significant consequences.

As deforestation degrades the forest, there will be decreased rainfall and a longer dry season because fewer trees create moisture in the atmosphere. As its own name implies, rainforests require large amounts of rain, and as precipitation decreases at some point, the area will no longer be able to support a rainforest ecosystem.[717]

The end result of deforestation is that more than 50% of the rainforest can be radically transformed into degraded savanna, where the vegetation is made up of more coarse grasses and scattered tree growth.[718] In addition to direct deforestation, other additional factors are impacting the rainforest. A warming climate and utilizing fire to clear areas of felled trees and weedy vegetation for pasture or agricultural crops also affect the forest's hydrological [water] cycle.[719] Together these forces are bringing the Amazon to a tipping point. In 2010 the World Bank released a report detailing the Amazon's dire situation and how it is close to this point of no return. Thomas Lovejoy, head of the committee responsible for this significant scientific investigation and world-renowned tropical biologist, warned,

> *The World Bank released a study that finally put the impacts of climate change, deforestation and fires together. The tipping point for the Amazon is 20 percent deforestation, and that is a scary result. The Amazon jungle is very close to a tipping point, and if destruction continues, it could shrink to one-third of its original size in just 65 years. The forest eventually converts to cerrado (the Brazilian savanna) after a lot of fire, human misery, loss of biodiversity and emission of carbon into the atmosphere.[720]*

The Amazonian south and southeast will receive much less rainfall making those areas more prone to fires. This will destroy the forest and further dry out the surrounding forest, reducing the Amazon's ability to produce rain. Clearing leads to more drought, leading to more fires and fewer trees, resulting in decreased rainfall and further drought. Extreme droughts have become more frequent and intense.[721] Once the tipping point is reached, it triggers a self-amplifying forest loss that pushes the Amazon into an irreversible downward slide. The World Bank report concluded,

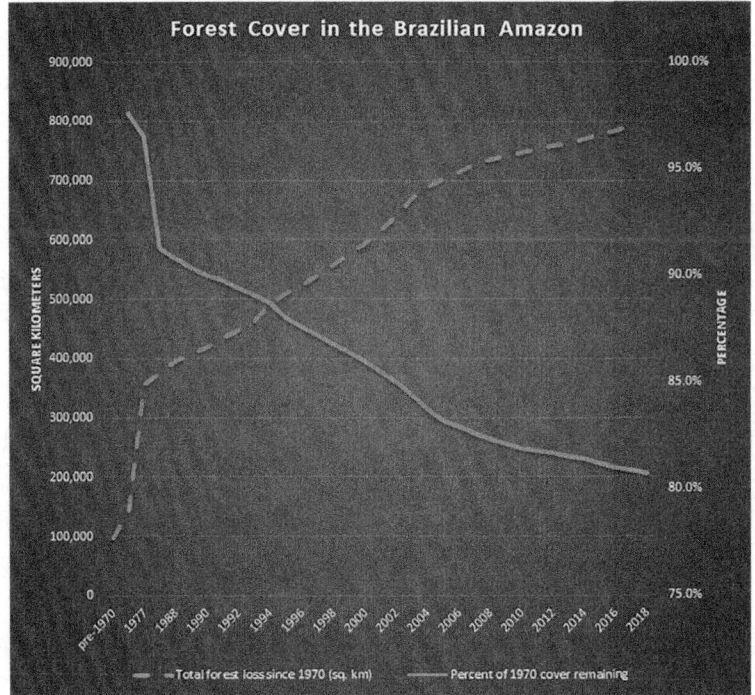

Amazon deforestation from 1970 to 2018.

For the Amazon as a whole, the remaining tropical forest will shrink to about three-quarters of its original area by 2025 and further to about only one-third of its original extension by 2075 as a result of these combined impacts of climate change, deforestation, and fire.[722]

In 2018, Lovejoy and Carlos Nobre, a member of the Brazilian Academy of Sciences, stated that once 20-25% of the Amazon is deforested, the

Amazon could "flip" to a non-forest ecosystem in eastern, southern, and central Amazonia.[723] He warns that if deforestation continues at current levels, it will be disastrous. The Amazon region could become drier and drier, unable to support healthy habitats or crops. Nobre stated,

> *Although we don't know the exact tipping point, we estimate that the Amazon is very close to this irreversible limit. Deforestation of the Amazon has already reached 20 percent, equivalent to 1 million square kilometers [386,000 square miles].*[724]

In 2019 Nobre declared that the critical tipping point for the Amazon had been reached. He urged for immediate and drastic action to stave off the global disaster of the Amazon's collapse.

> *We are no longer in a situation where the tipping point is on the horizon. It is here and now. If allowed to tip, it will affect continental climate and produce dieback in the south and east part of the central Amazon. That would represent an unconscionable release of carbon, loss of biodiversity and impact on the local people.*[725]

The Mato Grosso region in the southern Amazon, an area more than twice the size of California, is the most damaged region of the Amazon rainforest, which has already had 17% of its forest cleared. Mônica Carneiro Alves Senna and colleagues at the Federal University of Viçosa, Brazil, used computer models to simulate how the area would recover from various deforestation amounts. When 20% of the area was deforested, they found that the region became a dry, bare savanna and could not recover to its forested state even after 50 years.[726] Patrick Keys of the Stockholm Resilience Center in Sweden warns that Brazil's megacities of Rio de Janeiro and Sao Paulo, as well as Argentina's Buenos Aires, could also be vulnerable because much of their rainfall originates in the Mato Grosso region, where forests and grasslands are rapidly being replaced by corn and soy fields.[727]

What would be even more devastating is that if the forest itself is drawing in moisture across South America from the Atlantic as proposed by the "biotic pump" theory, then as it becomes more and more deforested, the force that drives moisture across the continent will significantly diminish. This would create a situation where it could turn the area not into a savanna but into a much drier desert.

> ...were the forest to disappear, then according to the [biotic pump] theory, moisture would no longer be sucked in and, given the natural fall-out rate of rainfall, some 600 kilometres [370 miles] from evaporation to precipitation, the land would dry out and in all likelihood turn to desert. Were that the case it would be a disaster of momentous proportions, not just dwarfing the likely changes resulting from global warming but indeed compounding them.[728]

In the Colombian Amazon, Leticia is about 2,500 kilometers (1,500 miles) west of the Brazilian coast and is currently a rainforest that receives 2,500 millimeters (98 inches) a year. If the moisture transportation system collapses, it could be altered to be as dry as Israel's Negev Desert, which only gets 10 centimeters (4 inches) of rain yearly.[729]

The tremendous amount of moisture pumped into the atmosphere by the billions of trees in the Amazon plays a significant role in regulating the earth's climate. If it is disrupted, the climate throughout the region and beyond will be significantly impacted.

> The importance of the Amazon rainforest in regulating not only South America's climate but also that of the entire world cannot be overestimated. Like the Earth's cryosphere, the Amazon and other rainforests are essential geographic features of the planet that help regulate the climate and provide habitat for unique wildlife. As with the melting polar regions, the loss of the Amazon to capitalist "resource development" will prove to be a self-destructive act for all of mankind.[730]

In 2015, much of Brazil was gripped by one of the worst droughts in its history. Massive reservoirs were dry, and water has been rationed in São Paulo, a megacity of 20 million people, in Rio and in many other places.[731] In October 2014, which is usually the beginning of the rainy season, it was drier than any time since 1930, leaving the volume of the water reserve system that supplies the state of São Paulo, Brazil, down to 5% of capacity.[732]

A Princeton study suggested that deforesting the Amazon could potentially contribute to drought in places as far away as California.[733] Other research indicates that recent droughts from Mexico to Texas have been linked to tree cutting in the Amazon. Specifically, deforestation of the Amazon was found to severely reduce rainfall in the Gulf of Mexico, Texas, and northern Mexico during the spring and summer seasons when water is crucial for agricultural productivity.

> *Associated changes in air pressure distribution shift the typical global circulation patterns, sending storm systems off their typical paths. And, because of the Amazon's location, any sort of weather hiccup from the area could signal serious changes for the rest of the world like droughts and severe storms.*[734]

Each year, the eight trillion tons of water that evaporate from Amazon forests[735] produce rainfall in the Andes Mountains in the west and agricultural areas in southern Brazil, Paraguay, and northern Argentina. Moisture from the Amazon reaches all the way up to the Midwestern United States during the time farmers are planting. Losing the Amazon rainforest could diminish rainfall across the Americas, with potentially severe consequences for farmers. Adrian Forsyth, a tropical ecologist who is the president and co-founder of the Amazon Conservation Association, noted,

> *There's this trillion-dollar subsidy of rainfall coming to agricultural and urban areas that people simply didn't know about until recently.*[736]

For decades economic and political forces have been steadily gnawing away at the Amazon, pushing it ever closer to tipping into an unstoppable downward spiral. Permissive land-use policies and cheap farm acreage have helped catapult Brazil into an agricultural superpower. Brazil is the world's largest exporter of soy, beef, and chicken, as well as a major pork and corn producer.[737] More international markets for Brazilian beef have opened, with exports steadily increasing.[738]

Agriculture is not only impacting the Amazon rainforest. Roughly the size of Mexico, the largest savanna in South America, the cerrado, is rapidly being converted to mega-farms to grow products such as soy to feed livestock and sugarcane to make biofuels. From 2008 to 2018, this habitat has lost 105,000 square kilometers (40,541 square miles) or about the size of South Korea.[739] The cerrado has seen about 50% of its native forests and grasslands converted to farms, pastures, and urban areas over the past 50 years. Every year some 51,800 square kilometers (20,000 square miles), an area the size of New Jersey, is cleared to make room for crops such as soy, wheat, and cotton.

Propelled by the worldwide growing desires for biofuels, meat, chicken, and pork, large portions of South America's natural environment are remade into massive mega-farms to feed these human appetites. The cerrado is being plowed under by Brazil's large-scale agribusiness and could disappear entirely by 2030[740], destroying 11,000 plant species, 199 species of mammals, and 837 species of birds. Deforestation in the cerrado is also changing Brazil's water cycle, further raising the likelihood of drought in the coming decades.

> *The fruits of converting the cerrado are fed into the maw of the global commodity market — soybeans to feed livestock in China, sugar cane to make ethanol to meet U.S. biofuel mandates, beef to feed the world's growing middle classes. The market — animated by our desires — moves so much faster than our learning curve. Our loyalties are to our own interests. Our affections rarely extend to scrubby woodlands and carnivorous plants. Our convictions don't run that deep underground.*[741]

There is an ever-increasing demand for Brazil's agricultural commodities worldwide and a push for large-scale transportation and energy infrastructure projects, including Amazon dams, roads, and railways.[742] Furthermore, despite cities like São Paulo already experiencing drought due to deforestation, the clear-cutting is likely to continue. Richard George, forest campaigner for Greenpeace U.K., explains,

> *Ironically, the drought is already affecting Brazil's agribusiness sector, a major cause of forest destruction in the Amazon. The risk, of course, is that instead of putting pressure on the government to stop deforestation, Brazil's powerful agribusiness lobby will demand the right to clear even more forest to make up for the declining yields.*[743]

Dr. Nobre and other climate experts are urging an immediate halt to deforestation, as well as large-scale planting of new forests, to bring the Amazon back to full health and stabilize its pivotal role in climate.[744] Yet, the upturn in resource conversion and degradation of the Amazon's natural habitat will likely gain further momentum. Moreover, with political and economic pressures, any environmental protections that exist are in danger of being rolled back, quickening deforestation. With the 2018 Brazilian election of Jair Bolsonaro as president, one group estimates a rapid increase in deforestation in the coming years. In addition, in 2019, the Brazilian government, under the leadership of Bolsonaro, unveiled plans to privatize the Trans-Amazonian Highway to complete and fully pave this 4,000-kilometer (2,485-mile) long road, which has already caused extensive deforestation.[745]

> *Based on an economic modeling approach that simulates the competition for land in order to meet a growing global demand for major commodities such as beef and soybeans, we estimate that, if environmental protections are removed by Brazil's next president, the average annual loss of primary forest in the Amazon will quickly rise to 25,600 square kilometers (9,884 square miles) per year, a figure similar to the deforestation rates measured at the beginning of the 2000s and an increase of 268 percent from 2017.*[746]

With this estimated acceleration in deforestation by 2021, the tipping point of 20% is quickly crossed, and by 2029 it exceeds 25% sending the Amazon into an unstoppable tailspin. Unfortunately, the situation could be even worse. A combination of expanding global demands for Amazonian products, Amazon forest fires, and drought may drive a more rapid dieback. By 2030, half the Amazon could be cleared or damaged, releasing 15 to 26 billion metric tons (16.5 to 28.6 billion tons) of carbon into the atmosphere.

Flattened Forests

Rising worldwide demands for biofuel and meat are creating powerful new incentives for agro-industrial expansion into Amazon forest regions. Forest fires, drought and logging increase susceptibility to further burning while deforestation and smoke can inhibit rainfall, exacerbating fire risk. If sea surface temperature anomalies (such as El Niño episodes) and associated Amazon droughts of the last decade continue into the future, approximately 55% of the forests of the Amazon will be cleared, logged, damaged by drought or burned over the next 20 years, emitting 15–26 Pg [1 petagram = 1 billion metric tons] of carbon to the atmosphere.[747]

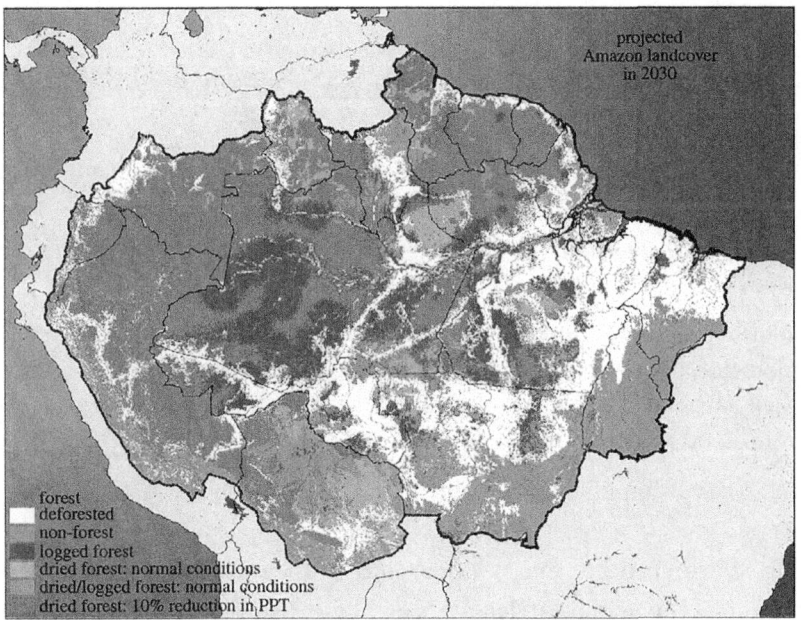

A map of Amazonia 2030, showing drought-damaged, logged and cleared forests assuming the last 10 years of climate are repeated in the future. (PPT = precipitation).

Forests absorb and store CO_2 from the atmosphere. That CO_2 is released back into the atmosphere when trees are cut and burned. The Amazon rain forest absorbs one-fourth of the CO_2 absorbed by all the land on Earth and is known as a carbon sink. However, because of deforestation, the amount absorbed today is 30% less than in the 1990s.[748] As the forest deteriorates and spirals down, a tremendous amount of stored carbon will be released into the atmosphere. So the forest would flip from a carbon sink to a carbon source amplifying the effects of a warming climate. Daniel Nepstad, executive director of the Earth Innovation Institute, noted,

> *There's this 90 billion-ton pool of carbon leaking out slowly with deforestation, and the potential for large belches of CO_2 going into the atmosphere through forest fire is very, very real. It's actually happening; it's not a hypothetical thing. Whether or not that locks us into a brand new climate in the Amazon [with half of the region dominated by grasslands instead of forest] remains to be seen, but it's a potential.*[749]

Tropical forests in the crosshairs

Tropical forests cover about 7% of the Earth's land surface yet produce 40% of its oxygen and are home to about half of all species on Earth.[750] Forests help moderate local climates keeping them cool. This effect is partly due to shading and the trees releasing moisture into the air through their leaves. This process extracts energy from the surrounding air, which cools it. This evaporation of hundreds of liters of water a day from a tree's leaves results in the forests' climate-cooling effects. Every 100 liters of water is equivalent to 70kWh [kilowatt hours] of cooling power or enough to power two 1,440 watt air-conditioning units per day.[751]

Deforestation is occurring not only in the Amazon but in tropical forests throughout the world. Africa and Asia have lost about 55% and 35% of their tropical forests, respectively.[752] Each year, an estimated 130,000 square kilometers (over 50,000 square miles) of tropical forests are demolished.[753] The deforestation rate had accelerated from 1980 to 1990 when the rate was 92,000 square kilometers (over 35,500 square miles) per year.[754] Approximately one-sixth of global carbon emissions are due to cleared or degraded forests.[755]

In 2017, slash and burn deforestation was primarily responsible for destroying an area the size of Italy. This was the second-worst year for tree loss since records began to be kept in 2001. A total of 292,000 square kilometers (113,000 square miles) was cleared from the Amazon, Congo, Indonesia, and Malaysia.[756] Over 10 years, the planet has lost 945,345 square kilometers (370,000 square miles) of natural forests or a little over the total size of Venezuela. The rate of loss of forest covering has doubled since 2003, and deforestation in tropical rainforests has doubled since 2008.[757] Frances Seymour of the World Resources Institute (WRI) said,

> *Tropical forests were lost at a rate equivalent to 40 football fields per minute. Vast areas continue to be cleared for soy, beef, palm oil and other globally traded commodities. Much of this clearing is illegal. We are trying to put out a house fire with a teaspoon.*[758]

In Indonesia and Malaysia, deforestation has been mainly due to substantial global demand for palm oil. The Indonesian island of Sumatra has been losing forests to palm oil cultivation faster than almost anywhere else on the planet.[759]

Off the coast of East Africa, the island country of Madagascar has the third-highest rate of biodiversity on Earth, after Brazil and Indonesia. Since humans' arrival 2,000 years ago, Madagascar has lost more than 90% of its original forest. Almost 40% of forest cover disappeared from the 1950s to 2000.[760] The Makira Forest, located in the northeast part of Madagascar, is one of the largest remaining contiguous tropical rainforest areas in Madagascar. Less than 10% of the Makira Forest remains. Causes of Madagascar's forests' destruction are from agriculture, uncontrolled wildfires, and lands burned for grazing.

> *Forests that once blanketed the eastern third of the island were degraded and fragmented, while endemic spiny forests have been diminished by subsistence agriculture, cattle grazing, and charcoal production. The central highlands have mostly been cleared for pasture, rice paddies, and eucalyptus and pine plantations. Each year as much as a third of the country burns, the result of fires set by farmers and cattle herders clearing land for subsistence agriculture. Meanwhile industrial miners from developed countries are tearing away at some of Madagascar's last remaining forest tracts.*[761]

Rosewood logging has affected tens of thousands of acres of protected rainforest. Much of the wood shipped to China to make wood products eventually sold in Europe and the United States. Illegal sapphire mines have also caused deforestation across Madagascar. Tens of thousands of Madagascans uproot trees and divert streams to find these gemstones to help lift them to survive in a country where 70% live in poverty.[762]

Deforestation is likely to reduce rainfall in many parts of the world while increasing it in others, considerably pushing up the global mean temperature. A case study of the effect of deforestation can be found in the Mau Forest in western Kenya, which is part of East Africa. It was once called the "water tower" because it stored rain during the wet seasons and pumped it out to the Rift Valley and Lake Victoria during dry months. From 1998 to 2013, approximately 200,000 hectares (770 square miles or larger than the size of the Hawaiian island of Maui) of the forest were converted to agricultural land.[763] Conversion of forest land into farms and settlements, illegal logging, and charcoal burning contribute to the destruction,[764] with 12,600 hectares (48 square miles) of forest lost every year.[765] Only 2% of the country is now forested when at one time half of the country was covered with jungles. From 1990 to 2010, East Africa's forest shrank by over 20%, from 107 million hectares down to 85 million hectares. That is a loss of 22 million hectares (84,900 square miles) or roughly the size of the state of Utah.[766]

Now the region suffers from severe drought, temperature extremes, and the formerly productive land has gone barren. The forest is the source of 11 main rivers feeding into five major lakes, including Lake Victoria, the source of the Nile.[767] As the rivers that flow from the forest are drying up, Kenya's harvest, tea industry, cattle ranching, hydroelectricity, lakes, and famous wildlife parks have suffered.[768] The deforestation at the heart of Kenya has triggered a "cascade of drought and despair in the surrounding valleys." In 2009 the normal rains failed. The breadbasket of Kenya, Narok County, became a barren dustbowl in April, the year's wettest month. Millions of cattle died, and the government declared a national emergency, with millions of Kenyans facing starvation.

In the Mau Forest, land surface temperatures measure 19°C (66°F) whereas, in the agricultural land, which until recently had been forest, it hovers close to 50°C (122°F). Kenya Sarah Higgins, a conservationist who runs the Little Owl Sanctuary for injured birds near Lake Naivasha east of the Mau Forest, has seen clear indications that the weather has changed with the forest's destruction.

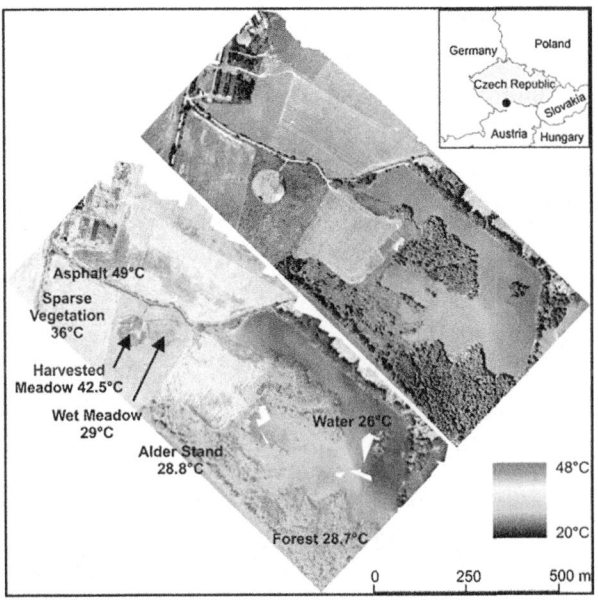

Surface temperature distribution in a mixed landscape with forest.

> When she started farming 30 years ago "we were almost guaranteed sufficient rainfall for our crops." Then came the destruction of the Mau Forest, and the area above and on either side of the farm was "denuded of trees and overgrazed, down to bare Earth. Our regular rainfall started to fail and we were seeing dry years, poor yields and more droughts."[769]

In February 2017, the Kenyan government declared a national drought emergency. Severe drought had dried up water resources in half of Kenya's 47 counties, and an estimated 3 million people lacked access to clean water.[770] Unicef reported in 2017 that 3.4 million people faced starvation in the country following a drought that ravaged the country that year.[771] Oxfam stated that the 2017 drought was "worse in a number of ways than in 2011, with some areas experiencing the failure of three rains in a row."

As a severe drought swept across East Africa, thousands of herders were forced to abandon their ancestral grazing lands and embark on a desperate search for water and pasture.[772] Dozens have been killed and injured in Kenya's drought-stricken Laikipia region as armed herders searching for scarce grazing land drove tens of thousands of cattle onto private farms and ranches from poor-quality common land.[773] Increased droughts due to deforestation exacerbated by climate change and population growth are increasing stress and conflicts throughout the area. More disasters loom on the horizon as deforestation of the Mau Forest continues.

As in the Mau Forest case, the loss of tropical forests worldwide will significantly impact the climate. Michael Wolosin and Nancy Harris of the World Resources Institute published a study that noted the global impact of tropical forest deforestation.

> *Tropical forest loss is having a larger impact on the climate than has been commonly understood. Deforestation contributes to warming and disrupts rainfall patterns at multiple scales. These changes will impact all of us, threatening agricultural productivity in the tropics and beyond... changes in rainfall driven by tropical deforestation combined with warmer temperatures could pose a substantial risk to agriculture in key breadbaskets halfway around the world in parts of the U.S., India, and China.*[774]

Altering land use to agriculture or pastures by destroying forests has a significant impact on the climate's warming. Not only does this human activity account for a rise in CO_2, but it also increases methane (CH_4) and nitrous oxide (N_2O), which are also greenhouse gases. These factors could account for as much as 40% of the warming in the climate since 1850.[775] This makes deforestation a critical component of the Paris climate agreement to keep the global temperature from rising less than 1.5°C (2.7°F) or at worst 2°C (3.6°F). The authors of a 2017 study noted that if land conversion changes continued unabated, temperatures would increase significantly even if all other human-caused climate-related activities, such as fossil fuel burning, were halted entirely. Temperatures could even exceed the upper agreed limit of 2°C (3.6°F).

> *Using a more realistic estimate of tropical-business-as-usual conversion of forests, we estimate about 1°C rise in temperature will result by 2100 solely due to LULCC (land use and land cover change), with a substantial probability of exceeding 2°C warming relative to pre-industrial temperatures even without non-LULCC (e.g., burning of fossil fuels) emissions.*

The Amazon rainforest is still the largest, pristine, and roadless wilderness in the world. Yet, an increasing number of people in the Amazon region have more investment money pouring in for economic development every year. The virtually singular drive for profit-generating growth fuels the ever-expanding road system that increases legal and illegal enterprises of ranching, agriculture, logging, mining, and more at the expense of the long-term viability of the rainforest.

Hacking down our future

The ongoing destruction of this ever-dwindling jungle is occurring just as it has happened throughout modern human history. Once people find resources they value, they lay waste to the environment to acquire them. Short-term profits and the political power derived from them almost always trump long-term thinking and sustainability. Deforestation destroys the habitat for an immense diversity of species and threatens the welfare of more than a billion people who rely on forests for their livelihoods. Large-scale deforestation is a global catastrophe in the making. It is primarily driven by the rampant consumer desires in developed countries for commodities such as beef, soy, gold, timber, and palm oil. Harrison Ngau, an indigenous tribesman from Sarawak, Malaysia, explains,

Flattened Forests

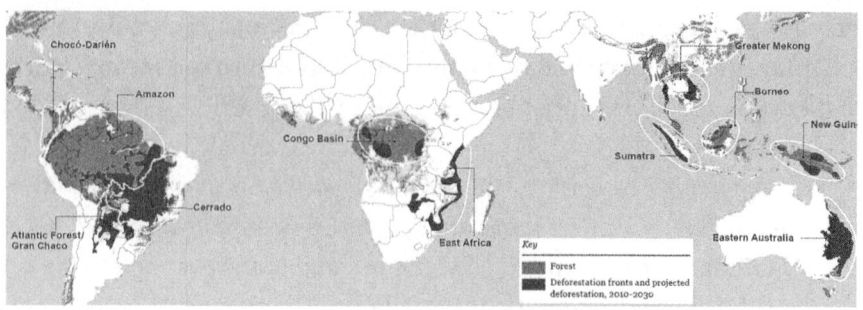

Hotspots of projected forest loss between 2010 and 2030.

> *The roots of the problem of deforestation and waste of resources are located in the industrialized countries where most of our resources such as tropical timber end up. The rich nations with one-quarter of the world's population consume four fifth of the world's resources. It is the throw away culture of the industrialized countries now advertised in and forced on to the Third World countries that is leading to the throwing away of the world. Such so-called progress leads to destruction and despair.*[776]

Human societies are often dysfunctional and destructive, believing that nature is simply a limitless resource to be consumed. Clearing forests may make profits for those who are doing it, but it impoverishes and imperils the planet as a whole over the long run. Reducing deforestation and replanting forests need to be prioritized in the Amazon and forests across the globe. A 2015 Living Forests report by the WWF (World Wide Fund for Nature) identified 11 major global deforestation hotspots.[777] These are places where the most extensive forest loss concentrations or severe degradation are projected to occur between 2010 and 2030 under business-as-usual scenarios and without interventions to prevent losses. According to Alfonso Cauteruccio, president of Greenaccord,

> *The forests represent the balance and stability within the ecosystem. However, each year we lose 16 million hectares [160,000 square kilometers/62,000 square miles] of forest. These are astonishing figures and they should be enough to start immediate action since forests produce oxygen, filter air, regulate the humility of the*

> *surrounding area, absorb enormous quantities of greenhouse gas and provide refuge and sustenance to local populations. Forests should be considered universal resources because they guarantee the equilibrium of the planet and because their protection means the protection of all humanity, but instead, they depend on the weak legislation of every country.*[778]

If the governments and people of the world do not get serious about bringing the forces of destructive development under control, these magnificent rainforests will continue to disappear decade by decade. Not only do forests need to be conserved and restored, but so do other vital natural ecosystems, including mangroves and peatlands. Without a serious effort to solve this ecosystem destruction problem, the planet will lose its most valuable climate stabilizing resources, vastly increasing the risk from a warming climate.

Reforestation removes CO_2 from the atmosphere and thus cools the climate. Clearing tropical forests and draining peatlands for agriculture, ranching, logging, mining, and increased settlements transfers CO_2 from these ecosystems to the atmosphere and heats the climate. In addition, the use of fire, which is often used in the tropics to quickly clear forest land, rapidly releases CO_2 as well as CH_4 and N_2O into the atmosphere, increasing the warming further.

The globalization of deforestation is far outpacing public understanding of the consequences of the resulting blowback. The Amazon is so massive that it has a substantial impact on the world's climate. No one knows the precise threshold when the Amazon will tip and enter a death spiral of unrecoverable deforestation. Distressingly, perhaps the tipping point has already been reached.

No one can fully foresee the cascading impacts as it collapses into savanna or desert. Yet, humankind proceeds on a reckless course forward toward impending devastation. Destroying the world's rainforests endangers our civilizations, for it is those very ecosystems that provide climate stability that enabled them to flourish. *By the time most of the public acts - it may be too late, with disastrous results for biodiversity, climate change, and all life on the planet.*

What you can do!

Tropical forests are being destroyed at an alarming rate. If current deforestation rates continue, all of our planet's original rainforest could be lost within decades, resulting in massive extinctions, accelerated climate change, and increased desertification. Here are some simple steps that you can take at a personal level to make a difference.

Avoid foods that cause deforestation – Many of the foods we eat are grown on deforested lands. For example, beef, soybean, and palm oil are principal drivers of deforestation in the Amazon basin. Fortunately, we can limit our contribution to these destructive industries. Reduce your meat intake, and buy your meat from local farms that use sustainable practices. Check your food product labels for soy or palm oil ingredients, and buy alternatives when possible. Choosing sustainably produced foods and products forces companies to change their practices.

Avoid products that cause deforestation – Choose products that are responsibly sourced or made from recycled materials. For example, mining for gold, mining for sapphires, and logging-threatened trees like ebony, mahogany, and rosewood, all of which drive rainforest destruction—research before you buy.

Buy sustainable wood products – Go paperless or reduce your use of paper products. Purchase paper products from sustainable sources, such as recycled paper or cardboard, tissues, paper towel, and napkins. Purchase eco-friendly office supplies, school and craft supplies, etc.

Choose used furniture – Consider purchasing or repurposing used furniture you already own and refurbishing when needed. Even if the piece requires some work, such as a new coat of paint or needs to be upholstered, you can create a piece that is uniquely yours while saving you money.

Use fewer paper products – Support companies committed to reducing deforestation and harvest wood from sustainably sourced forests. This ensures that any damage to the surrounding ecosystem can be monitored and managed.

Plant Trees – Plant native trees, where possible, near your home or in your community. Join organizations that are committed that are working to stop tropical deforestation and are engaged in replanting trees.

Flattened Forests

Few problems are less recognized but more important than the accelerating disappearance of the earth's biological resources. In pushing other species to extinction, humanity is busy sawing off the limb on which it is perched.
– Paul R. Ehrlich

When I consider that the nobler animal have been exterminated here - the cougar, the panther, lynx, wolverine, wolf, bear, moose, dear, the beaver, the turkey and so forth and so forth, I cannot but feel as if I lived in a tamed and, as it were, emasculated country... Is it not a maimed and imperfect nature I am conversing with?
– Henry David Thoreau, The Journal 1837-1861

In no way does civilized man so quickly revert to his former state as when he is alone with the beasts of the field. Give him a gun and something which he may kill without getting himself in trouble, and, presto! He is instantly a savage again, finding exquisite delight in bloodshed, slaughter, and death, if not for gain, then solely for the joy and happiness of it. There is no kind of warfare against game animals too unfair, too disreputable, or too mean for white men to engage in if they can only do so with safety to their own precious carcasses.
– William T. Hornaday, 1887

The great buffalo slaughter

When Europeans first arrived on the North American continent, approximately one-third of it was covered with massive bison herds, more widely known as American buffalo.[779] These immense herds literally blackened the plains, and it appeared to the new arrivals to early America that the whole country was one mass of buffalo. In 1871 Colonel Richard Dodge of the United States Army reported on a great southern herd as he rode in a light wagon 34 miles between two forts in Arkansas. That single massive herd was estimated to contain approximately four million buffalo.

> *The great herd on the Arkansas through which I passed could not have averaged, at rest, over fifteen or twenty individuals to the acre, but was, from my own observation, not less than 25 miles wide, and from reports of hunters and others, it was about five days in passing a given point, or not less than 50 miles deep. From the top of Pawnee Rock, I could see from 6 to 10 miles in almost every direction. This whole vast space was covered with buffalo, looking at a distance like one compact mass.*[780]

Buffalo was so pervasive that those at the time thought it impossible that such an enormous number of animals could ever be in danger of extermination. Some Native American tribes believed that buffaloes were continually issued from the Earth and that the supply was inexhaustible. However, within a few years, the once vast southern herd would face near-total annihilation.

As Europeans arrived in North America and began to migrate across the continent, many common regional animal species were systematically eradicated. East of the Mississippi River from 1730 to 1830, animals, including the buffalo, were hunted down primarily for food and skins, helping to fuel the rapid settlement expansion of the fledgling United States. In 1830, there was an ever-increasing desire for "buffalo-robes," as buffalo-dressed skins were termed. With this rising demand for the buffalo's flesh and hides, the era of the systematic slaughter of the buffalo began. By 1852 buffalo had entirely disappeared east of the Mississippi River.[781] At that time, with the rapidly dwindling herds, Dr. Leidy predicted that "the day is not far distant when it [the buffalo] will become quite extinct."

By 1869 the newly completed transcontinental railroads divided the vast areas of the great herds of buffalo in the American West. The three great railways, the Kansas Pacific, the Atchison, Topeka and Santa Fé, and the Union Pacific, provided easy and inexpensive transportation for buffalo hides, significantly increasing the buffalo market.[782] Tourists shot buffalo from the windows of trains for sport.[783]

Long-range rifles were another technological advance that greatly aided in the slaughter. The Sharps 40-90 or 45-120 and Remington were the favorite weapons of the buffalo-hunter. Unfortunately, these two advances and the buffalo's inability to comprehend the human threat they faced made them easy targets for the ever-expanding carnage.

> *The building of three lines of railway through the most populous buffalo country there came a demand for robes and hides, backed up by an unlimited supply of new and marvellously accurate breech-loading rifles and fixed ammunition. And then followed a wild rush of hunters to the buffalo country, eager to destroy as many head as possible in the shortest time.*[784]

A buffalo hide yard in Dodge City, Kansas, in 1874. Hunters who arrived in the spring of 1874 noted the scarcity of buffalo north of the Arkansas River; vast herds had been decimated in the hunts of 1872 and '73.

Buffalo became a booming industry, and hundreds of thousands of buffalo were slaughtered, quickly decimating the massive herds.[785] For a time, buffalo tongue was regarded as a delicacy and was often the only part extracted, leaving the rest of the animal to rot on the open plains. Tongues were purchased at 25 cents each and sold in the markets farther east at 50 cents. Until people got tired of them, buffalo tongues were in considerable demand, and hundreds, if not thousands, of barrels of them were shipped east from buffalo country.

Known as "still-hunting," a hunter would secure a position within 100 to 250 yards of his game. With the hunter hidden from view, the animals were easily picked off one by one without any risk to the hunter. It was not unusual for hunters to kill 60, 70, 80, or more buffalo a day to feed the frenzied desire for their hides. Often the buffalo was stripped of its skin with the remaining meat untouched. Whole plains were covered with the decaying buffalo remains.

> ...thousands are still killed annually merely for so-called "sport," no use whatever being made of them; thousands of others of which only the tongue or other slight morsel is saved; hundreds of thousands of others for their hides, which yield the hunter but little more than enough to pay him for the trouble of taking and selling them; while many more, though escaping from their wouldbe captors, die of their wounds and yield no return whatever to their murderers. Of the hundreds of thousands that for the last few years have annually been killed, probably less than a fourth have been to any great extent utilized.[786]

In just three years, from 1872 to 1874, hunters killed 4 million buffalo. Of those killed, 3 million were only for their hides.[787] In 1872 Colonel Dodge commented on the wholesale slaughter that was occurring.

> During this autumn, when riding some thirty to forty miles along the north bank of the Arkansas River to the East of Fort Dodge, there was a continuous line of putrescent carcasses, so that the air was rendered pestilential and offensive to the last degree.

And in 1873, he commented,

> Where there were myriads of buffalo the year before, there was now myriads of carcasses. The air was foul with a sickening stench, and the vast plain, which only a short twelvemonth before teemed with animal life, was a dead, solitary, putrid desert.[788]

Exterminate

William Frederick Cody earned fame as "Buffalo Bill" because of his accomplished skill in shooting buffaloes with a rifle from the back of a galloping horse. In 1867 he entered into a contract with the Kansas Pacific Railway to deliver all the buffalo meat to feed the laborers engaged in building the railroad that would pass through western Kansas. In eighteen months, Buffalo Bill reportedly killed 4,280 buffaloes, an incredible average of 7 to 8 per day.[789]

An estimated 40 to 80 million native peoples, referred to as Indians, lived throughout the Americas before the Europeans' arrival.[790] However, when natives were inadvertently introduced to epidemic disease from contact with the new arrivals, their populations were devastated. From 1492 to 1650, there maybe have been as much as a 90% population reduction leaving only 5.6 million natives. Later with the steady and constant migration of white men into the territories of the West, native peoples were restricted to increasingly narrower lands.

The United States government had a strategy of ridding the Plains of buffalo and Indians. Major-General Phillip Sheridan had the task of forcing Native Americans from the Great Plains and onto reservations.[791] In 1867, a member of the United States Army was said to have given orders to his troops to "kill every buffalo you can. Every buffalo dead is an Indian gone." In 1868, General Sherman wrote in a letter to General Sheridan that as long as buffalo roamed parts of Nebraska, "Indians will go there. I think it would be wise to invite all the sportsmen of England and America there this fall for a Grand Buffalo hunt and make one grand sweep of them all." In a letter back to General Sherman, General Sheridan wrote, "make them poor by the destruction of their stock, and then settle them on the lands allotted to them." In 1875, General Sheridan urged that medals with a dead buffalo on one side and a discouraged Indian on the other side be created for anyone who killed a buffalo.[792]

For thousands of years, the indigenous people of America relied on buffalo for everything essential that they needed for survival. By decimating the hunting grounds they relied on, the native peoples were pushed to a state of desperation and increasingly dependent on government assistance. Eventually, the buffalo had disappeared from that entire region, leaving the Blackfeet Indians on the verge of starvation. Some of the Cree, Chippewyan, and other native peoples starved to death. By 1877 only 300,000 natives were left living within the United States.[793]

The buffalo's increased slaughter was aided by some native peoples in trade for luxuries such as canned provisions, fancy knickknacks, firearms, ammunition, or a pint of whiskey. Amongst the Crows, who were liberally provided for by the government, horse racing was a common pastime, and the stakes were usually dressed buffalo robes.

> *In 1876, some 3 to 4 million buffalo killed on the Plains supplied hides and bones for robes and fertilizers. Three thousand hides were loaded onto each boxcar and 350 boxcars went east. In a space of 10 to 15 years, buffalo were removed from the Plains and the remaining Plains tribes relocated to reservations.*[794]

1892: bison skulls await industrial processing at Michigan Carbon Works in Rogueville (a suburb of Detroit). Bones were used processed to be used for glue, fertilizer, dye/tint/ink, or were burned to create "bone char" which was an important component for sugar refining.

By the 1870s, humans had annihilated not only buffalo but much of the indigenous wildlife. Animals other than the buffalo, such as elk, moose, deer, pronghorn, and mountain sheep, were slaughtered with the utmost recklessness. Wolves, too, had in a considerable measure been exterminated over much of the buffalo range. By the close of the hunting season of 1875, the great southern buffalo herd had ceased to exist. The land once teaming with life had been radically transformed.

> *Not many years ago, the region we traversed was swarming with buffaloes; now their skulls whitening on the plain, and the deep worn grass-grown tracks which traverse the prairies in all directions are the only evidence of their former existence. Not a single buffalo was seen during the journey, and very little of large game of any kind, only a few antelopes or cabri, one moose and one red deer. Foxes, wolves, badgers, skunks, minks, and beavers were seen or heard occasionally.*[795]

> *...the total killed [Bison] between 1870 and 1875 cannot have been less than about two and a half million annually. The effect of this destruction upon the already terribly thinned herds has been most marked, and if continued at a proportional rate, will unquestionably in a few years exterminate the race.*[796]

By 1894 the American Buffalo, once was estimated to have numbered as many as 50,000,000, was practically eradicated from North America.[797] By the close of the nineteenth century, the buffalo population had probably reached its low of about 800 animals.[798] The eradication of the American buffalo was equivalent to killing the entire human population of modern-day Spain, leaving only the passengers on a single Airbus A380 passenger jet alive.

From the brink of extinction, the American Buffalo still endure today. Approximately 500,000 live in North America,[799] although only about 20,000 make up wild herds in national parks and private reserves.[800] The other 96% have been hybridized with cattle genes and are raised commercially for meat and hides. About 70,000 buffalo are butchered each year to supply meat to a relatively small trendy market. That number is minuscule compared with the demand for beef, which results in 125,000 cattle slaughtered a day.[801]

The wanton killing of the iconic American buffalo is one egregious example of a mass slaughter of animals, but it is far from the only case.

Billions to zero

During the early years of Colonial North America, observers reported that the sky was darkened by vast flocks of passenger pigeons. These flights often continued from morning until night and could last for several days. In 1813 John Audubon remarked on an enormous number of birds that passed overhead along the Ohio River banks for three days.

Pigeon Shoot Historic Illustration.

He calculated the flock to be made up of over 1.1 billion pigeons.[802] In 1847 an enormous multitude of passenger pigeons flew over Hartford, Kentucky. A local ornithologist calculated that the mass of birds was numbered at an astonishing 2.23 billion.[803] The estimated total number of passenger pigeons that lived in America when Europeans arrived was 3 to 5 billion.[804]

After the American Civil War (1861-1865) came technological developments that set in motion the destruction of not only the buffalo but also the passenger pigeon. The national expansions of the railroads and the telegraph enabled a commercial pigeon industry to explode.

> *The professionals and amateurs together outflocked their quarry with brute force. They shot the pigeons and trapped them with nets, torched their roosts, and asphyxiated them with burning sulfur. They attacked the birds with rakes, pitchforks, and potatoes. They poisoned them with whiskey-soaked corn.*[805]

The growth of major Eastern and Midwestern cities in the mid-1800s drove the demand for meat, with passenger pigeons being an abundant and easily hunted source. Cookbooks were published with recipes for passenger pigeons to be potted, broiled, roasted, smoked, salted, pickled, or stuffed into pies. Passenger pigeon became an elegant dish in households and restaurants.[806]

> By 1855, 300,000 pigeons per year were hunted in the Midwest and sent to New York City as a source of cheap meat for the expanding human population. At the height of massive hunts to supply eastern markets in the 1870s, an army of 500-1,000 professional hunters, or "pigeoners" used telegraph and rail to track and pursue flocks to well-known roosting sites. The "great killing" in Michigan in 1878 yielded 300 tons of passenger pigeons, packed 55 dozen to a barrel, and shipped by rail to New York.[807]

Single moving birds were attached as a lure to a stool to attract hundreds of birds to be captured. This is the origin of the term "stool pigeon," which meant a person acting as a decoy. By the 1860s and 1870s, the markets had exploded with multimillions of birds being shipped east from Midwest counties, quickly decimating flocks.[808] By the mid-1890s, wild flock sizes numbered in the dozens rather than the once-massive millions or billions.

By 1900, all passenger pigeons had been killed in the wild. By 1914 the last one died in captivity at the Cincinnati Zoo.[809] The bird that once darkened the very sky with their multitude was hunted out of existence, victimized by the fallacy that no amount of exploitation could threaten a creature so bountiful.

The great hat craze

During the same period in the late 1800s, a widespread fashion craze of adorning women's hats with bird feathers swept the United States and other Western countries. To satisfy this fashion trend, plume hunters began to slaughter egrets, herons, eagles, condors, falcons, and other birds by the millions to supply milliners (women's hat makers) with this widely desired frill. In 1900 roughly 80,000 people, most of whom were women, were employed in the millinery business in New York City.[810]

As more women entered the middle class in the 1800s, a widespread market developed for bird feathers used in haute couture [high fashion], fueled by magazines such as Harper's Bazaar. Great snowy egrets, with their long feathery plumes known as "aigrettes," were most prized, but roseate spoonbills and shore birds were also routinely killed by plume hunters, the former for their colored feathers, the later for their long billed carcasses, which, stuffed whole, adorned women's hats.[811]

Woman with an entire bird in her hat, circa 1890. Late-Victorian and Edwardian fashions led to the deaths of several hundred million birds in the days before state, national, and international laws stepped in to help prevent the extinction of many of them.

At the start of the fad, plumes brought a few dollars an ounce, but at the height, an ounce of feathers was worth more than an ounce of gold.[812] In 1903 prominent dealers in New York City offered $32 an ounce for beautiful plumes.[813] In 1910 long white back feathers of egrets, aigrettes, sold for as much as $80 an ounce in New York City.[814] In 1910 gold was sold for about $21 per ounce[815], and a New York City painter who made 50 cents an hour working an average 44-hour workweek [816] needed to labor more than 3½ weeks to earn the same as an ounce of these feathers was worth. In the peak days of plume hunting, 400 or 500 birds were worth about $10,000.[817]

The egrets' brilliant white plumage became more prominent during mating season and was in exceptionally high demand among milliners. Hunters killed and skinned the mature birds, leaving orphaned hatchlings to starve or be eaten by crows endangering the entire egret population.[818] Consumer fashion demand and the desire for quick profits drove a relentless slaughter resulting in an estimated 5 million North American birds from 50 species killed each year.[819] According to The Audubon Magazine in 1888,

> *Within the past few years, the destruction of our birds has increased at a rate which is alarming. This destruction now takes place on such a large scale as to seriously threaten the existence of a number of our most useful species. It is carried on chiefly by men and boys who sell the skins or plumage to be used for ornamental purposes – principally for the trimming of women's hats, bonnets, and clothing. These men kill everything that wears feathers.*[820]

William T. Hornaday estimated that there had been a 46% decline in bird numbers in 28 states, including the District of Colombia and Indian Territory, just in the period 1883-1898.[821] Rosetta spoonbills, snowy egrets, great white herons, and short-tailed hawks nearly vanished from Florida. Wild flamingos and the lime-green-and-carmine Carolina parakeet were hunted to extinction.[822] The reddish egret was almost exterminated and slowly recovered after a single breeding pair was found in 1938.[823] Nearly 95% of Florida's birds had been killed by plume hunters, all in the name of fashion.[824]

> *Americans and Europeans even in urban areas, women and children found a decent supplemental income in stoning birds to death or killing them with pea-shooters, stringing them up, and selling them to hat-makers.*[825]

In the early 1900s, the tireless work of Doctor Pearson of the National Association of Audubon Societies eventually won a hard-fought battle over the milliners, as well as the indifference of lawmakers and the public, to make it illegal to wear or sell plumes in the United States.[826]

> *But back in 1910, when the National Association of Audubon Societies and Doctor Pearson began the drive to make the selling illegal, 16,000 milliners rose up to declare such a law beyond all bounds of reason. They had $20,000,000 invested in feathers and plumes. They said it was only a bunch of cranks and sentimentalists and schoolteachers who wanted to interfere with a free people's legitimate business.*

Doctor Pearson and a small contingent of activists, including Harriet Hemenway and her cousin Minna Hall, educated the public, and eventually, the outlawed aigrette went out of style.[827] This allowed the egret to slowly increase their population, bringing them back from the brink of extinction.[828]

Massacre on the high seas

Human ingenuity in exterminating large numbers of animals also extended to the oceans. Whales are mammals of the open sea, and for hundreds of years, people hunted them for their oil to fuel lamps and candles as well as using the oil to lubricate machinery.[829] Between 1712 and 1899, it is estimated that 300,000 sperm whales were killed globally by crews on sailing vessels.[830]

In the 1840s, there were more than 700 whaling ships on the world's oceans. More than 400 vessels called New Bedford, Massachusetts, their home port, and wealthy whaling captains built luxurious houses in the best neighborhoods. Because of whale oil, New Bedford became known as the "City that Lit the World."[831] Herman Melville, who set sail on a whaling ship from New Bedford in January 1841, would later publish Moby Dick ten years later.

> *At its peak, the cetacean [whale] killing business employed 70,000 people and became the fifth largest business in the United States. Two thirds of the world's whaling fleet sailed from U.S. At its peak, in the 1850s, it produced more than six million gallons of whale oil a year.*[832]

When we think of whaling, we primarily think of old wooden sailing vessels and the men that would go out on longboats to hunt whales with hand-thrown harpoons. Yet, whaling continued and greatly expanded well into the 20th century.

> *The opening of vast new whaling grounds, linked with a rising whale oil price, brought an enormous expansion in the fleet. In 1927/28 there were seventeen floating factories with sixty-one catchers in the Antarctic; in 1930/31 this had risen to forty-one floating factories and two hundred catchers, with a rise in whale oil production from 733,000 barrels to 3,400,000.*[833]

Grytviken whaling station on South Georgia Island during the First World War. It has long been abandoned.

As with the other animals hunted to extinction or near extinction, technological advances allowed humans to quickly overexploit. According to Quentin R. Walsh of the United States Coast Guard in 1938:

> *Some of the larger factory vessels with their capacity of over 2,500 barrels of oil per day capture more in two days than the original floating factories of 1904 were able to carry away with them in an entire season. One modern factory ship can take more whales in one season than the entire American whaling fleet of 1846 which number over 700 vessels.*[834]

During the late 1800s, steam-powered ships replaced the old sailing boats. By the early 1900s, new diesel engines[835] allowed for greater efficiency. Large floating factories became accomplished at hunting and processing whales, traveling worldwide to kill vast numbers of whales with exploding harpoons. Technology and innovation created an industry that could virtually catch, kill, and quickly process any whale in any ocean.

> *...with the aid of diesel engines and exploding harpoons, twentieth-century whalers matched the previous two centuries of sperm-whale destruction in just over 60 years.*[836]

From 1900 to the middle of 1962, the same number of sperm whales had been killed as had been killed during the 18th and 19th centuries. Astonishingly, this feat was then repeated between 1962 and 1972. During the entire 20th century, nearly 2.9 million large whales were killed all over the world.[837] In terms of sheer biomass, twentieth-century whaling is perhaps the most massive hunt in human history. While hunting for whales, the whalers also slaughtered walruses, ducks, cod, polar bears, and reindeer.

> Was the hunting of whales the biggest in history? Assuming an average hunted adult whale weighed 100,000 pounds means 100,000 * 2,900,000 = 290,000,000,000 total pounds and assuming an average adult American buffalo weighed 1,400 pounds means 1,400 * 50,000,000 = 70,000,000,000 total pounds. So by weight, the slaughter of whales was probably about 4 times larger than that of the American buffalo.

...whalers captured roughly 2.4 million kilograms [5.3 million pounds] of non-whale meat over more than 71,000 days at sea—roughly 34 kilograms [75 pounds] per day. The vast majority of their prey were walruses, caught for both food and tusks. Between the 1850s and 1860s, the number of captured walruses rose by 500 times, before collapsing. This fits with other historical evidence showing that in the mid-to-late 19th century, American ships killed as many walruses as currently survive on the planet.[838]

During the 1950s, '60s, and '70s, the number of whales and the size of those whales caught by most whaling nations declined. The International Whaling Commission (IWC) was set up to regulate whale stocks attempting to properly manage whaling. In 1982 the commission passed a global moratorium.[839]

As a result of whaling, many whale populations had been reduced to small fractions of their once natural abundance. For example, Southern Ocean blue whales are estimated to be less than 1% of their original numbers.[840] Sperm whales have been driven down to one-third of their pre-whaling population. Several whales are currently listed as endangered, including the right whale, bowhead whale, blue whale, fin whale, sei whale, humpback whale, and sperm whale. The North Atlantic right whale hovers on the brink of extinction, with only about 450 whales remaining.[841] With only 100 reproductively mature females left alive, marine biologists warn that this species could be extinct by 2040.[842]

Wild America decimated

Numerous other animals have seen their numbers decimated over the centuries. By the end of the 1800s, white-tailed deer populations vanished from most of the United States. Their numbers plummeted from between 24 and 33 million in the late 1500s to about half a million. Elk was estimated to number around 10 million across North America had their numbers hunted down to fewer than 41,000. Pronghorn antelope, also known as the American antelope, were once as numerous as the buffalo (50,000,000) were too hunted down to just 26,700 by 1924.[843] When Europeans arrived in the 1500s, the gray wolf numbered as many as 2 million across North America.[844] Their present-day numbers have fallen to less than 20,000.[845]

Like these historical examples of the past, massive animal depopulations, possibly ending in extinction, continue right up to the present day. Today many iconic large animal populations are facing increased human pressures causing their numbers to plummet. The human footprint is increasing across the globe, negatively affecting species and regions through hunting, habitat degradation, expanding agriculture, human settlement encroachment, and livestock competition.

Lions

Lions have historically lived over most of Africa, southern Europe, and the Middle East, to northwestern India. In 1950, there may have been 400,000 in the wild, and by 1975, perhaps only 200,000.[846] Over time, lion populations have been fragmented and are now confined to scattered populations in sub-Saharan Africa and a remaining population in India's Gir forest. [847] The vast majority of wild lions are gone, with their populations declining approximately 43% over the past 21 years from 1993-2014.[848]

Indiscriminate killing in defense of human life and livestock, habitat loss, and reduction in the animals they hunt are the main reasons for population decrease. An emerging threat is the African and Asian trade in bones and other body parts of lions for traditional medicine. According to estimates, 23,000-39,000 lions remain in all of Africa.[849] Scientists projected a 67% chance that the number of lions in Central and West Africa will drop by half by 2035.[850] U.S. Fish and Wildlife Service director Dan Ashe warned the African lion could go extinct by 2050.[851]

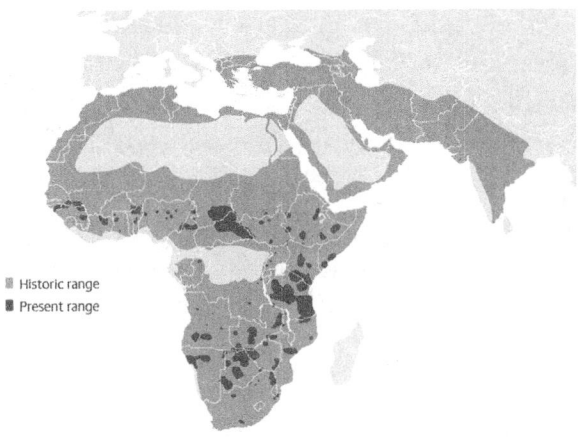

Current and historic distribution of lions: Historically, lions lived across Africa, southern Europe, the Middle East, all the way up to Northwestern India. Today their habitat has been reduced to a few tiny pockets of the original area.

Giraffes

The giraffe population is also plummeting. Giraffe numbers have declined over the last 30 years (1985-2015), falling from 157,000 down to 97,500.[852] By 2020, their numbers had dropped to under 68,300, a decrease of 56% since 1985.[853] Habitat loss, land conversion for human development, and illegal hunting are factors driving down their numbers.

> "Whilst giraffes are commonly seen on safari, in the media, and in zoos, people – including conservationists – are unaware that these

majestic animals are undergoing a silent extinction," said Julian Fennessy, co-chair of the IUCN's [International Union for Conservation of Nature] giraffe and okapi specialist group. "It is timely that we stick our neck out for the giraffe before it is too late," he said.[854]

Rhinos

Throughout most of the 20th century, the Black Rhino was the most numerous of the world's rhino species, which at one time could have numbered around 850,000. However, relentless hunting of the species and clearance of land for settlement and agriculture reduced their numbers so that by 1960 only an estimated 100,000 remained. Between 1960 and 1995, large-scale poaching caused a dramatic 98% collapse in their population.[855] South Africa is home to approximately 80% of the world's rhinos, where rhinos' recent slaughter has drastically escalated. In 2007 poachers killed 13 rhinos, and in 2014 poachers had killed 1,215, which is a 9,000% increase.[856] Today black rhino is critically endangered, with only about 5,000 left worldwide or a 99.4% decline from their once abundant numbers.[857]

The main threat for rhinos is poaching for the international rhino horn trade. Rhino horn is used in Chinese medicine as a hangover cure and used to make ornately carved handles for ceremonial daggers called jambiyas. Rhino horn is now more valuable than cocaine, heroin, or gold. A horn can fetch between $25,000 and $60,000 per kilogram ($11,360-$27,270 per pound),[858] with a single horn worth as much as £250,000 (over $325,000).[859] Rhino horn is the feathery aigrettes of their day, driving a thriving worldwide criminal organization to obtain this endangered commodity. Kingpins recruit poachers who are usually uneducated, poor people who are often simply desperate for an income to do their dirty work.

Conservation efforts have allowed a small rhino population to rebound in 2010, but they are still critically endangered. Unfortunately, poachers use military-style weapons, vehicles, and helicopters to pursue elephants and rhinos, frustrating wildlife rangers' efforts to stop them. Because of the spectacular profits, transnational organized crime-backed poaching is an enormous global threat facing rhinos and other wildlife. In the past decade, poachers are estimated to have cost the lives of about 1,000 wildlife rangers.[860]

Cheetahs

Cheetahs, the fastest land animals achieving 112 kph (70 mph), also face extinction. An assessment in 2017 determined that a little over 3,500 adult cheetahs are alive in an extensive South African area larger than the size of France.[861] Only a year earlier, their population was estimated at 7,100, which had already declined 90% since the turn of the 20th century. Cheetahs have disappeared from 91% of their historic habitat.[862] At the end of 2018, there were only about 6,600 cheetahs left in the wild.[863]

Cheetahs face increasing pressures from humans through dwindling habitats, loss of prey, and illegal wildlife trade. The Asiatic cheetah is on the brink of extinction, with only a population of about 50 left alive in Iran. Cheetahs are hunted and killed by local sheep and goat herders because cheetahs will occasionally kill and eat one of their animals. They are also frequently killed crossing highways.[864]

Pangolins

Pangolins, which look like armored anteaters, are the most trafficked mammal in the world. Hunted for its meat and scales, from the 1960s to the early 2000s, their population has declined in China by about 90%.[865] 100,000 pangolins are killed each year[866], with an estimated 1.1 million slaughtered by poachers between 2006 and 2015 in tropical Asia and sub-Saharan Africa.[867]

> Pangolins are the most trafficked mammal in the world. Each year 100,000 pangolins are killed which is equal to about a dozen every hour.

Other slaughters

Many other animal populations are plummeting. More than 100,000 elephants were killed by poachers from 2010 to 2012. Central Africa has lost 64% of its elephants in just a decade.[868] At the start of the 20th century, there were an estimated 100,000 tigers in the wild.[869] In the present day, there are only 2,154 to 3,159, which is at least a 96.8% decrease.[870] A 2018 study in PLOS Biology shows that all charismatic and iconic animals are vulnerable, endangered, or critically endangered. They are all on a downward trajectory whose endpoint is extinction.[871]

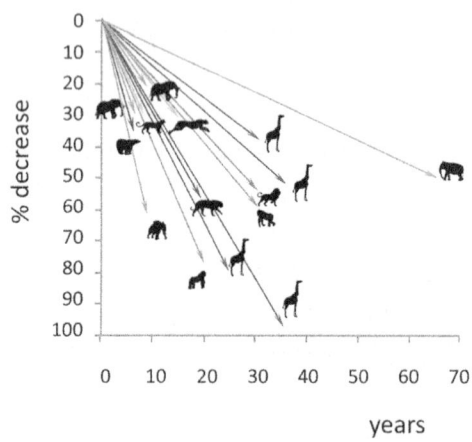

Recent, dramatic declines of the most charismatic animals. Time, but not date, is taken into account, explaining why all trajectories have the same origin. Long, steep lines indicate a large decline at a high rate. Icons represent populations. Wolf is not represented and 4 subspecies of giraffes are represented.

The declines are tigers: over 55% in the last 20 years; African lions: 54% over the last three decades; African elephants: over 20% over less than 10 years; savannah elephants: over 30% between 2007 and 2014; Central African forest elephants: 62% between 2002 and 2011; Asian elephant: over 50% in 65 years; giraffes: 38% in the last 30 years; Masai giraffes: 52% in 35 years; reticulated giraffes: 80% in 25 years; Nubian giraffes: 97% in 35 years; leopards: over 30% in 8 years; cheetahs: over 30% in the last 15 years; southern Beaufort Sea polar bears: 63% between 2004 and 2010; Grauer's gorillas: 77% in less than 20 years; Western lowland gorillas: nearly 60% in 30 years.

Bushmeat

The considerable reduction of animals is partly due to a massive bushmeat trade. Bushmeat, also known as wild meat, has long been a traditional food source for many rural people. Still, as more roads have been built into remote areas, it has allowed for large-scale commercial hunting, leaving forests, savannahs, grasslands, and deserts devoid of wildlife.[872] Like the American buffalo and the passenger pigeon of yesteryear, many animals are being eaten into extinction spurred by human encroachment and technology. Bats, pangolins, chimpanzees, gorillas, clouded leopards, and many others are on the menu.[873] At just one market in West Africa, over 9,000 primates are killed and sold every year.[874]

Wealthier nations exacerbate or even drive the problem by inflating demand and prices for meat, trophy, medicinal, and ornamental wildlife products. Bushmeat is harvested for sports hunters as well as novelty food for wealthy tourists in high-end restaurants. The illegal trade in elephant ivory, rhino horn, and other highly valued products threatening Africa's wildlife are dwarfed by the multi-billion dollar international bushmeat trade involving hundreds of species.[875]

> "There are plenty of bad things affecting wildlife around the world, and habitat loss and degradation are clearly at the forefront, but among the other things is the seemingly colossal impact of bushmeat hunting," said Prof David Macdonald, at the University of Oxford. "The number of hunters involved has gone up, and the penetration of road networks into the remotest places is such that there is no refuge left. So it becomes commercially possible to make a trade out of something that was once just a rabbit for the pot. In places like Cameroon, where I have worked, you see flotillas of taxis early in the morning going out to very remote areas and being loaded up with the [bushmeat] catch and taken back to towns."[876]

In 2008 more than 260 tons of wild meat was estimated to be smuggled in personal baggage into just a single European airport – the Charles de Gaulle Airport in France.[877] In 2011 the Center for International Forestry

Research estimated 6 million metric tons (6.6 million tons) of bushmeat is annually taken just from the Congo and Amazon basins.[878] More than 70% of the world's largest species are being pushed towards extinction primarily due to killing these animals for meat and body parts.[879]

As wealthier nations expand their fishing fleets into the waters of poor countries, it forces the local population to turn to bushmeat to survive. Tuna are among the main fish species targeted by foreign vessels. Since China has exhausted its fisheries by the mid-1980s, its distant-water fishing fleet of nearly 2,600 fishing vessels is fishing off the coast of West Africa. Their mega trawlers with their mile-long nets sweep up virtually every living thing, resulting in fishing stocks that are plummeting.[880] Each year their vessels pull three million tons of fish from the sea.[881]

> *"Foreigners complain about Africa migrants coming to their countries, but they have no problem coming to our waters and stealing all our fish," said Moustapha Balde, 22, whose teenage cousin drowned after his boat sank in the Mediterranean.*[882]

With declining fish stocks, an increasing number of Africans are compelled to hunt wild animals for sustenance, including lions, leopards, hyenas, monkeys, hippos, giant hogs, and antelopes. Hunting is the primary cause of a reported 50% decline in Gabon's apes over 20 years.[883]

Mainly, due to the bushmeat trade, Kenya has lost about 50% of its wildlife in recent decades.[884] Bushmeat hunting, primarily for food and medicinal products, drives a global crisis putting 301 terrestrial mammal species at risk of extinction.[885] Each year, rural people in Africa consume some one million metric tons of bushmeat from wildlife, equivalent to four million cattle. Elephants, gorillas, and other large forest mammals may become extinct in central Africa by midcentury if hunting meat to feed hungry human populations continues at the current rate.[886]

Worldwide wildlife slaughter

Exorbitant prices for rare items, driven by trivial human desires and superstitions, have led to the poaching of diminishing and endangered species. The shallow lust for the exotic energizes the expanding illegal wildlife trade, which is estimated to be worth $23 billion,[887] making it the fourth-largest unlawful market in the world behind drugs, guns, and human trafficking.[888] In 2019, officials seized 750 ivory tusks hidden in tree trunks. The tusks came from 325 slaughtered elephants with an estimated worth of £6.5million ($8.3 million).[889]

> For a gangster, these animals are like bundles of cash lying almost unprotected in the wilderness. This is a profit-hungry global crime conducted by some of the same ruthless and violent groups that traffic drugs and guns... poachers slice off the faces of live rhinos to steal their horns; militia groups use helicopters to shoot down elephants for their tusks; factory farmers breed captive tigers to marinate their bones for medicinal wine and fry their flesh for the dinner plate; bears are kept for a lifetime in tiny cages to have their gall bladders regularly drained for liver tonic.[890]

The lure of these considerable profits to feed human trivialities provides organized crime the incentive to perpetuate this global butchery. Illegal exploitation of natural resources has become the top way criminal and lawless militia groups worldwide fund themselves. Obliterating the natural environment and the illicit exploitation and taxation of gold, oil, minerals, diamonds, and other natural resources have overtaken drug trafficking, kidnapping, and ransom, which have long been the traditional way such groups fund themselves.[891] According to Interpol Secretary General Jürgen Stock,

> The huge volume of illicit money being generated through the exploitation of natural resources is of great concern. Criminal networks and their activities fuel violent conflict which in turn undermines the rule of law.

Shark fin soup costs upwards of $100 per bowl[892] and is seen in Asian cultures as a sign of respect, honor, and appreciation to the guests during weddings, banquets, and important business deals.[893] The population of reef manta rays, majestic cousins of the shark, has dropped by 90%[894] as they are hunted down for their gills. Their gills are sold for $75 per pound as a fraudulent Chinese medicine treatment to increase the amount of breast milk, detoxify the blood, cure chickenpox, heal tonsillitis and clear a smoker's lungs.[895]

> *You can buy a permit to shoot an endangered Tibetan antelope for US $35,000 or an argali wild sheep for $23,000. Deer antlers, musk, bones, and other parts of wild animals are used in Chinese medicine, and many animals, including blue sheep and wild yak, are being poached by hunters to supply meat markets in China.[896]*

In China and across Southeast Asia, the dried tiger penis is believed to be potent sexual medicine for men. Tiger penis soup can command up to $300 a bowl, and a whole tiger penis costs as much as $5,000.[897] A report compiled with the Norwegian risk analysis centre RHIPTO and the Global Initiative Against Transnational Organised Crime, 85% of the global trade in tigers occurs within the European Union.[898] Chinese pangolin scales can sell on the black market for over $3,000 a kilogram ($1,360 a pound). Pangolin scales are used to make coats[899] and are also roasted and chewed as part of a health fad.[900]

> *A dead tiger costs $5,000, while a live tiger costs 10 times more. A baby tiger costs $3,200 while tiger bone costs $2,000. Its penis costs $1,300 while its remains may sell as high as $70,000 in China and its skin, $35,000. Snake venom can also make an illegal wildlife trader very rich in a short period. A litre costs $215,175. Bear bile costs $200,000 per pound. The bladder of the totoaba fish costs $200,000 in China. Gorillas cost $400,000. The scales of Pangolin, the most hunted and trafficked mammal, cost $3,000 per kilogram. Polar bear skin can cost up to $9,000. If you think these are hard to get, what about tortoises that cost $10,000 in Madagascar. Orangutan costs $45,000. Ivory sells at $850 per kilogramme.[901]*

The totoaba is a marine fish that is endemic to the Gulf of California off Mexico's coast. It is being hunted because its bladder is prized in Chinese traditional medicine. It is believed to rejuvenate the skin and heal a host of ailments, from arthritis pain to discomfort during pregnancy.[902] People in Asian cultures also use the swim bladder in a soup called fish maw.[903] Not only is the critically endangered totoaba being hunted, but the vaquita marina is also being wiped out. The vaquita, the world's smallest porpoise, gets caught in the same kind of net used to capture the totoaba. According to scientists, only 30 of the vaquita remain. The totoaba bladder is considered the "cocaine of the sea," fetching as much as $20,000 in China. The lure of big profits has driven armed poachers, which are believed to be funded by drug cartels, to turn the Gulf of California into a battleground.

Primates are also increasingly being targeted. Monkey hands sell for £20-£30 ($26-40) each, and monkey skulls can fetch £100 ($130) on the Internet. According to Detective Constable Sarah Bailey at the Metropolitan Police of London,

> *The animals are shot out of trees by hunters, the young sold into the pet trade, or their skeletons sold for £250 ($330) a time. Older ones are dismembered, with some "shot to order."*

In Asia, the wealthy elites that consume illegal animal products overlap with the political and commercial establishment. The criminal entrepreneurs who run the supply lines can conceal their identity behind front companies and hide their enormous profits by using the same secretive offshore jurisdictions exploited for tax avoidance by multinational corporations. These illegal transnational organizations are plundering the natural world, driving forward the collapse of wildlife across the globe.

The sixth mass extinction

Although life around us seems relatively eternal and constant from a human perspective, over the last 3.5 billion years of Earth's history, approximately 99% of species have become extinct.[904] Species extinction is a natural phenomenon usually balanced by the evolution of new species. Yet throughout our planet's history, there have been various significant events that cause a much larger number of species to become extinct over a short period. Five mass extinctions have swept across the Earth in prehistoric times, which have exterminated species in large numbers. The most famous event that wiped out the dinosaurs is formally known as the Cretaceous-Tertiary (K–T) extinction.

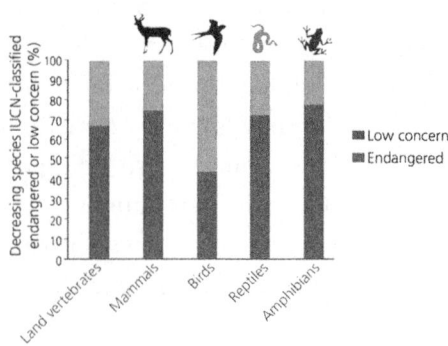

The percentage of decreasing species classified by IUCN as "endangered" (including "critically endangered," "endangered," "vulnerable," and "near threatened") or "low concern" (including "low concern" and "data-deficient") in terrestrial vertebrates. This figure emphasizes that even species that have not yet been classified as endangered (roughly 30% in the case of all vertebrates) are declining. This situation is exacerbated in the case of birds, for which close to 55% of the decreasing species are still classified as "low concern."

The relatively recent rapid loss of species and populations indicates that a sixth mass extinction event is underway on Earth. Over the past 500 years, humans have triggered an average 25% decline in Earth's animal species populations.[905] Ecosystems are now losing species at rates only seen in previous mass extinction events, with extinction rates between 100 and 1,000 times higher than pre-human levels.[906] Monopolizing resources, fragmenting habitats, pollution, introducing invasive species, spreading pathogens, killing species directly, and changing the global climate is creating a radically transformed world.

> ...there are clear indications that losing species now in the 'critically endangered' category would propel the world to a state of mass extinction that has previously been seen only five times in about 540 million years. Additional losses of species in the 'endangered' and

'vulnerable' categories could accomplish the sixth mass extinction in just a few centuries.[907]

According to the IUCN (International Union for Conservation of Nature) Red List of Species, as of 2012, 712 species of animals and 89 species of plants have been documented to have become extinct in modern times. However, local and global extinctions probably occur unnoticed because the approximately 1.2 million species currently named and cataloged represent only about 14% of the species estimated to exist on Earth.[908] In other words, researchers haven't been able to document most of the species on earth, much less track their population levels and possible extinction. As of 2019, the Red List report has almost 30,000 different species at risk of extinction. According to Lee Hannah, a climate change biologist at Conservation International,

> *The numbers are just horrendous, that's totally frightening. We've had a lot of great progress, we've got national parks, community conservancies, a lot of great conservation going on around the world, and these numbers tell us that it's just not enough.*[909]

Local extinctions are occurring all over the world. Lake Victoria is one of the African Great Lakes, and it is under assault from overfishing, pollution from industrial and agricultural sources, over-harvesting, and land clearing. A 2018 survey found that 76% of species in the lake are threatened with extinction.[910] Ugandan scientists and researchers from the International Union for Conservation of Nature and Natural Resources noted the lake's "biodiversity is being decimated."

> *Pollution has already contributed to the extinction of one of the world's rarest aquatic wildlife, the finless Porpoise (River Pig) of the Yellow River. Likewise, in the last two decades, the Yangtze sturgeon – source of the freshwater caviar and the Yangtze dolphin (Baoji) have become extinct.*[911]

According to a new report released by 16 conservation groups worldwide, nearly one-third of all freshwater fish species are threatened by extinction.[912]

> 'The World's Forgotten Fishes,' says that 80 freshwater species — which make up more than half of all the world's species — have already been declared extinct, with 16 disappearing in 2020 alone. Migratory populations have declined by more than three-quarters since the 1970s, while populations of larger species, weighing more than 60 pounds, have fallen by an even more "catastrophic" 94 percent."

Land use has been ranked as the most intensive driver of terrestrial environmental change in the present century. Each year, an estimated 13 million hectares (over 50,000 square miles) of tropical forests are destroyed, causing the loss of 14,000–40,000 species.[913] According to the IUCN, 1,169 of the world's 4,556 assessed terrestrial mammals (approximately 26%) are listed as threatened with extinction.[914] Figures approaching 30% extermination of all species by about 2050 are not unrealistic.[915]

Although species extinctions are obviously of great concern, declines in the number of populations in local communities and changes in species composition in that community can significantly impact the ecosystem. It's not just the large iconic animals like lions and elephants whose numbers have declined. Across Europe, bird populations have decreased by over 420 million over the past 30 years.[916] The house sparrow has fallen by 147 million or 62%, the starling by 53%, and the skylark by 45%. Mountain hare numbers on moorlands in parts of the eastern Highlands in Scotland have declined by 99% since the 1950s.[917]

Strong evidence indicates that globally moths and butterflies have declined in abundance by 35% over the last 40 years.[918] Research in Germany has shown a 75% decline in its flying insects, figures since matched by Dutch, and some data from England.[919] This threatens the global food supply because an estimated 75% of all the world's food crops rely on pollination provided by these insects for free.[920]

> Up to a third of all species of vertebrates are now considered threatened, as are 45% of most species of invertebrates. Among the vertebrates, amphibians are getting clobbered, with 41% of species in trouble, compared to just 17% of birds—at least so far. The various orders of insects

suffer differently too: 35% of Lepidopteran [butterflies and moths] species are in decline (goodbye butterflies), which sounds bad enough, but it's nothing compared to the similar struggles of nearly 100% of Orthoptera species (crickets, grasshoppers, and katydids).[921]

According to a 2019 study in the journal Biological Conservation, the authors examine "the dreadful state of insect biodiversity in the world," with 50% of insect species in rapid decline and over 30% threatened with extinction.[922] Habitat change and pollution are the main drivers of this looming global catastrophe. In particular, the intense use of industrial, agricultural methods over the last 60 years is the root cause of this problem. The widespread and relentless use of synthetic pesticides is a primary driver of insect losses in recent times. The authors' conclude that,

> Unless we change our ways of producing food, insects as a whole will go down the path of extinction in a few decades. The repercussions this will have for the planet's ecosystems are catastrophic, to say the least, as insects are at the structural and functional base of many of the world's ecosystems.[923]

A United Nations panel on biodiversity in 2019 reported one million species are currently threatened with extinction. Most of those are insects that comprise two-thirds of the earth's species.[924] According to a Living Planet Index report from 1970 to 2014, there has been a 58% overall decline in the numbers of fish, mammals, birds, and reptiles worldwide, potentially reaching 66% by 2020.[925] By 2020 the updated report found from 1970 to 2016, the decline had already reached 68%.[926] Species population declines are especially disastrous in the tropics, with South and Central America suffering an 89% decline during the same period.[927] World Wide Fund for Nature (WWF) conservation scientist Martin Taylor commented:

> This is definitely human impact, we're in the sixth mass extinction. There's only been five before this, and we're definitely in the sixth. Governments (need) to take action to halt the slow death of the planet because it isn't just affecting wild species it's affecting us too. This is a threat to our future as a species, what we're doing to the planet. We only

have one planet if we screw it up, then we're gone.

The rapid extinction is due to habitat loss, overexploitation of resources, pollution, and climate change. Wetlands, lakes, and rivers were the worst hit since 1970, seeing an 81% decrease in their species population. Marco Lambertini, director-general of WWF, commented,

> *The richness and diversity of life on Earth is fundamental to the complex life systems that underpin it. Life supports life itself, and we are part of the same equation. Lose biodiversity and the natural world and the life support systems, as we know them today, will collapse.*[928]

Human-caused biological annihilation

Since the dawn of civilization, humankind has caused the loss of 83% of all wild mammals and 50% of plants. Humans, and what humans generally like to eat, have primarily replaced wild animals. Livestock, mainly cattle, and pigs, make up 60% of mammals, 36% are humans, with only 4% being wild animals. Farmed poultry, such as chickens, makes up 70% of all birds, with only 30% wild.[929] A 2017 study found that the advance of the sixth mass extinction is further underway and more severe than previously feared. In their paper, the authors abandoned the dispassionate tone usually used by scientists and used the term "biological annihilation" to describe the dire situation.

> *...while the biosphere is undergoing mass species extinction, it is also being ravaged by a much more serious and rapid wave of population declines and extinctions... Population extinctions, however, are a prelude to species extinctions, so Earth's sixth mass extinction episode has proceeded further than most assume... When considering this frightening assault on the foundations of human civilization, one must never forget that Earth's capacity to support life, including human life, has been shaped by life itself. When public mention is made of the extinction crisis, it usually focuses on a few animal species (hundreds out of millions) known to have gone extinct and*

projecting many more extinctions in the future. But a glance at our maps presents a much more realistic picture: they suggest that as much as 50% of the number of animal individuals that once shared Earth with us are already gone, as are billions of populations.[930]

The authors note that the main drivers of biological annihilation are habitat destruction, overhunting, toxic pollution, invasion by alien species, and climate change. The authors pointed to the leading causes of the problem as "human overpopulation and continued population growth, and overconsumption, especially by the rich." Study author and Bing Professor of Population Studies at Stanford University Paul R. Ehrlich noted:

The serious warning in our paper needs to be heeded because civilisation depends utterly on the plants, animals, and microorganisms of Earth that supply it with essential ecosystem services ranging from crop pollination and protection to supplying food from the sea and maintaining a livable climate.[931]

We are basically annihilating the life on our planet -- and that is the only known life ... in the entire universe. It's life that shaped the planet, that made it possible for us to live here. It's life that still makes it possible for us to live here. (If) we don't have the diversity of other organisms, we're done.[932]

Human beings are taking an increasingly substantial toll on the Earth's biodiversity, stamping out other forms of life for profit and pleasure. We live in a world already radically transformed by human activity, and humanity's global footprint is expanding and continuing to adversely affect species across the planet.

The destructive environmental impact by civilization has occurred since before modern times but has become considerably more evident and severe in recent centuries. The world now faces a sixth mass extinction event whose causes are well connected to stressors such as habitat loss, overexploitation, invasive species, and climate change. Environmental fragmentation can be more than sufficient to cut species off from food, water, and mates, which allows them to thrive. This can cause an irreversible downward population spiral. As more wildlife is exterminated and habitats are destroyed, planetary ecosystems will fall apart, and extinctions will snowball.

The dark history of humankind shows a total lack of empathy for animals that live on the same planet. When there is financial gain, even for the most trivial tastes or fashions, humans will kill the animals that are the source of these frivolities. Negligent greed, or the sheer pleasure of killing another living form of life, can ultimately allow for the carnage. As William T. Hornaday observed in 1887, humans find "exquisite delight in bloodshed, slaughter, and death." To the hunter, the buffalo, passenger pigeon, whale, lion, elephant, rhino, pangolin, shark, or others are not living animals but merely a source of profit or something to satisfy a malicious lust to murder. This leaves a paltry few living animals of various iconic species in a few zoos and mostly besieged wild preserves, creating a comfortable fiction that wildlife is being preserved when the bitter truth is the opposite.

The fundamental nature of humanity is one in which human beings generally feel entitled to overtake and inhabit every square inch of land on the planet. The human race's relentless proliferation is squeezing and decimating every other form of life on the Earth – some to extinction and others to the brink. With the rapid expansion of civilization and technology, no living thing on the planet is safe from humankind. Animals of particular use are domesticated and then slaughtered yearly by the billions, mostly without even the remotest thought or compassion. Humans have radically changed and continue to change the face of the world, transforming it into a planet primarily made up of humans, cattle, chickens, pigs, wheat, corn, and soy.

The current massive degradation of habitat and extinction of species is taking place on a catastrophically short timescale. Earth's rapidly declining resources and the escalating demand from an ever-increasing human population will undoubtedly result in an apocalyptic blowback for humanity's lack of thought, concern, and empathy. Ultimately there will be a very high price to pay for the decimation of most of the only known life in the universe.

Regardless of the answers, the rapid biodiversity loss suggests that we cannot afford much delay before choosing solutions to these environmental stressors. Only bold changes will substantially diminish the imminent possibility of humans destroying much of the world's wildlife to the point of functional or global extinction. *It will take overcoming a nature of destructive hubris and heartless greed to prevent the extermination of the other forms of life on the planet and ultimately ourselves.*

What you can do!

We are on the verge of a Sixth Mass Extinction, with species being lost up to 1,000 times faster than expected because of human activities. With such a massive global problem, it seems there is little that any one person can do. Yet, here are some simple steps you can take personally to make a difference.

Preserve nature – In your local area, do all you can to make it a more nature-friendly world. Plant an organic garden. Don't think of insects as pests but as part of your natural space outside. Encourage bees, butterflies, and other insects by planting trees, bushes, and wildflowers. Refer to organic gardening books and websites to learn how to control insects naturally. By not using pesticides and chemicals to make that "perfect" lawn, you reduce toxins that you and your children and pets are exposed to. Studies of occupational exposure to agricultural pesticides, such as 2,4-D and glyphosate, have positively correlated with certain cancers.[933]

Love nature – Reconnect with nature and instill a love and reverence for nature into our children. Teach them the wonders that can be found in nature. Unfortunately, humanity has become disconnected from nature and how living in balance with nature is vital to the survival of the human race.

Choose Sustainability – Many of the things we purchase and use, such as food, palm oil, coffee, clothing, wood, gold, precious gems, and more, come from places worldwide where rare and threatened species live. Endangered species lose their habitats because of the demand for unsustainable products that increase deforestation. To help reduce your impact, be aware of where products are sourced before purchasing them, and when possible, buy used instead of new. For example, buying at used clothing and furniture stores helps ensure that you aren't contributing to additional habitat destruction.

Go organic – Insect populations are collapsing all over the planet. Over the last 60 years, intensive agriculture is the root cause of insect biodiversity destruction and is leading to insect extinction.[934] Over 1 billion pounds (the weight of over 300,000 cars) of pesticides are used in the United States each year, and approximately 5.6 billion pounds (the weight of around 1.7 million cars) are used worldwide.[935] Insecticides have a devastating impact on bees, butterflies, and other beneficial insects. If insect life continues to collapse, our planet will be in a dire state of affairs. Also, in many developing countries, it has been estimated that as many as 25 million agricultural workers worldwide experience unintentional pesticide poisonings yearly.

Avoid all products that use toxic ingredients, including pesticides. Instead, support your local organic farmers and grow your own food organically. Organically grown and local food is not only better for you; it's better for the community and your health. Your health improves with increased nutrients, omega-3 fatty acids, and lower toxic metals and pesticide residue.[936]

Avoid products derived from threatened species – We may not often notice because our local environment seems unchanging, but our natural world is being destroyed bit by bit. If we don't make changes, not only will many species vanish, but we imperil our own existence. Therefore, avoid any products that are harming the natural world.

Many products are still available such as shark fin, pangolin scales, ivory made from rhino horns or elephant tusks, totoaba, monkey skulls, and many others. Traditional Chinese medicines that use endangered animals should be avoided. Hunting and selling these is pushing the Earth ever closer to mass extinction events. Huge illicit profits are being made by exploiting natural resources that fuel criminal networks and fund violent conflicts worldwide. Avoid all these products and condemn their trade and sale.

Think where your seafood comes from – Worldwide, life in our seas is being decimated by overfishing. Also, overfishing off the coast of poor countries can push the people in those areas to resort to bushmeat to survive, further endangering our natural world. Read the chapter "Fished Out" to find out more.

Exterminate

CONSUMERISM GONE MAD

In a nation that was proud of hard work, strong families, close-knit communities, and our faith in God, too many of us now tend to worship self-indulgence and consumption. Human identity is no longer defined by what one does, but by what one owns. But we've discovered that owning things and consuming things does not satisfy our longing for meaning. We've learned that piling up material goods cannot fill the emptiness of lives which have no confidence or purpose.
– Jimmy Carter, 1979

The greatest blessings of mankind are within us and within our reach. A wise man is content with his lot, whatever it may be, without wishing for what he has not.
– Seneca

Nothing is sufficient for the person who finds sufficiency too little.
– Epicurus

Happiness

We all want to be happy with a life of satisfaction and a feeling of general well-being. It's something that has been sought after by humanity throughout history. "Life, Liberty and the pursuit of Happiness" is even enshrined in the United States Declaration of Independence as an "unalienable" right.

In Western civilization, this quality of life for many was undreamed of only a century ago. Before the early twentieth century, people longed for decent housing, clean water, sanitation, safe and hygienic workplaces, child labor laws, schools for their children, shorter workdays, and other basics to make life healthy and enjoyable. As Western societies have progressed, technological advancements and improved social structures changed life from this once abysmal drudgery and struggle to happier and more stable circumstances for millions of people.

One of the spinners in Whitnel Cotton Mill. She is 51 inches high, and has been in the mill for one year. Sometimes she works at night; runs 4 sides and earns 48 cents a day. When asked how old she was, she hesitated, then said, "I don't remember", then confidentially, "I'm not old enough to work, but do just the same." Out of 50 employees, ten are children about her size. Whitnel, N.C, December 1908.

Since that time of wondrous transformation, there has been an explosion in the availability of consumer products. People are blasted with an almost continual onslaught of advertisements in an attempt to persuade them that buying a wider variety of different products will bring more joy into their lives. The goal is to convince the consumer that the newer, bigger television with more features, the latest fashion trend, the more expensive car, the ever-expanding oversized house, the better cell phone, that latest technological gadget, or the extravagant meal at a trendy restaurant will provide that ultimate fulfillment.

In the United States, the average child sees between 50 and 70 commercials a day on television, and the average adult sees 60 minutes of advertisements and promotions a day.[937] About $150 billion per year are spent to embed consumer advertisements in every conceivable space. The message is clear. You can have a "good life" by making lots of money and spending it on products that claim to make all of us happy, loved, and esteemed.[938]

The preoccupation with consumption as a path to happiness had its roots in stimulating economic profits. American business leaders knew that people could be convinced that, however much they had, it would never be enough.[939] President Herbert Hoover's Committee report on the economy published in 1929 exemplified the attitude that the public had to be manipulated to create the desire for more products to spur economic growth.

> "Wants are almost insatiable. One want satisfied makes way for another... We have a boundless field before us; there are new wants that will make way endlessly for newer wants, as fast as they are satisfied... by advertising and other promotional devices, by scientific fact finding, by a carefully predeveloped consumption, a measurable pull on production has been created... it would seem that we can go on with increasing activity."[940]

Austrian-American Edward Bernays is known as the father of public relations, a term he invented because it sounded more agreeable than referring it to as propaganda. In the early 1900s, he became an expert in influencing consumer behavior by what he called the "engineering of consent." He pioneered the mass marketing of fashion, food, soap, cigarettes, books, and a multitude of other consumer products.[941] In his seminal work, Propaganda, he stated his belief that,

> *The conscious and intelligent manipulation of the organized habits and opinions of the masses is an important element in democratic society... We are governed, our minds are molded, our tastes formed, our ideas suggested, largely by men we have never heard of... In almost every act of our daily lives, whether in the sphere of politics or business, in our social conduct or our ethical thinking, we are dominated by the relatively small number of persons... who understand the mental processes and social patterns of the masses. It is they who pull the wires which control the public mind.*[942]

Bernays and others left behind a terrible legacy. They used psychology to create an insatiable consumer demand that drives the throwaway-culture-based economic system that we live with today.

> Advertising may be described as the science of arresting the human intelligence long enough to get money from it." – Stephen Leacock

We live in a society where we are frequently seduced into buying a great deal of meaningless junk, which generally doesn't bring true happiness but does create lots of waste and debt. Today many Americans have been widely convinced that they need more to be happy, and they work long hours to fulfill these mostly artificially induced desires.

Most advertising is repetitive and misleading, sowing anxieties often by "subtle cinematic and psychological techniques concealed from the consumer."[943] Advertising strategies not only promote a particular product but also result in the growth of status restlessness and material consumption as natural and permanent aspects of society. The modern phrase "keeping up with the Joneses" represents the anxiety of failing to meet the same socio-economic status as your neighbors. That expression originated from an early twentieth-century comic strip of the same name.

> *Buy things for every occasion. Buy things to celebrate. Buy things to mourn. Buy things to keep up with the trends. Buy things while you're buying things, and then buy a couple more things after you're done buying things. If you want it — buy it. If you don't want it — buy it. Don't make it — buy it. Don't grow it — buy it. Don't cultivate it — buy it. We need you to buy. We don't need you to be a human,*

we don't need you to be a citizen, we don't need you to be a capitalist, we just need you to be a consumer, a buyer.[944]

Today many own twice as much compared to Americans in the 1950s.[945] Then, it was typical for a family to have one bathroom or two or three growing children to share a bedroom. The average American house size has more than doubled since that time and now averages 2,349 square feet (218 square meters), where each person not only has their own television but each person also often has a private bathroom.[946]

Keeping Up with the Joneses comic strip is a domestic comedy following a family of social climbers, the McGinises, that ran from 1913 to 1938.

Yet, with all this greatly expanded material wealth, there has been no increase in well-being. Ironically, there has actually been an increase in depression and psychological problems. According to Hope College psychologist David G. Myers, Ph.D.,

> ...our becoming much better off over the last four decades has not been accompanied by one iota of increased subjective well-being.[947]

Researchers are reporting that continually buying more can damage relationships and self-esteem and increase anxiety and depression. The stress of working to afford all of these material things takes away time from what research shows genuinely makes people happy, like family, friendship, and engaging work.[948] Socrates, the classical Greek philosopher, hailed as one of the most influential thinkers of all time, believed happiness didn't come from external rewards or accolades but from the private, internal success people bestow upon themselves.[949] His philosophy and studies show what is essential to happiness, and contentment is the polar opposite of today's pervasive buy your way to happiness mindset.

Indeed, to those who are poor or working many hours to make ends meet, more money to afford a comfortable life with adequate housing, food, and clothing makes people more secure and happy. So money is a contributing factor to happiness, but only to a certain degree. Beyond that point, there are diminishing returns to where the excesses become unsatisfying. Research consistently shows that the more people value materialist aspirations and goals, the lower their happiness and life satisfaction. They have also been found to experience fewer pleasant emotions daily.[950] Yet materialism is widespread. Daniel Gilbert, a Harvard psychology professor, explains,

The secret of happiness, you see, is not found in seeking more, but in developing the capacity to enjoy less.
– Socrates

> *If it were the case that money made us totally miserable, we'd figure out we were wrong. But it's wrong in a more nuanced way. We think money will bring lots of happiness for a long time, and actually, it brings a little happiness for a short time.*[951]

Influencers on YouTube, Instagram, and other social media platforms flaunt their large, beautiful, luxurious, expensive homes and lavish lifestyles. They promote a culture of consumption, wealth, and privilege that affects the consumption habits of millions who aspire to live this unattainable and unsustainable lifestyle.

Excessive consumerism acts very much like an addictive drug. A purchase of a new, and often unneeded thing, provides a quick burst of euphoria, which quickly fades, being replaced with debt, the extra work hours required to pay it off, and the resulting time away from friends and family.

> *Exhausted by long hours working and commuting, people begin to wonder what happened to real happiness. Advertisers are there with the answer: You just need to spend still more on plastic surgery, antidepressants, or a new car.*[952]

The unfulfilled promise of bliss by acquiring more possessions is rarely realized because people are inundated with the continual "more is better" message. People become stuck on a psychological treadmill, confident that they would fill their internal void and be happier if only they make just a few more of the right purchases. This is often followed by an emotional letdown as the promise fails to deliver. Then the cycle repeats. In the end, homes are cluttered with unneeded or even unused purchases, which only amplifies the stress, leaving people feeling even more anxious, helpless, and overwhelmed.[953]

Planet trash

On top of not fulfilling its promise of happiness, excessive materialism has immense societal and ecological burdens. We generally never consider where our possessions come from, what impact manufacturing them has on the world and the people who make these goods, or where they end up when we are done with them.

> *When people are under the sway of materialism, they also focus less on caring for the Earth. As materialistic values go up, concern for nature tends to go down. Studies show that when people strongly endorse money, image, and status, they are less likely to engage in ecologically beneficial activities like riding bikes, recycling, and reusing things in new ways.*[954]

Through massive public relations campaigns, industries have fueled binge buying by tapping into primal feelings of fear and desire. Our modern, intense need for material things has fueled not only jobs and economic benefits but also a disposable culture of waste. The estimated amount of garbage humans in the world throw away each year is a staggering 2.1 billion metric tons,[955] equivalent to the weight of 4,200 Burj Khalifas, the world's tallest building, or equal to over 1.4 billion cars. In developed countries, people produce about an average of 2.6 pounds (1.1 kilograms) of trash a day. Americans make significantly more trash at 4.4 pounds (2 kilograms) of waste per day.[956]

The Burj Khalifa is the tallest building in the world. It has 163 floors, is 829.8 (2,722 feet) meters tall, and weighs 500,000 metric tons

In a nation of nearly 324 million people, that amounts to more than 700,000 tons of garbage produced daily — enough to fill around 60,000 garbage trucks. The EPA estimates that Americans generated about 254 million tons of garbage in 2013.

More than half the world's population lives without access to regular trash collection. Trash simply piles up in unregulated or illegal dumpsites, which hold over 40% of the world's garbage.[957] In poor countries, waste is often dumped in low-lying areas and land adjacent to slums. Poorly collected or improperly disposed of waste can have a tremendous negative impact on the environment. Potentially infectious medical and hazardous materials can be mixed with garbage, harmful to waste pickers and the environment. Environmental threats include contamination of surface water and groundwater and air pollution from the burning of waste that is not properly collected and disposed of.

Dumpsites are a global health emergency. The World Health Organization and others calculated that in 2012 exposures to polluted soil, water, and air resulted in an estimated 8.9 million deaths worldwide

— 8.4 million of those deaths occurred in low-and middle-income countries. By comparison, HIV/AIDS causes 1.5 million deaths per year and malaria and tuberculosis fewer than 1 million each. More than 1 in 7 deaths are the result of pollution.[958]

In addition to the toxic environment resulting from the excessive materialism of the world's wealthy nations, millions of modern-day slaves are used to make items to satiate the demand for cheap products. The global slavery index attempts to measure the scope and scale of slavery on the planet. Their 2018 report indicates that $354 billion of "at-risk" products were imported by the G20 countries.[959] The United States is the worst offender importing $144 billion, three times the second-worst offender, Japan, at $47 billion.

> As of 2020, there are 20 members of the G20: Argentina, Australia, Brazil, Canada, China, the European Union, France, Germany, India, Indonesia, Italy, Japan, Mexico, Russia, Saudi Arabia, South Africa, South Korea, Turkey, the United Kingdom, and the United States.

"The prevalence of modern slavery is driven through conflict and oppression, but it's also derived in more developed countries by consumer demand," said Fiona David, executive director of research at Walk Free, the organisation that produces the global slavery index.[960]

The top five products at risk of modern slavery imported into the G20 are laptops, computers, and mobile phones at $200.1 billion; garments at $127.7 billion; fish at $12.9 billion; cocoa at $3.6 billion, and sugarcane at $2.1 billion. As of 2016, 40.3 million people exist as modern-day slaves, with three-quarters being female. While many slaves are found in impoverished countries, 403,000 people lived in conditions of modern slavery in the United States, a prevalence of 1.3 victims of modern slavery for every thousand in the country.[961]

Wasted food

One thing that is wasted in enormous amounts across the world and that which sustains life is food. Even though there are those in the world that are hungry or starving, there is still a massive amount of food wasted daily. Nearly 800 million people worldwide suffer from hunger. At the same time, about one-third of food is squandered,[962] or about 1.6 billion metric tons (1.76 billion tons) a year, equal to the weight of over 4,800 Empire State Buildings, with a value of about $1 trillion. If this wasted food were stacked in a 20-cubic meter bin, it would fill 80 million of them, enough to reach all the way to the moon and encircle it once. The wasted food is enough to feed the hungry twice over.[963]

> *At least 1.3 billion tons of food is lost or wasted every year — in fields, during transport, in storage, at restaurants, and in markets in industrialized and developed countries alike. In rich countries alone, some 222 million tons of food [are] wasted, which is almost as much as the entire net food production of sub-Saharan Africa.*[964]

There is a vast "farm to the fork to the landfill" problem in the United States, with 40% of food going to waste and the uneaten food ending up rotting in landfills.[965] Roughly 50% of all produce is thrown away, weighing 60 million tons annually, worth $160 billion.[966] Families throw out approximately 25% of the food and beverages they buy.[967] One analysis found that students discarded roughly 40% of their fruits and 60 to 75% of their vegetables from their school lunches.[968]

It's not only the food that is wasted but also all of the water, fertilizer, pesticides, seeds, fuel, and land squandered to grow it. Globally, each year the uneaten food utilizes as much water as the entire annual flow of the Volga River, Europe's most voluminous river.[969]

> *About one-quarter of produced food is lost along the food-supply chain. The production of this lost and wasted food accounts for 24 percent of the freshwater resources used in food-crop production, 23 percent of total global cropland area, and 23 percent of total global fertilizer use.*[970]

In the United States, the Environmental Protection Agency (EPA) estimates that more food reaches landfills and incinerators than any other single material in our everyday trash, accounting for about 21% of the waste stream.[971] Food production uses up 10% of the total energy budget, 50% of the land, and 80% of all freshwater consumed.

Growing food releases hundreds of millions of pounds of pesticides into the environment each year, leading to water quality impairment in the nation's rivers and streams. The energy used from agriculture, transportation, processing, food sales, storage, and preparation to produce the 133 billion pounds of food that retailers and consumers throw out in the United States yearly is massive. It is equivalent to more than 70 times the amount of oil lost in the Gulf of Mexico's Deepwater Horizon disaster.[972]

Contributing to this food waste culture is the increase in American food portion sizes over the past 30 years, which has helped expand food waste and waistlines. The average restaurant portion today is more than four times as big as it was in the 1950s. A cup of soda is now six times as large. Burgers and a portion of fries have both tripled in size. A chocolate bar is now over 1,200% larger than it was in the early 1900s.[973] From 1982 to 2002, the average pizza slice increased by 70% in calories, the standard chicken caesar salad doubled in calories, and the average chocolate chip cookie quadrupled. Portion sizes can be 2 to 8 times larger than the USDA (United States Department of Agriculture) or FDA (United States Food and Drug Administration) standard serving sizes.[974]

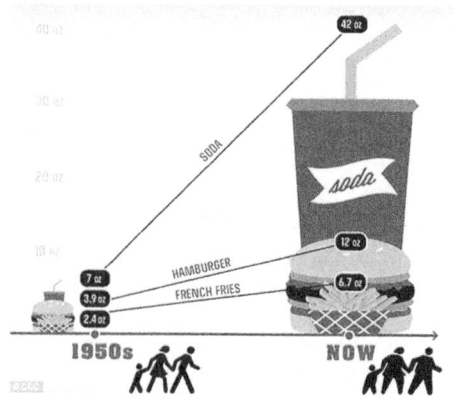

Out of control: The Centers for Disease Control and Prevention (CDC) released this graphic as part of its The New (Ab)normal campaign. It shows just how meal sizes have increased in the past 60 years.

Not only have food portions increased, but the bowls, glasses, and plates have grown to match. In America, the surface area of the average dinner plate has increased by 36% since 1960.[975] In supermarkets, the number of larger sizes has increased 10-fold between 1970 and 2000.[976] Larger packages in grocery stores, larger portions in restaurants, and larger kitchenware in homes cause an unconscious perceptual shift that makes people think it is appropriate to eat more than they should.[977]

Since the Great Depression of the 1930s, with its breadlines and food rationing, many in the United States now have easy access to vast amounts of inexpensive food twenty-four hours a day. Not surprisingly, with such an abundance of readily available food, American adults have also increased in size.

As of 2018, 42.4% of Americans are now obese. Another 30.7% of adults were categorized as overweight, and 9.2% had severe obesity.[978] This means that 82.3% of American adults are overweight or heavier. Those with obesity have a higher increased risk of overall mortality, high cholesterol, diabetes, heart disease, stroke, many types of cancer, depression, anxiety, body pain, and difficulty with physical functioning.[979] As of 2019, 1,408,647 Americans died from heart disease, cancer, and stroke, or on average 3,850 every day.[980] Childhood obesity has reached epidemic levels in developed countries. In the United States, 18.5% of children and adolescents are obese, or nearly 1 in 5.[981] Medical costs due to obesity are now $190 billion per year comprising over 21% of all healthcare costs in the country.[982]

With increasing portions, Americans are wasting more of the food they buy. The average American wastes 10 times as much food as someone in Southeast Asia and 50% more than Americans did in the 1970s.[983] Unfortunately, with increasing prosperity comes a generally indifferent attitude towards readily available and easily accessible food. Since, for many people, food represents a small portion of many Americans' overall budgets, food waste has become relatively unimportant. From an economic perspective, the more food consumers waste, the more profits are to be had by the food industry. Any waste downstream in the food supply chain means more sales and profit for anyone upstream.

American consumers have also come to expect a level of quality in the produce they purchase. A great deal of produce is culled and thrown away if it does not meet specific, unrealistic, and stringent cosmetic standards of size, color, and weight, as well as being free from blemishes.[984]

> *A large tomato-packing house reported that in mid-season, it can fill a dump truck with 22,000 pounds of discarded tomatoes every 40 minutes. And a packer of citrus, stone fruit, and grapes estimated that 20 to 50 percent of the produce he handles is unmarketable but perfectly edible.*

Supermarkets have, in recent years, started running their produce departments like beauty pageants. They say they are responding to customers who expect only ideal produce.[985] Unfortunately, the demand for "perfect" fruit and vegetables means many are discarded and left in the field to rot, even though blemishes do not necessarily affect freshness or quality. This demand for flawless-looking fruits and vegetables results in as much as half of the food grown in the United States being thrown away.[986]

> *"It's all about blemish-free produce," says Jay Johnson, who ships fresh fruit and vegetables from North Carolina and central Florida. "What happens in our business today is that it is either perfect, or it gets rejected. It is perfect to them, or they turn it down. And then you are stuck."*[987]

The food that is discarded is rarely composted and instead ends up in landfills. Food waste slowly decomposes and releases methane gas, which is considered a greenhouse gas. According to the EPA, methane causes 21 times as much warming as an equivalent mass of carbon dioxide over 100 years. In the first 20 years after its release, the intensity is much more severe. During that period, methane is 84 times as potent a greenhouse gas as carbon dioxide.[988] In the United States, landfills account for about 20% of the total methane emissions.[989] A report from the United Kingdom estimates that the amount of greenhouse gases emitted from food scraps rotting in landfills is equivalent to the emissions of one-fifth of all the country's cars. If food waste were a country, it would be the third-largest producer of greenhouse gases in the world, after China and the United States.[990]

Fast fashion

The modern culture of waste is also reflected in cheap clothes. In many industrialized countries, cut-rate clothes are available everywhere. People can select from endless piles of inexpensive jeans, shirts, and other articles of clothing at retail and fast-fashion chain stores. Like food, apparel has, in no small measure, lost any real value. According to the Bureau of Labor Statistics, in 1901, clothing accounted for 14% of Americans' total discretionary expenditures. By 1960 it had decreased to 10.4%, and then by 2013, it dropped to 3.1%.[991]

Because so much clothing can be cheaply purchased, some of it even remains unworn in closets and drawers with the sales tags still attached. In a study of women over the age of 18, the majority admitted that they had never worn at least 20% of the items in their wardrobe.[992] A survey of 2,000 women found that most clothes were only worn seven times, and a third considered clothes "old" after just being worn three times.[993] One in seven said they were strongly influenced by a social media culture that made it unacceptable to be pictured twice in the same outfit.

> *In wealthy countries around the world, clothes shopping has become a widespread pastime, a powerfully pleasurable and sometimes addictive activity that exists as a constant hum in our lives, much like social media. The internet and the proliferation of inexpensive clothing have made shopping a form of cheap, endlessly available entertainment—one where the point isn't what, or even whether, you buy, but the act of shopping itself.*[994]

Fast and low-cost fashion feeds the immediate gratification process. Each new purchase fires up excitement in the pleasure center in a shopper's brain with virtually no downside because of such low prices.

Stores such as H&M, Forever21, and Zara are only too happy to feed this desire with an almost continuous stream of something new to look at and buy while simultaneously creating enormous profits for the people at the top of these companies. The disposable clothing industry creates demand and then continuously churns out massive quantities of cheap clothes. Companies like H&M fuel binging on fashion by supplying cheap items such as women's T-shirts for $5.99 and boys' jeans for $9.99.[995]

Glossy fashion magazines eagerly declare what fashion trends "are in and out" for every year. At one time, there were the fashion seasons of spring, summer, fall, and winter. Today the fashion industry creates a buzz about 52 "micro-seasons" per year. These new fashion trends that come out every week allow the fashion industry to sell vast quantities of garments as quickly as possible.[996]

Along with the fashion industry push, the advent of digital technology shopping has become incredibly easy and more pleasurable. According to the Urban Land Institute, a survey of millennials (also known as Generation Y) found that 50% of men and 70% of women consider shopping a form of entertainment.[997] In 1991 the average American bought 34 pieces of clothing a year. By 2007 that number had doubled to 67 items a year or about 1 new piece of clothing every 5 days.[998]

Fashion is a huge business. In 2012 the global garment market was estimated to be worth $1.7 trillion, employing an estimated 75 million people in 2014.[999] China exports 30% of the world's fast fashion, with Americans buying approximately 1 billion garments each year or about 4 pieces of clothing for every United States citizen.[1000]

By definition, fast fashion is a phenomenon in the fashion industry where clothes are designed to be made as quickly and cheaply as possible. These quick and generally poorly made clothes are designed to last as long as the short fashion trend remains. Fast fashion companies profit from a small markup from the hundreds of millions of garments sold each year. If the outfit quickly falls apart, that only serves to drive the next sale.

> *"You see some products, and it's just garbage. It's just crap,"* says Simon Collins, dean of fashion at Parsons The New School for Design, on NPR. *"And you sort of fold it up and you think, yeah, you're going to wear it Saturday night to your party — and then it's literally going to fall apart."*[1001]

This trend of cheap and disposable clothing has become a severe problem for the environment, not only in its production but also in the massive quantities thrown out. The Council for Textile Recycling estimates that Americans throw away 32 kilograms (70 pounds) of clothes and other textiles each year. Approximately 10.5 million tons of garments end up in American landfills annually. Each year over 80 billion pieces of clothing are produced worldwide, with 75% ending up in landfills or incinerated.[1002] Every second, the equivalent of one garbage truck of textiles is dumped into a landfill or incinerated.[1003]

Secondhand stores receive so much excess clothing that they only resell about 20% of it.[1004] Pietra Rivoli, a professor of international business at the McDonough School of Business of Georgetown University, notes that "there are nowhere near enough people in America to absorb the mountains of castoffs, even if they were given away."

Every purchased article of clothing has an impact on the planet. The founder of the apparel company that bears her name, Eileen Fisher, has called the clothing industry "the second largest polluter in the world, second only to oil."[1005] The fashion industry is responsible for 10% of humanity's carbon emissions, producing more than all international flights and maritime shipping combined.[1006]

The fashion industry is the second-largest consumer of the world's water supply.[1007] In 2015 alone, the fashion industry consumed 79 billion cubic meters of water, enough to fill 32 million Olympic-size swimming pools.[1008] That figure is expected to increase by 50% by 2030.

Cotton, which is used in 40% of all clothing worldwide, is grown on just 2.4% of the world's cropland, yet it accounts for 24% of global sales of insecticides and 11% of pesticides.[1009] More chemical pesticides are used for the growth of cotton than any other crop.

Cotton is a water-intensive crop taking 2,700 liters (713 gallons) of water to make just one t-shirt[1010] or as much as a person would drink throughout three years. The average pair of jeans takes 7,000 liters (1,849 gallons) of water to make, with about 2 billion pairs made every year,[1011] that's 14 trillion liters (3.7 billion gallons), or enough to fill 5,600 Olympic-size swimming pools.[1012] It takes 20,000 liters (5,290 gallons) of water to produce just one kilogram (2.2 pounds) of cotton.[1013]

Only about 1% of cotton grown worldwide is organic.[1014] While organic is much more sustainable than conventional cotton, it still uses vast amounts of water and may even be dyed with chemicals and shipped globally.

The process of dyeing clothes uses 1.7 million metric tons of various chemicals.[1015] Alkylphenols, phthalates, brominated and chlorinated flame retardants, azo dyes, organotin compounds, perfluorinated chemicals, chlorobenzenes, chlorinated solvents, chlorophenols, pentachlorophenol (PCP), short-chain chlorinated paraffins, cadmium, lead, mercury, and chromium are some of the hazardous chemicals used in clothing.

According to the American Apparel & Footwear Association, the United States, the world's largest apparel market, imports over 97% of clothing and 98% of shoes purchased.[1016,1017], Globalization has pushed the manufacturing of cheap clothing to the world's poorest parts with few, if any, labor laws or environmental regulations.

In Jakarta, Indonesia, the Citarum River is known as the most polluted river in the world. Unregulated factory growth since the 1980s has choked the Citarum with both human and industrial waste. Plastics and other debris float on top of the toxic water filled with dyes and chemicals, including lead, arsenic, and mercury produced by the over 200 textile factories that line the river.[1018] The typically untreated dye wastewater is discharged directly into nearby rivers, where it eventually reaches the ocean.[1019] More than a half-trillion gallons of fresh water are used in the dyeing of textiles each year.

Made from petrochemicals, polyester and nylon are not biodegradable and flood the earth with plastic microfibers, contaminating the food chain. Clothing made from polyester is estimated to take 200 years to break down in a landfill.[1020] Clothes release half a million metric tons of microfibers into the ocean every year, equivalent to more than 50 billion plastic bottles.[1021]

Waste products from a garment factory in Dhaka, Bangladesh, spill into a stagnant pond.

The manufacturing of synthetic fabrics is an energy-intensive process that takes enormous amounts of crude oil to make. An estimated 70 million barrels of oil are used each year to produce the virgin polyester used in fabrics.[1022] A study by the Massachusetts Institute of Technology calculated that over 706 billion kilograms of greenhouse gas can be attributed to polyester production for use in textiles in 2015.[1023] Those greenhouse gases are equal to the weight of over 471 million cars, over 2,100 Empire State Buildings, and equivalent to the annual emissions of 185 coal-fired power plants. The production of synthetic materials also emits a large amount of nitrous oxide, a greenhouse gas, during manufacturing. The impact of one pound of nitrous oxide on global warming is 310 times that of the same amount of carbon dioxide.[1024]

Cheap clothing has been made possible because of the virtual slave labor in poor countries such as Bangladesh. More than four million people work in Bangladesh's garment industry, accounting for about 80% of its foreign trade. Ever decreasing clothing prices driven by fast-fashion chains in the West keep workers' wages at levels as low as $68 a month or about $2.25 a day.[1025]

Between 20 to 60% of clothing production is sewn at home. Millions of workers in poor countries are "hunched over, stitching and embroidering the contents of the global wardrobe... in slums where a whole family can live in a single room."[1026] Children can work in harsh conditions for

10 hours a day to earn a dollar.[1027] Former child factory worker Nazma Akter founded the Awaj Foundation, which fights for labor rights in Bangladesh. Akter noted that the situation for workers had not improved even after the Rana Plaza disaster in 2013, where 1,130 people died when a run-down eight-story textile factory complex collapsed.

> *They have higher production targets. If they cannot fulfill them, they have to work extra hours but with no overtime. It is very tough; they cannot go for toilet breaks or to drink water. They become sick. They are getting the minimum wage as per legal requirements, but they are not getting a living wage... There is too much work pressure, and they have not enough food, and they suffer malnutrition. They spend most of their youth in the garment industry for multinational retailers, and then they have to retire at 40 when their health is ruined.*[1028]

Major brands such as Gap, Marks & Spencer, and Adidas rely on Cambodian sweatshops to stitch their clothing. Garment prices are low in the West because workers, primarily women, will sew for about 50 cents an hour.[1029] Although the legal working age is 15 years old, girls as young as 12 drop out of school and labor in Cambodian factories. According to UNICEF and the International Labor Organization, an estimated 170 million children worldwide are currently working in the clothing industry.[1030] That number of children is equal to more than half the entire population of the United States.

Electronic waste

As technical innovation has advanced and global trade has increased, the worldwide quantity of electronic devices has exploded. Electronics are now incorporated into the daily lives of most of the people on the planet. In 2007 Apple's iPhone worldwide sales were 1.4 million. By 2015 sales had increased by over 16,400%, reaching a whopping 231.2 million.[1031] Apple is only part of an overall massive market for smartphones. Sales for smartphones have skyrocketed from 122.3 million in 2007 to over 1.5 billion in 2017.[1032] Falling prices now make electronic devices affordable for most people in the world while encouraging early equipment replacement or new acquisitions in wealthier countries. Thousands of people sometimes wait for hours for the latest electronic gadget.[1033] In 2020 there were 3.5 billion smartphone users.[1034]

Enormous amounts of modern gadgets sold – from smartphones to computers to motorized toothbrushes to greeting cards that play music to robot vacuums – are eventually thrown away. If a device breaks down, slows down, or a newer model is available, people often discard electronics and replace them.

> *Electronics have always produced waste, but the quantity and speed of discard has increased rapidly in recent years. There was a time when households would keep televisions for more than a decade. But thanks to changes in technology and consumer demand, there is hardly any device now that persists for more than a couple of years in the hands of the original owner.*[1035]

Discarded televisions, smartphones, solar panels, refrigerators, and many other devices are classified as electronic waste, also known as e-waste. In 2016, 44.7 million metric tons (49 million tons) of e-waste were generated.[1036] By 2020 that number increased to 53.6 million metric tons (59 million tons).[1037] The current rate of e-waste is equivalent to throwing out 1,000 laptops every single second.[1038]

> *In 2016 the world generated e-waste - everything from end-of-life refrigerators and television sets to solar panels, mobile phones, and computers - equal in weight to almost nine Great Pyramids of Giza, 4,500 Eiffel Towers, or 1.23 million fully loaded 18-wheel 40-ton trucks, enough to form a line from New York to Bangkok and back.*[1039]

In 2016, the worldwide e-waste average was 6.1 kilograms (13.5 pounds) per person or 24.5 kilograms (54 pounds) for a family of four. In the United States or Canada, the figure was 3.3 times higher. This surging global trend makes e-waste the fastest-growing part of the world's domestic waste stream.[1040] The carbon impact of the entire Information and Communication Industry (ICT), including personal computers, laptops, monitors, smartphones, and servers, is rapidly increasing. In 2007 ICT represented 1% of the carbon footprint and had tripled by 2018 and is set to exceed 14% by 2040.[1041]

It's not only that electronics are trashed, but all the valuable materials initially mined and procured that go along with them. Metals such as gold, silver, copper, platinum, and palladium that were thrown out with the e-waste were worth $64.61 billion in 2016.[1042] Ruediger Kuehr, head of the United Nations University's Sustainable Cycles Programme, noted that only 20% of e-waste goes into the official collection and recycling schemes.[1043]

> *Smartphones are particularly insidious for a few reasons. With a two-year average life cycle, they're more or less disposable. The problem is that building a new smartphone–and specifically, mining the rare materials inside them–represents 85% to 95% of the device's total CO_2 emissions for two years. That means buying one new phone takes as much energy as recharging and operating a smartphone for an entire decade.*[1044]

Cobalt is used to build rechargeable lithium-ion batteries used in smartphones and other electronic devices. Most of the world's global cobalt production comes from the Katanga region in the Democratic Republic of Congo (DRC). The DRC is home to the world's largest-known cobalt reserves and is responsible for about two-thirds of its cobalt production.[1045] In 2017, 64,000 metric tons of cobalt were mined, far ahead of the world's second-largest cobalt producer that year, Russia, with 5,600 metric tons. Some of the cobalt is extracted by artisanal (or informal) miners.[1046] According to UNICEF, an estimated 40,000 boys and girls work digging for cobalt for electronic devices. Some artisanal miners use chisels and other hand tools to excavate holes tens of meters deep, while others handpick rocks rich in cobalt ore at the surface. What a 13-year-old boy named Arthur, who was a miner from age 9 to 11, said is reminiscent of what Western children experienced a century or so ago,

I worked in the mines because my parents couldn't afford to pay for food and clothes for me. Papa is unemployed, and mama sells charcoal.

Children endure long hours of up to 12 hours a day working at the mines, hauling back-breaking cobalt loads of between 20 and 40 kilograms (44 to 88 pounds) for $1-2 per day. Many children are frequently ill, inhaling cobalt dust, which can cause hard metal lung disease, a potentially fatal condition. In addition, some children are killed in tunnel collapses, while others are paralyzed or suffered life-changing injuries from accidents.[1047]

Worker at an e-waste recovery site in Guiyu, Guangdong Province, China.

Most e-waste ends up in technology graveyards in poor countries in West Africa and Asia, causing significant environmental pollution and health risks for local populations. Around 41,500 metric tons (45,745 tons) of used electronic items were shipped into Nigeria from Europe and the United States hidden inside cars, buses, and trucks. Large quantities of photocopiers, smartphones, desktop computers and laptops, kettles, irons, air conditioners, power generators, microwave ovens, washing machines, and electric cookers often end up in open dumpsites.[1048] Electronic products are made up of hundreds of different materials and contain toxic substances such as lead, mercury, cadmium, arsenic, and flame retardants. Once in a landfill, these toxic materials seep into the environment, contaminating land, water, and air. Also, devices are often dismantled in primitive conditions, with those who work at these sites suffering frequent bouts of illness.[1049]

About 70% of globally generated electronic waste ends up in China, primarily through illegal channels.[1050] Guiyu, in Guangdong Province, China, is possibly the world's largest e-waste disposal site. In small workshops and in the open countryside, thousands of men, women, and children take apart old computers, monitors, printers, DVD players, photocopying machines, telephones and phone chargers, music speakers, car batteries, and microwave ovens, stripping them down to the smallest components.[1051] Hundreds of thousands of people here have become experts at dismantling the world's discarded electronics.

> They use primitive methods that leave them exposed to environmental hazards. For example, circuit boards and other computer parts are burned, individually, over open fires to extract metals. This smelting process releases large amounts of toxic gases into the air. Plastics are graded by quality, and other parts are burned to separate plastic from scrap metal. After this thorough dismembering, any remaining combustibles are left to burn in open fires, filling the air with the acrid stench of plastic, rubber, and paint.[1052]

The workers at Guiyu are poor migrants, and they and their children live and work near the smoldering waste that creates clouds of toxic smoke. The groundwater in Guiyu is undrinkable. Streams are black and pungent and choked with industrial waste. Streams that were tested were found to have acid baths leaching into them, with the water so acidic that it was found to disintegrate a penny after a few hours.[1053]

> Over 95% of the e-waste is treated and processed in the majority of urban slums of the country, where untrained workers carry out the dangerous procedures without personal protective equipment, which are detrimental not only to their health but also to the environment.[1054]

Globalization

While globalization seems like a great thing because we can get our products from all over the world, the environmental and health costs are enormous. Your foreign-made creation likely traveled halfway around the world in a container ship fueled by the dirtiest of fossil fuels. These ships burn fuel, which is made from the dregs of the oil refining process. Marine heavy fuel, also known as "bunker fuel," is pitch black, thick as molasses, and has up to 2,000 times the sulfur content of diesel fuel used in the United States and European automobiles. These ships leave behind a trail of potentially lethal chemicals of sulfur and smoke linked to breathing problems, inflammation, cancer, and heart disease.[1055] According to a Natural Resources Defense Council (NRDC) report, one container ship operating along the coast of China emits as much diesel pollution as 500,000 new Chinese trucks in a single day.[1056]

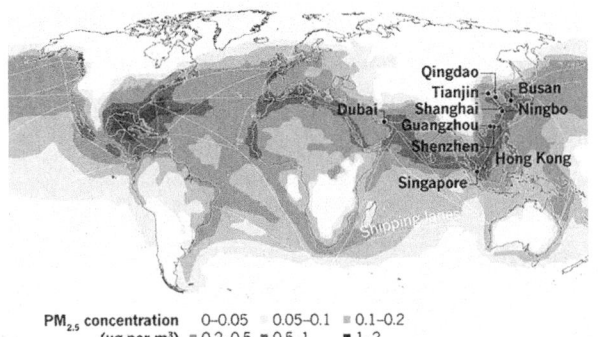

Concentrations of air pollution along global shipping routes. Until recently, most of the world's 10 busiest ports — nine of which are in Asia — had no limits on emissions. PM$_{2.5}$ refers to atmospheric particulate matter (PM) that have a diameter of less than 2.5 micrometers, which is about 3% the diameter of a human hair.

As populations increase and global economies expand, the amount of cargo that ships transport has vastly increased. From the 1970s to 2016, shipping volumes have increased by 300%, making cargo ships one of the world's largest sources of air pollution. Ship-related health impacts include approximately 400,000 premature deaths from lung cancer and cardiovascular disease and 14 million childhood asthma cases annually.[1057] However, as increasing evidence indicates the enormous risks of burning bunker fuel, the global maritime industry is overhauling cargo ship fuel supplies. Starting January 1, 2020, the International Maritime Organization (IMO) required ships to use fuel no more than 0.5% sulfur, down from 3.5%. This is projected to roughly prevent 150,000 premature deaths and 7.6 million childhood asthma cases each year, although it will still result in 250,000 premature deaths and 6.4 million childhood asthma cases.[1058]

Shipping traffic can also be a significant source of tiny plastic particles that end up in our oceans.[1059] Binders, which are part of marine paint, are applied to the hulls to protect the ships and slowly chip away, resulting in significant ocean pollution. Ships leave microplastic pollution in the water much as tire wear particles from cars are on land.

> *In the European Union alone, several thousand tons of paint end up in the marine environment every year. With potentially harmful consequences for the environment: coatings and paints used on ships contain heavy metals and other additives that are toxic to many organisms. These antifouling components are used to protect ships' hulls from barnacles and other subaquatic organisms and are constantly rubbed off by the wind and waves.*[1060]

Throwaway society

In 1955, LIFE magazine published an article titled "Throwaway Living. Disposable items cut down household chores." The new promise was that as a modern society, we didn't have to be bothered with the mundane chores of cleaning items when those items can just be thrown out. That notion has vastly escalated ever since that time. Our ever-expanding "throwaway living" is characterized by rampant consumerism and the ever-increasing trend to easily discard things and buy something new. With consumers primarily interested in price and quality, environmental and social costs are rarely considered. Industries interested in increasing sales and providing for their customers' wants also have little incentive to consider those costs. Advertisers who help compel people to purchase things they don't necessarily need are also oblivious or uncaring to these costs.

1955 LIFE magazine article – "Throwaway Living." The objects flying through the air in this picture would take 40 hours to clean-except that no housewife need bother. They are all meant to be thrown away after use.

As affluence increases and prices fall, often through the use of cheap labor in countries with lax or nonexistent labor and environmental laws, the actual value of food, clothing, electronics, and other products is lost. With the actual costs of what people buy hidden, it propels the notion that things are disposable. The primarily invisible costs continue to build and magnify with industries, media, and governments rarely interested in exposing the enormous downsides to a world economic system dependent on growth, waste, and exploitation. Yet this perpetual expansion system of consumption and waste is simply unsustainable and leaves a massive wake of human and environmental destruction. According to Story of Stuff creator Annie Leonard,

> *It is a linear system, and we live on a finite planet. You cannot run a linear system on a finite planet indefinitely. Too often, the environment is seen as one small piece of the economy. But it's not just one little thing, it's what every single thing in our life depends upon.*[1061]

Shopping is a way that people often search for themselves and their place in the world. Because people often conflate the search for self with the search for stuff, they ultimately never really find true happiness. The good news is that sustainable happiness is achievable. It could be available to everyone, and it doesn't have to cost the planet. It turns out that we don't need to use up and wear out the world in a mad rush to produce the stuff that is supposed to make us happy. We don't need people working in sweatshop conditions to manufacture cheap stuff to feed an endless appetite for possessions. It begins by assuring that everyone can obtain a basic level of material security, and it requires an understanding that excessive acquisition of things isn't the path to happiness.

Mindfulness – the focused awareness of the present moment, which can be cultivated through meditation and contemplative practice – may be an effective remedy to empty or compulsive consumption. The research shows that a sense of meaning and deep satisfaction isn't achieved through mass accumulation; instead, it comes from other sources. *We need loving relationships, thriving natural and human communities, meaningful work opportunities, and a few simple practices, like gratitude and mindfulness. With that definition of sustainable happiness, we really can have it all.*

What you can do!

The human legacy of waste and greed destroys the environment, ruins people's health, enslaves others to make cheap products while not increasing happiness for those who are overconsuming. Here are some simple steps that you can take at a personal level to make a difference.

Avoid advertising – Watching television and other media, you are bombarded with advertisements. All these commercials are attempting to influence you to purchase products that you often don't need. Turn off the television and avoid ads as much as possible. Anytime you encounter an advertisement, be aware that they are designed to manipulate you to buy their product. Don't succumb to their often false promises of happiness. Once you are aware of the manipulation, you will be far less susceptible to their influence, which will be better for you and the planet.

Decrease food waste – When food is wasted, all the resources that went to producing, manufacturing, packaging, and transporting it are also squandered. By being aware, only buying the amount you need and cutting down that waste will reduce the amount of land and resources used to grow the food. This will reduce the fertilizers, pesticides, and everything associated with raising that food. As a result, water and land quality improve. Starting your own organic food garden gives you a greater appreciation for food and further reduces waste. Consider creating a compost bin for left over food scraps. If you go to a restaurant, bring a glass or other container to pack away leftovers and take them home.

Avoid buying new clothing – New clothing takes enormous amounts of resources to produce. There are massive quantities of used clothing in consignment shops that are often more unique and generally of superior quality. If you buy new clothes, consider the most sustainable and environmentally friendly materials, such as hemp and bamboo. Maintain your quality clothing by learning how to sow and donate anything you don't need anymore.

Donate or dispose of electronics properly – If the electronic device is still working, consider donating it to a school, library, or daycare. If the product is at the end of life, check with e-Stewards or other organizations to properly dispose of it. Always consider repairing instead of replacing if possible.

Donate or dispose of white goods – Whitegoods are large domestic appliances, such as refrigerators or washing machines. They contain significant amounts of metal, plastic, insulating material, refrigerant gases, and other non-renewable and valuable material. Some charities or furniture reuse organizations accept donations of white goods, and many offer collection services. Recycling keeps these materials in the economy. It also helps prevent toxic substances such as flame-retardants from entering the environment. Check your area for home collection services, or often shops will collect your unwanted electricals when they deliver a new one.

Use online local buy-sell sites – Many times, items you can no longer use can be utilized by someone else. Reusing items reduces the environmental cost of making new things and the environmental cost of disposing of them. There are online websites that promote local buy, sell, and swap online.

Join a local repair café – There are thousands of local repair cafés around the world. Repair cafes are places where volunteers in the community help you repair household electrical and mechanical devices, computers, bicycles, clothing, and other items. It's fun, and you can learn how to fix your devices instead of disposing of them.

Avoid novelty electronics – Music playing greeting cards, light-up shoes, LED glasses, smart belts, smart floss dispensers, Bluetooth-enabled "smart forks," a device that monitors if you need to buy more eggs — These are some of the largely needless novelties and gadgets that are available to purchase. They are part of an ever-growing number of products that are of little value but add to the massive amount of e-waste generated. Avoiding these products will reduce pollution and save you money.

Buy the essentials – Too frequently, we buy what we really don't need filling our homes with clutter. Taking the time to think about what you really need and avoiding impulse buying will reduce stress on the planet, improve your life quality, and save you money.

Avoid keeping up with the Joneses – Consumer buying habits on tech has been to dispose of our electronics as soon as it's perceived to be "old." In the world of tech, that change happens at a very rapid rate. We need to ask ourselves if we are really getting the full lifespan and potential out of our products? Do we really need the newest, smallest, fastest, latest and greatest, or are we merely suffering from a fear of missing out or not keeping up with others?

The latest fashion trend, the new upgraded technology, or other trendy products impact the environment and cost you money. While a particular product might make you feel better for a short while, in the end, it probably really won't, while leaving you and the planet more impoverished. So carefully consider purchases and why you are buying something. If buying something is to maintain the same lifestyle as your neighbors or peers, it might be wise to reconsider.

Find happiness – Happiness comes from within, not from the latest snacks, electronic gadgets, clothing trends, automobiles, or anything else you can buy. You don't need to compete with your neighbors and friends to get the biggest television or fanciest car. Instead, engage in life by spending time with your family and friends, finding hobbies that interest you, reading and learning, working on your health, volunteering to help those less fortunate than you, creating a piece of art, and involving yourself with many others beneficial activities. Once you break away from the materialistic mindset, you may realize just how trapped you may have been. You can find real happiness, spend less money, and help decrease the enormous stress on our small world.

TOXIC WORLD

*The planet was being destroyed by manufacturing processes,
and what was being manufactured was lousy, by and large.*
– Kurt Vonnegut, Breakfast of Champions

*When the last tree is cut, the last fish is caught, and the last river is
polluted; when to breathe the air is sickening, you will realize, too late,
that wealth is not in bank accounts and that you can't eat money.*
– Alanis Obomsawin

*Don't it always seem to go
That you don't know what you've got
'Till it's gone
They paved paradise
And put up a parking lot.*
– Joni Mitchell, Big Yellow Taxi

Paradise lost

It was the end of spring in 1534 when French explorer Jacques Cartier began to explore the Saint Lawrence River. While still far from shore, he and his crew encountered sea birds swarming all over one rocky island. They were in such huge numbers that Cartier exclaimed, "All the ships of France might load a cargo of them without once perceiving that any had been removed." Around the island, a hundred times more birds flew than were on the island itself. The numbers were "so great as to be incredible unless one has seen them."[1062]

When Europeans arrived in North America, they stood in amazement at the natural bounty they saw. The skies were filled with unimaginable numbers of birds. Coastal waters, rivers, and lakes swarmed with an unbelievable amount of fish, which left even experienced seamen to declare, "the abundance of Sea-Fish are almost beyond believing."[1063] The forests and grasslands were teeming with an unfathomable amount and diversity of wild animals. In the mid-1600s, Nicolas Denys reported it as a country "where there is so great an abundance of game of all kinds that it is astonishing."[1064] These early arrivals believed they had discovered a type of paradise,[1065] and compared to the relatively sparse wildlife of present-day North America, the cornucopia of life would be startling. The predominantly unspoiled land was also filled with a massive assortment of plant life. The coasts, lakes, and rivers were largely pristine and pure. The air was clean and fresh.

Before the Europeans began to settle in the Americas, there lived a large native population numbering in the tens of millions. These natives had advanced cultures that had, to some degree, already impacted the environment through extensive earthworks, roads, fields, and settlements. Unfortunately, their societies would suffer an era of mass extermination from warfare, enslavement, famine, and European diseases with the arrival of early European explorers, starting with Christopher Columbus in 1492.[1066]

These initial European explorers had inadvertently brought infectious diseases that the native populations had never before encountered.

Diseases such as influenza, smallpox, and measles caused widespread death among the natives, hastening the local population's collapse.[1067] From 1492 to 1650, their numbers were decimated, with approximately 90% of their population dying.[1068] Some estimate that the native population declined from 54 to 61 million people, down to a minimum of 6 million by 1650.[1069]

After this enormous human calamity, the vastly depopulated environment reverted to a more natural state. The native societies that had existed were virtually erased from the landscape and would only be rediscovered by scientists centuries later. During this time, about 65 million hectares

> After the massive Native American depopulation, the regeneration of forests, woody savanna and grassland allowed for a large intake of carbon from the atmosphere which contributed to a decline in CO_2 of 7-10 parts per million (ppm) between 1570 and 1620.

(251,000 square miles or slightly smaller than the modern state of Texas) of land in the Americas previously altered by the natives reverted back to the forest.[1070] This considerable increase in plant life utilized more CO_2 and actually caused atmospheric levels to drop. This decrease in atmospheric CO_2 was enough to be a factor in chilling the planet during the 1500s/1600s and was known as the "Little Ice Age." During this period, the Thames River in London would regularly freeze over.[1071]

The "Great Dying" of the Americas' Indigenous people left the land largely unspoiled and ripe for Europeans to explore, exploit, and inhabit. With abundant natural resources and ever-evolving and improving technologies, the new settler populations rapidly increased in size and expanded across the continent. After 1750, and especially after the start of the Industrial Revolution in the early 1800s, resources were more intensively exploited, and the environment's modification escalated, continuing to the present day. With the aid of scientific innovations, the once primarily unspoiled wilderness was also fundamentally changed into a modern world undreamed only a few centuries ago, filled with incredible achievements, such as plumbing, electricity, roads, automobiles, planes, phones, computers, and more.

Since the 1950s, not only has the growth of human populations accelerated, but there has been an enormous increase in the use of new materials such as plastics, pharmaceuticals, pesticides, industrial fertilizers, and a vast quantity of other chemicals. These remarkable industrial, societal, and technological changes have proceeded at different rates across the globe, bringing an incredible standard of living to millions but also simultaneously causing the death and suffering of millions of others. As technological changes accelerate across a finite-sized planet, it is becoming increasingly apparent that there are massively disastrous human and environmental costs for these modern advances.

Water is life and death

Water is essential for the existence of plant and animal life. An adult human can live for more than three weeks without food but can only survive at most a week or so with no, or very limited, water intake.[1072] For many, a simple turn of a faucet or the twist of a bottle cap provides clean drinkable water. Conversely, if this indispensable life-sustaining elixir is contaminated, it can be transformed into a source of illness or even death.

Filthy and deadly drinking water is something that was commonplace before the turn of the twentieth century. Indeed, vast numbers died throughout the world from waterborne diseases like dysentery, typhoid, and cholera. During the American Civil War (1861 to 1865), the Union army lost over 186,000 men to infectious disease, double the number killed in combat. Nearly half died from typhoid and dysentery.[1073] At the National Cemetery in Andersonville, Georgia, records show that 75% of soldiers from Indiana buried there died from diarrhea, dysentery, and scurvy (a disease caused by a deficiency of vitamin C).[1074]

Cholera is a disease of poor sanitation and hygiene. The illness is usually transmitted through water or food contaminated with human and animal waste. Lack of knowledge in basic hygiene and primitive or nonexistent sanitation resulted in countless deaths and an almost unbroken series of cholera pandemics. From 1817 to 1923, six major cholera pandemics swept across the globe.[1075] Between 1817 and 1860, cholera claimed an estimated 15 million lives in India. In Ireland, the 1849 cholera outbreak is estimated to have killed as many people as during the Irish Famine of 1845 to 1849.

> *In North America, 3,500 people (5.5% of Chicago's population) died of cholera in 1854, with up to 150,000 Americans dying of cholera between 1832 and 1860... London's epidemic in 1852-1854 killed 10,738... By 1866, the outbreak reached North America, causing up to 50,000 deaths... The 1883-1887 epidemic claimed 250,000 lives in Europe and, in spreading, killed at least 50,000 in America, 267,890 in Russia, 120,000 in Spain, 90,000 in Japan, 60,000 in Persia, and more than 58,000 in Egypt.[1076]*

Deaths from contaminated water continue to the present day. The seventh cholera pandemic occurred from 1961 to 1975. Each year 3 to 5 million people are infected with cholera.[1077] In 1991 an outbreak in Peru spread across the continent, killing 10,000 people.[1078] In the 2008 Zimbabwe epidemic, the more than 90,000 suspected cholera cases resulted in over 4,000 deaths. In the 2010 Haiti epidemic, more than 8,500 of the nearly 700,000 infected people lost their lives.[1079] The disease occurs in at least 47 countries, resulting in 95,000 deaths each year.[1080] Advanced sanitation systems have made Europe and North America cholera-free for decades. Unfortunately, this is not true for other parts of the world, with 2.2 billion people lacking access to safely managed drinking water services and 4.2 billion people lacking adequate sanitation services.[1081]

There are 4 billion cases of diarrhea per year worldwide, resulting in 1.8 million deaths, over 90% of them (1.6 million) among children under five.[1082] Meaning that a child under the age of five dies every 20 seconds from water-related diseases. According to a United Nations Environment Programme (UNEP) report, polluted drinking water claims more lives than all forms of violence, including war.[1083] In developing countries, one in three hospital cases may be due to water contamination.[1084] In India alone, a third of a million children under five still die each year from diarrhea.[1085]

In India, the Ganges River starts high in India's Himalayas, flowing 2,500 kilometers (1,550 miles) and eventually discharges into the Bay of Bengal.[1086] This highly polluted river provides water to over 400 million people. The river runs through 50 major Indian cities, which generate some 3 billion liters (800 million gallons) of sewage every day, only a fraction of which is treated before dumping it into the river.[1087] So much water is now being taken from the river for drinking water, agriculture, and industry that some of the once free-flowing parts of the river are now sluggish or even stagnant during the dry months. The river has been transformed into little more than an open sewer.

For Hindus, the Ganges River is very sacred. Bodies are burned on pyres made of high beds of logs on the bank of the river. When the fires die down, workers throw the ashes into the river. Some 32,000 human corpses are cremated here each year, with up to 300 metric tons (330 tons) of half-burnt human flesh released into the Ganges.[1088]

> *Though untreated human waste is the single largest contributor of pollution to the Ganges, industrial waste from leather, paper, sugar, and brass factories along the river banks contributes to some 20 percent of the toxic flow of the Ganges. And though the government banned the practice of dumping the dead into the Ganges over a decade ago, thousands still slip under the cloak of darkness to set their loved ones afloat. The bodies are then left to bloat and decompose into the river bed, only exacerbating the pollution problem.*[1089]

The Yamuna River is the Ganges largest tributary, which snakes through densely populated Delhi in India. With the famous Taj Mahal built on its banks, the river was described in the 16th century by the first Mughal emperor as "better than nectar."[1090] Regrettably, the river that has sustained civilization in Delhi for at least 3,000 years, and the sole water source for more than 60 million Indians today, has been transformed into a mostly lifeless river filled with toxic chemicals and sewage. Like many other polluted rivers, the Yamuna River's sewage problems stem from a lack of developed infrastructure. Just 55% of Delhi's population is connected to the sewage system, and of that sewage, only 48% receive treatment. Untreated wastewater accounts for 80% of water pollution in the Yamuna.[1091]

Raw sewage spills directly into the Yamuna River at the northern edge of New Delhi. The chemical waste from factories manufacturing leather goods, dyes and other goods floats down the river in white blocks of what looks like icebergs of detergent.

The presence of fecal coliform bacteria in aquatic environments indicates that the water has been contaminated with human and animal fecal material. The stretch of the river which flows through Deli contains a concentration of 1.1 billion fecal coliform bacteria per 100 milliliters of water. The standard for bathing is 500 coliform bacteria per 100 milliliters making the river over 2 million times higher than that standard.[1092] Environmentalist Manoj Misra of Yamuna Jiye Abhiyaan, a group that campaigns to clean up the river, notes,

> *That is the reason why this stretch of the Yamuna is called dead. Because there is no life here. There cannot be life here. There's nothing here.*[1093]

> *Yamuna is not a river anymore. It's a collection of 18 drains flowing into it, carrying a toxic cocktail of sewage, chemicals, detergents, industrial waste, and excreta.*[1094]

The Yangtze is the longest river in China, with one-third of China's population living near its banks and relying on it for freshwater. For the 20 million residents of Shanghai, the Yangtze River is the only drinking source. Yet, the river receives more waste than any other in China. With 10,000 chemical businesses dumping a toxic cocktail of heavy metals, fertilizers, pesticides, and weed killers into the Yangtze,[1095] it has become one of the most polluted rivers in the world.[1096] According to a World Wildlife Federation report, over the past 50 years, pollution levels in the Yangtze River have increased by 73%. Each year 25 billion tons of waste (equal to the weight of over 68,000 Empire State Buildings), including 42% of China's sewage and 45% of its industrial waste, gets dumped into the Yangtze.[1097]

In 2013, 60 miles outside of Shanghai, more than 16,000 dead pigs were found in tributaries that supply the city's drinking water. In Jiaxing, a city close to Shanghai, 7.7 million pigs are raised for food annually. Each year, farmers dump up to 300,000 pig carcasses into the rivers. One tributary, the once clean Jiapingtang River, now has dead piglets floating on it and has been described as "inky black, covered in a slick of lime green algae, and it smells like a blocked drain."[1098] Greenpeace East Asia estimates that 320 million people in China are without access to clean drinking water.

Human communities, industries, and agriculture are significant sources of water pollution. Globally, 80% of municipal wastewater is discharged untreated into waterways. Industries are responsible for dumping heavy metals, solvents, toxic sludge, and other wastes into water bodies each year.[1099] Two million tons of untreated sewage and industrial and agricultural wastes are discharged every day into rivers and seas, spreading disease and damaging ecosystems.[1100]

Population growth, economic development, and climate change are placing increasing pressure on the Earth's water resources. According to microbiologist Professor Joan Rose, laboratory director in water research at Michigan State University, water worldwide is quickly worsening.

> *"There are 7 billion people, and most of their waste is going into water. The water quality of lakes, rivers, and coastal shorelines around the world is degrading at an alarming rate. There has been a great acceleration since the 1950s of human and animal popula-*

tions, water withdrawals, pesticide, and fertilizer use. But at the same time, there has been a deceleration, or shrinkage, in wetlands." [1101]

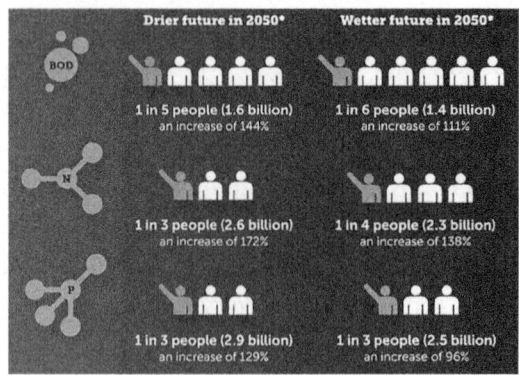

In 2050, more people will be at high risk of water pollution due to increasing BOD, nitrogen and phosphorous.

According to a new study conducted by the International Food Policy Research Institute, water quality is projected to rapidly deteriorate over the next several decades into 2050.[1102] Biochemical Oxygen Demand (BOD) is a measurement of the amount of dissolved oxygen (DO) that is used by aerobic (oxygen requiring) microorganisms when decomposing organic matter in water. High BOD levels can indicate contamination with human and animal waste. Nitrogen (N) and phosphorus (P) are essential nutrients for sustaining plant life and the health of aquatic ecosystems. However, too much of these nutrients cause algae to grow faster than usual, killing other marine life by depleting oxygen.

Human activities contribute significant amounts of BOD, N, and P in global waterways. Globally, 1 in 8 people are at high risk of water pollution from BOD; 1 in 6 people are at high risk of N pollution, and 1 in 4 people are at high risk of P pollution. Based on either a drier or wetter climate scenario, these risks are all expected to get worse for vast amounts of the world's population. By 2050, there could be increases in pollution risk from BOD by 111-144%, N by 138-172%, and P by 96-129%. This escalating water quality deterioration crisis will pose a significant threat to aquatic environments and the people who depend on them.

Industrial agriculture

As many populations have grown more affluent, it has driven the desire for and increased consumption of calorie-rich foods, increasing obesity and associated health problems. In addition, people's diets that were once mostly plant-based have shifted their diets to contain higher portions of meat, eggs, dairy, and oils. These require more land and other resources, increasing humanity's environmental footprint. Today, animal agriculture accounts for 70% of all agricultural land use and 30% of the planet's land surface.[1103]

The desire to live the Western lifestyle includes increased consumption of meat. From 1970 to 2011, the world's population doubled (x 2.12) from 3.3 billion to 7 billion,[1104] yet, at the same time, the total number of livestock has more than tripled, from 7.3 billion units in 1970 to 24.2 billion units in 2011 (x 3.32).[1105]

> *The livestock sector is growing and intensifying faster than crop production in almost all countries. The associated waste, including manure, has serious implications for water quality. In the last 20 years, a new class of agricultural pollutants has emerged in the form of veterinary medicines (antibiotics, vaccines, and growth promoters [hormones]), which move from farms through water to ecosystems and drinking-water sources.*[1106]

Industrial animal agriculture concentrates large numbers of animals in relatively small confined areas. This negatively impacts the environment, particularly water quality, as animal feces also contain pathogens, drug residues, hormones, and antibiotics. Meat from these industrial farms contributes to unsustainable agricultural production and to water-quality degradation. The overuse and misuse of agricultural chemicals, water, and drugs designed to increase productivity have resulted in higher pollution loads in the environment, including rivers, lakes, aquifers, and coastal waters.[1107]

With increasing agricultural land use, countries have increasingly used synthetic pesticides to manage pests. Today, pesticide production is a multibillion-dollar industry, with the global market worth more than $50 billion per year.[1108] Approximately 20,000 commercial pesticide products, including insecticides, herbicides, fungicides, and rodenticides, are commercially available. More than 1.1 billion pounds of these products (equal to the weight of over 330,000 cars) are used in the USA each year. An estimated 5.2 billion pounds (equal to the weight of over 1.5 million cars) of these chemicals are used globally.[1109]

Acute pesticide poisoning causes significant human mortality worldwide, especially in developing countries where poor farmers often use highly hazardous pesticide formulations. According to the United Nations, every year, 200,000 people die due to chronic exposure to agricultural chemicals, with the majority being farmers in the developing world.[1110] In some countries, pesticide poisoning even exceeds fatalities from infectious diseases.

> *Not only is human health directly at risk, early in 2016 one-hundred (100) nations met in Kuala Lumpur to discuss an alarming development, a rapidly accelerating loss of the world's pollinators, such as bees, flies, beetles, moths, butterflies, wasps, ants, birds, and bats among others, as a result of global pesticide contamination, climate change, and habitat loss. In some instances, 75% of a pollinator population species has already been wiped-out. That alone meets the definition of "an extinction event."*[1111]

Since the world's natural pollinators are an absolute necessity for pollinating 75% of global food crops, their extinction risks mass human starvation. At the National University in Río Negro, Argentina, Lucas Garibaldi noted, "We know wild insects are declining, so we need to start focusing on them." Without such changes, the ongoing loss is destined to compromise agricultural yields worldwide.[1112]

Since 1950, more than 140,000 new chemicals and pesticides have been created. Of these, the 5,000 produced in the highest volume are ubiquitous in the environment and are responsible for nearly universal human exposure and their resultant negative health impacts. Yet, less than half of these high-production volume chemicals have undergone any testing for safety or toxicity.[1113]

> *Chemicals and pesticides whose effects on human health and the environment were never examined have repeatedly been responsible for episodes of disease, death, and environmental degradation. Historical examples include lead, asbestos, dichlorodiphenyltrichloroethane (DDT), polychlorinated biphenyls (PCBs), and the ozone destroying chlorofluorocarbons.*

Prescription drugs

Lakes, rivers, and streams worldwide are also being flooded with thousands of tons of over-the-counter and prescription drugs. Many drugs such as analgesics, antibiotics, anti-platelet agents, hormones, psychiatric drugs, and antihistamines have been found at levels considered dangerous for wildlife. Worldwide yearly consumption of diclofenac, a nonsteroidal anti-inflammatory drug that goes by the brand name of Voltaren, equals 2,400 metric tons (2,645 tons) or the weight of over 1,600 cars. Hundreds of metric tons of diclofenac remain in human waste, with only about 7% filtered out by treatment plants, 20% absorbed by ecosystems, and the remainder ending up in the oceans. Between 70% and 80% of the thousands of tons of antibiotics consumed by humans and farm animals end up in natural environments.[1114]

> *There's a good chance that if you live in an urban area, your tap water is laced with tiny amounts of antidepressants (mostly SSRIs like Prozac and Effexor), benzodiazepines (like Klonopin, used to reduce symptoms of substance withdrawal) and anticonvulsants (like Topomax, used to treat addiction to alcohol, nicotine, food and even cocaine and crystal meth).*[1115]

The typical American medicine cabinet is full of unused and expired drugs, with only a fraction of those drugs correctly disposed of. By flushing medications down the toilet or dumping them down the drain, consumers become responsible for a percentage of the pharmaceutical and personal care products that wind up the waterways. According to MarineSafe, there may be as many as 82,000 chemicals polluting our marine environments, just from personal care use.[1116]

A study by the Citizens Campaign for the Environment found that contrary to public health guidelines, over 50% of the medical facilities in nearby Suffolk County in New York State flush their unused medicines down the toilet.[1117] Tested wastewater downstream from pharmaceutical manufacturing facilities has been found to have high concentrations of drugs. In some cases, concentrations were a thousand times higher than the safe levels for wildlife.[1118]

Drugs that have been consumed can also end up in water sources. Because our bodies only process a fraction of the medications taken, the remainder gets flushed out through urine and feces. These chemicals enter the waterways through wastewater treatment plants, which currently have no way to remove antidepressants, antimicrobials, NSAIDS (Nonsteroidal Anti-inflammatory Drugs), caffeine, and contraceptive hormones.

> *A study conducted by the U.S. Geological Survey in 1999 and 2000 found measurable amounts of one or more medications in 80% of the water samples drawn from a network of 139 streams in 30 states. The drugs identified included a witches' brew of antibiotics, antidepressants, blood thinners, heart medications (ACE inhibitors, calcium-channel blockers, digoxin), hormones (estrogen, progesterone, testosterone), and painkillers.*[1119]

While there is some uncertainty about the human health effects of the small amounts of medications consumed from our water supply, there's quite a bit of evidence for pharmaceuticals in the water affecting aquatic life. A study examining the fish collected in the upper Niagara River found that the fish consume and accumulate pharmaceutical pollution.[1120] Antidepressants were found at 20 times the levels in fish brains than in the water they live in. This indicates that the fish are accumulating these drugs in their bodies.

The study found that the major pollutants in the water were the antidepressants citalopram [Celexa], paroxetine [Paxil], sertraline [Zoloft], venlafaxine [Effexor], and bupropion [Wellbutrin], and their metabolites norfluoxetine and norsertraline, as well as the antihistamine diphenhydramine [Benadryl], all of which were found in the fish to differing degrees.[1121]

Studies have shown that antidepressants and their metabolites can influence fish behavior, including predator avoidance, feeding behavior, growth, and reproduction. In a preliminary investigation at the University of Idaho, fathead minnows were placed in water spiked with a combination of SSRIs (Selective Serotonin Reuptake Inhibitors, e.g., Prozac) and anticonvulsants. After swimming in the contaminated water for 18 days, the minnows exhibited 324 genetic alterations associated with human neurological disorders.[1122]

Victoria Braithwaite of Penn State University worries that these sorts of environmental changes could trigger the collapse of an entire fish population or even seriously disturb the biodiversity of the Great Lakes, which is the largest freshwater ecosystem in the world.[1123] Nearly half the United States population is on at least one prescribed medication.[1124] Antidepressant use has increased by 65% between 1999 and 2014, with about one in every eight Americans over the age of 12 using antidepressants.[1125] As antidepressant use continues to grow, the amount entering the environment can only follow.

Three species of vultures in South Asia have been pushed to the brink of extinction due to unintentional poisoning by diclofenac, which is used widely in livestock to relieve fever and lameness. The drug caused acute kidney failure in vultures that ate the carcasses of animals recently treated with it.[1126] Widespread use of diclofenac in cattle in South Asia has been linked to the deaths of millions of vultures that ate carcasses containing the drug, causing some populations to decline by more than 99% since the 1990s.[1127] As a result, the Indian government banned veterinary diclofenac in 2006. As a result, vulture population decline had slowed or ceased in some parts of the Indian subcontinent, although wild populations still remain precariously low.[1128]

Forever plastic

Microplastics are not only found in our oceans but in our lakes, rivers, streams, and our sources of drinking water. In 2017 researchers found that 83% of the globe's tap water contains microscopic plastic fibers.[1129] A glass of tap water from the United States has 4.8 fibers, whereas a European glass contains 1.9 fibers.[1130]

Plastic particles were also found in bottled water. Researchers at the State University of New York tested 259 bottles from 11 different bottled water brands from 9 different countries. Microplastic contamination was found in 93% of the bottles tested. One bottle contained more than 10,000 microplastic particles per liter.[1131] Some of this microplastic contamination of bottled water may arise from the industrial process of bottling the water and the packing material itself. Some fragments could also be entering the water when the cap's seal is broken to open the bottle. The water tested was polluted with polypropylene, nylon, and polyethylene terephthalate particles, ranging from the width of a human hair to a red blood cell size.

> *The results varied not just between brands, but between individual bottles. One bottle of Nestlé Pure Life tested at 10,390 particles per liter, the highest level of plastic out of any sampled,- though most of the brand's bottles tested much lower and one had as few as six particles. Bisleri (5,230), Gerolsteiner (5,160), and Aqua (4,713) all had at least one bottle with a high concentration of plastic particles.* [1132]

Microplastics are not only in our water but everywhere in the environment. In our homes, they originate from clothing, carpets, and any number of consumer products. A 2018 study estimated that an average dinner plate could be accumulating roughly 100 pieces of microplastics from household dust that drifts through the air. In total, an average person could be eating 70,000 bits of microplastic every year.[1133] These tiny bits can also attract toxic pollutants, which are consumed with microplastics.

Research shows that high levels of microplastics are set free from infant-feeding bottles during formula preparation.[1134] Polypropylene infant-feeding bottles can release up to 16 million microplastics and trillions of smaller nanoplastics per liter. Oceania, North America, and Europe were found to have the highest possible exposure levels corresponding to 2,100,000, 2,280,000, and 2,610,000 particles/day.

New information is showing just how much our entire planet is becoming contaminated with plastic. A research team found tiny microplastic fibers, beads, and shards are deposited in the environment delivered like dust by wind and rain.[1135] An estimated 1,000 tons of plastic – the equivalent of 120-300 million plastic bottles – falls upon national parks and protected wilderness areas in the western United States each year.

Lead researcher Janice Brahney of Utah State University was "shocked" at the amount of microplastic her team uncovered. She noted if their work was extrapolated to the entire United States, it would equal 22,000 tons of plastic raining down, equal to the weight of over 13,000 cars. Gregory Wetherbee, a research chemist of the USGS (United States Geological Survey), noted,

> *I think the most important result that we can share with the American public is that there's more plastic out there than meets the eye. It's in the rain, it's in the snow. It's a part of our environment now.*[1136]

Evidence strongly indicates that we breathe in an ever-increasing amount of microplastics wafting through our indoor and outdoor environments. The plasticizers and dyes and other toxins that plastics attract, along with the plastics themselves, could be causing breathing problems in much the same way as traditional air pollution from fossil fuels. According to Professor Frank Kelly, an expert in environmental health from King's College London,

> *If you can breathe them [microplastics] in, they could potentially deliver chemicals to lower parts of our lungs, maybe even across into our circulation in the same way we worry about vehicle emissions.*[1137]

A study in 1998 had already detected plastic fibers in the lungs. These threads can stay there for prolonged periods and possibly for life. Inhaled fibers were found in 87% of lungs and 97% of malignant lung specimens examined. This may pose a health risk by causing chronic inflammation resulting in lung diseases similar to that of the asbestos mineral fibers' inhalation, which is widely recognized. According to the study,

> *Although the physical and chemical properties of cellulosic and plastic fibers and asbestos differ greatly, all three fiber types resist biodegradation. Accordingly, it would be reasonable to postulate that inhaled biopersistent [tending to remain inside a biological organism] cellulosic and plastic fibers, particularly those that contain mordants, dyes, and various chemicals, may contribute to different pulmonary diseases, including lung cancer.*[1138]

In 2020 when researchers analyzed tissues in all the major filtering organs of the body.[1139] They found evidence of plastic contamination in tissue samples taken from the lungs, liver, spleen, and kidneys of donated human cadavers. The researchers said that anyone who wants to avoid ingesting plastic is out of luck. There is too much plastic in the environment, continuously being ground down into microscopic particles for anyone to prevent contamination.

In the United States, the average person produces a half-pound of plastic waste every day.[1140] In 2018, worldwide plastics production increased to around 359 million metric tons (about the weight of 1,084 Empire State Buildings or over 239 million cars), with massive amounts ending in our environment. This amount will continue to expand and contaminate the environment affecting all animal and human life.

Unfresh air

We all appreciate clean, fresh air. People often flock to parks and other green, open spaces on sunny days to relax and enjoy nature. Yet, dirty truck exhaust, smoke from coal-fired power plants, residential wood heating, and other sources cause the deteriorating air quality in many places in the world. A massive, grey, and yellow-tinged smog can even be seen from space engulfing large parts of China, blanketing all of the capital, Beijing.[1141]

Ground-level ozone, commonly known as smog, occurs when emissions from combusting fossil fuels react with sunlight. Soot is the dirty, smoky part of that stream of exhaust made of invisible particle pollution. Collectively, millions of these particles blur the spread of sunlight and can be seen as a dirty haze.

$PM_{2.5}$ refers to atmospheric particulate matter (PM) with a smaller than 2.5 micrometers diameter, about 1/3 the diameter of a red blood cell. These tiny and light particles are produced by large-scale human activities such as fossil fuel burning in factories, power plants, diesel- and diesel and gasoline-powered motor cars, trucks, and equipment. Other sources include burning wood in residential fireplaces and wood stoves or wildfires. They tend to stay longer in the air than more massive particles, increasing the likelihood of humans and animals inhaling them into their bodies. In addition, because of their tiny size, these particles can penetrate deep into the lungs where they get trapped. Ultrafine particles are smaller than 0.1 microns in diameter and are small enough to pass through the lung tissue into the bloodstream, circulating like the oxygen molecules themselves.[1142]

In the winter of 2015 in Beijing, China, the concentration of $PM_{2.5}$ reached 666 micrograms per cubic meter, which is 26 times the recommended maximum of 25 micrograms.[1143] In 2017, large swathes of North India were cloaked in a hazardous fog of toxic pollution, which reached 23 times the recommended maximum.[1144]

Studies have found a close link between exposure to fine particles and premature death from cardiovascular and lung diseases. According to the American Heart Association, exposure to $PM_{2.5}$ over a few hours to weeks can trigger cardiovascular disease-related mortality and nonfatal events. These particles are known to worsen the risk of stroke and play a part in initiating cardiovascular disease by causing systemic inflammation.[1145]

Longer-term exposure over the years increases cardiovascular mortality risk and reduces life expectancy by several months to a few years.[1146] In the United States, between 7,500 and 52,000 people meet early deaths because of small particles resulting from power plant emissions.[1147] Globally, this fine particle air pollution has a massive health impact and is responsible for many premature deaths.

> We estimated that long-term exposure to ambient fine particle air pollution (PM$_{2.5}$) caused 4.2 million deaths and 103.1 million lost years of healthy life in 2015, representing 7.6% of total global mortality, making it the fifth-ranked global risk factor in 2015. Exposure to ozone was responsible for an additional 254,000 deaths.[1148]

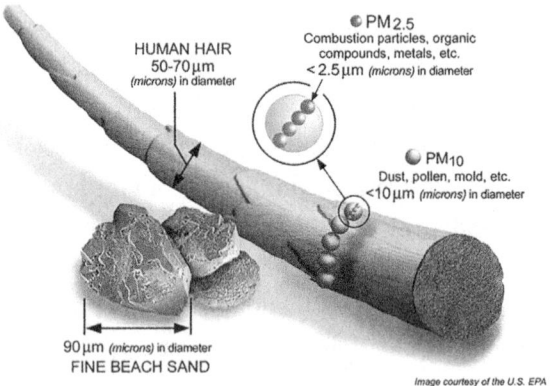

PM stands for particulate matter (also called particle pollution): the term for a mixture of solid particles and liquid droplets found in the air. Some particles, such as dust, dirt, soot, or smoke, are large or dark enough to be seen with the naked eye. Others are so small they can only be detected using an electron microscope.

Overall, air pollution was found to have had the most significant pollution impact on people worldwide, with dirty air accounting for around 6.5 million premature deaths in 2015.[1149] In Africa alone, polluted air could be prematurely killing 712,000 people a year compared with approximately 542,000 from unsafe water, 275,000 from malnutrition, and 391,000 from unsafe sanitation.[1150] Annual deaths from air pollution across Africa increased by 36% from 1990 to 2013. Unless there are significant changes, the number of deaths worldwide due to air pollution will likely increase by more than 50% by 2050.[1151]

The Fukushima disaster

On Friday, March 11, 2011, a magnitude 9.0 earthquake hit northeastern Japan.[1152] The quake was centered 130 kilometers offshore Honshu Island and presented as a rare complex double quake with a severity lasting around 3 minutes.[1153] The earthquake's force moved the landmass of Japan a few meters east, and the local coastline subsided half a meter. An area of the seafloor also extending 650 kilometers north-south moved around 10-20 meters horizontally.[1154] The widespread shifts in the land caused a series of tsunami waves that devastated many of the country's coastal areas. The tsunamis inundated 560 square kilometers of coastal land resulting in a recorded 19,500 deaths. Coastal ports and towns were demolished, with over a million buildings partially destroyed or completely collapsed.[1155] The tsunami also caused a major nuclear incident at Fukushima Daiichi Nuclear Power Plant.[1156]

The Fukushima Daiichi Nuclear Power Station is located in Futaba and Ohduma, 250 kilometers north of Tokyo city in Japan. The first unit of the nuclear station was commissioned in 1971. The station has six boiling water reactors, which together have a power generation capacity of 4.69 gigawatts.[1157] One gigawatt can generate power to supply about 250,000 homes.[1158]

Nuclear power plants generate electricity by using controlled nuclear fission chain reactions (i.e., splitting atoms) to heat water and produce steam to power turbines. Nuclear is often labeled a "clean" energy source because no greenhouse gases (GHGs) or other air emissions are released from the power plant.[1159]

Even though no greenhouse gases are released, nuclear power plants create waste, most notably high-level radioactive waste in the form of uranium fuel. This fuel is used in a nuclear power reactor, and once it is no longer efficient in producing electricity, it is called "spent." Spent fuel is highly radioactive and thermally hot, requiring remote handling and shielding. Nuclear reactor fuel contains ceramic pellets of uranium-235 inside of metal rods. Before the fuel rods are used, they are only slightly radioactive and can be handled without special shielding.

During the fission process, uranium atoms split, creating energy, which is used to produce electricity. The fission creates radioactive isotopes of lighter elements such as cesium-137 and strontium-90. These isotopes are called "fission products" and account for most of the heat and radiation in high-level waste. Some uranium atoms capture neutrons produced during fission. These atoms form elements like plutonium, which is heavier than uranium. However, they do not generate as much heat or penetrating radiation as fission products do, although they take much longer to decay. These heavier or "transuranic" wastes account for most of the radioactive hazard remaining in high-level waste even after 1,000 years.[1160] Radioactive isotopes eventually decay to harmless materials; however, Strontium-90 and cesium-137 have half-lives of about 30 years (half the radioactivity will decay in 30 years). Plutonium-239 has a half-life of 24,000 years.[1161] High-level wastes are hazardous because they produce fatal radiation doses during short periods of direct exposure.

Radioactive decay is where the emission of energy is in the form of ionizing radiation. Radioactive decay of particles occurs in unbalanced atoms called radionuclides. When it decays, a radionuclide transforms into a different atom – a decay product. The atoms will keep changing to new decay products until they reach a stable state and are no longer radioactive.[1162]

At the time of the earthquake, Fukushima Nuclear Power Station had eleven reactors operating at 4 nuclear power plants. Once affected by the earthquake, these reactors were shut down, cutting the main power to the cooling systems. While having survived the quake, the power plant was not equipped to withstand the flood due to the tsunami, which damaged backup power generators. With power gone, the cooling systems failed in three reactors within the first few days of the disaster.[1163] Their cores overheated, leading to partial meltdowns of the fuel rods. Some reactors had complete meltdowns, while others were also leaking.[1164]

The molten nuclear fuel accumulated at the bottom of the vessel, forming holes as the bottom melted. The steam from evaporated water formed hydrogen and oxygen, which is a highly volatile mix. This reacted with the zirconium alloy of the fuel rods.[1165] A hydrogen explosion occurred on March 12 in Reactor 1. To control this, up to 8,000 liters per hour of water were pumped into the reactor. However, due to the wrecked vessel in the basement, it completely evaporated or drained away.[1166] Reactor 2 had a complete meltdown, and Reactor 3 also exploded. Authorities discovered that at Reactor 2, highly radioactive water escaped and into the ocean through a manhole.

The accident released radionuclides cesium-134, cesium-137, strontium-90, and Iodine-129 into the air, topsoil, and the ocean next to the plant, exposing marine life to radioactive materials. While many of the exposed marine organisms remain around Japan, several species are highly migratory and swim across the North Pacific to the West Coast of North America. Two migratory types of fish are the Pacific Bluefin tuna and albacore tuna, with both cesium-134 and cesium-137 being detected in these species. However, the levels of radiation were below levels that are considered cause for concern for public health.[1167]

Ten years after the incident, chemist and environmental scientist Satoshi Utsunomiya of Japan's Kyushu University and colleagues analyzed 31 particles from the fallout of Reactor 1. He discovered that there were larger cesium-bearing particles with higher levels of activity than before. These new radioactive particles were found 3.9 kilometers northwest of the power plant, with the highest cesium-134 and cesium-137 activity documented. The team also found that the characteristic compositions and textures of the particles differed from those previously seen in the Reactor unit 1 fallout.[1168]

There is further work needed to determine the impact on wildlife living around the Fukushima Daiichi facility. For example, filter-feeding marine mollusks have previously been found to be susceptible to DNA damage and necrosis on exposure to radioactive particles. According to Utsunomiya, the half-life of cesium-137 is around 30 years, so the activity in the newly found highly radioactive particles has not yet had significant decay. Therefore, they will remain radioactive in the environment for many decades to come.[1169]

According to a Greenpeace report, areas are still contaminated with radioactive cesium despite an enormous decontamination program. On average, only 15% has been decontaminated. For example, in the case of Namie (5-30 kilometers north-northwest of the nuclear plant), of the 22,314 hectares that make up the municipality, only 2,140 hectares have been decontaminated – just 10% of the total. Much of Fukushima prefecture is a mountainous forest which may make it difficult to easily be decontaminated.[1170]

Radioactive releases measured in 2020 are dominated by radiocesium. However, other isotopes were also released. For example, radioactive strontium-90 was released and is a threat due to bone-seeking radionuclide, which, when ingested, concentrates in bones and bone marrow, increasing the risk of contracting cancer. The greatest danger from strontium-90 comes from the enormous amount at the Fukushima site, particularly in the melted reactor fuel cores in reactor units 1-3. Smaller amounts of strontium and other radionuclides are present in the 1.23 million tons of contaminated tank water stored at the site. These are not without significant risks. Yet, the government is planning to discharge this polluted water into the Pacific Ocean.[1171]

Considering this event occurred 10 years ago, global contamination is still of grave concern. Since the 2011 disaster, cooling water has constantly been escaping from the damaged primary containment vessels into the basements of the reactor buildings. Additional cooling water has been pumped into the reactors to cool the melted fuel remaining inside them to make up for the loss.[1172]

On February 13, 2021, a powerful 7.3 magnitude earthquake struck off Japan's east coast province of Fukushima. The wrecked Fukushima nuclear plant operator says cooling water levels fell in two reactors, indicating possible additional damage.[1173] Increased leakage could require more cooling water to be pumped into the reactors, which would result in more contaminated water being treated and stored in huge tanks at the plant. The plant's operator, Tokyo Electric Power (TEPCO), says its storage capacity of 1.37 million tons will reach capacity by the summer of 2022.[1174]

On April 13, 2021, Japan announced it will release 1.25 million tonnes of contaminated-treated wastewater into the Pacific Ocean.[1175] Trial releases could begin in 2 years. Neighboring countries, environmental groups, and fishery organizations have condemned the move, citing the vast amount of water involved. In addition, marine scientists have expressed concern about the possible impact of this action on marine life and fisheries.[1176] TEPCO and government officials say tritium, a radioactive material that is not harmful in small amounts, cannot be removed from the water.[1177]

Nuclear power had been expected to play an even more significant role in Japan's future in the context of the government's 'Cool Earth 50' energy innovative technology plan in 2008. The Japan Atomic Energy Agency (JAEA) modeled a 54% reduction in CO_2 emissions (from 2000 levels) by 2050, leading to a 90% reduction by 2100. Following the Fukushima accident in October 2011, the government sought to significantly reduce nuclear power's role. However, in 2014 the new government adopted the 4th Basic (or Strategic) Energy Plan, with a 20-year perspective, declaring nuclear energy as a key base-load power source which will continue to be utilized to achieve stable and affordable energy supply and to combat global warming.[1178]

Like all industries and energy-producing technologies, the use of nuclear energy results in waste products. Nuclear power creates radioactive wastes such as uranium mill tailings, spent (used) reactor fuel, and other radioactive wastes. These materials can remain radioactive and dangerous to human health for thousands of years.[1179]

There are three types of nuclear waste, classified according to their radioactivity: low, intermediate, and high levels. The majority of waste (90% of total volume) is composed of lightly contaminated items, such as tools and work clothing, and contains 1% of the total radioactivity. By contrast, high-level waste mainly comprises used nuclear fuel that has been designated as waste from nuclear reactions – it accounts for 3% of the total volume of waste but contains 95% of the total radioactivity.[1180]

Spent reactor fuel assemblies are highly radioactive and, initially, must be stored in specially designed pools of water. The water cools the fuel and acts as a radiation shield. Spent reactor fuel assemblies can also be stored in specially designed dry storage containers. An increasing number of reactor operators now hold their older spent fuel in dry storage facilities using special outdoor concrete or steel containers with air-cooling. In the United States, there is currently no permanent disposal facility for high-level nuclear waste.[1181]

Where radiation is emitted into the environment, the plant workers and emergency teams are most at risk of high radiation exposure, leading to radiation sickness or acute radiation syndrome (ARS). Symptoms of ARS include skin burns, vomiting, diarrhea, and possibly even coma. The cause of death in most cases of ARS is damage to the bone marrow, which leads to infection and internal bleeding. In addition, high exposure to ionizing radiation damages DNA, causing cancers and genetic mutations that can be transmitted to future generations. Radiation poisoning also impacts surrounding wildlife and domestic animals.[1182]

Nuclear reactors and power plants risk will always pose some risk of an uncontrolled nuclear reaction that could result in widespread air, water, and land contamination. In addition, there will always be a concern for reactor safety in an uncontrollable world of natural hazards such as hurricanes and earthquakes, human error, mechanical failure, or design flaws that can trigger the release of radioactive contamination.[1183]

Fracking

Since 2008, the United States has become the world's leading oil and gas producer, mainly due to advances in hydraulic fracturing.[1184] Hydraulic fracturing, or fracking, is a technique designed to recover gas and oil from shale rock.[1185] Simply put, fracking is the process of drilling down into the earth and using a high-pressure water mixture directed at the rock to release the gas inside. Next, water, sand, and chemicals are injected into the rock at high pressure, which allows the gas to flow out to the head of the well."[1186]

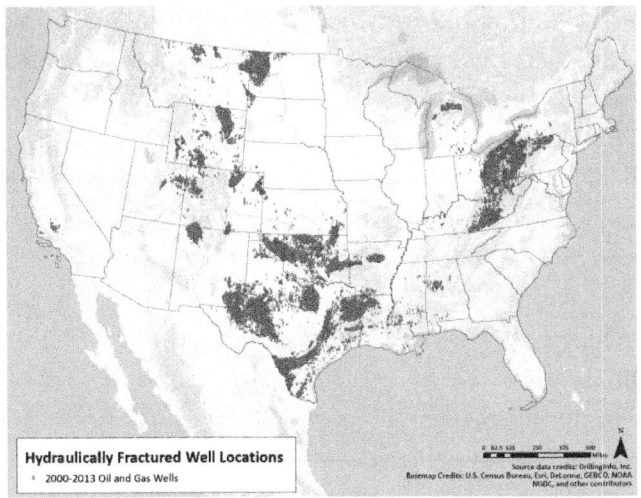

Locations of approximately 275,000 wells that were drilled and likely hydraulically fractured between 2000 and 2013.

Fracking is not a new technology; it dates back to the 1860s, with modern-day hydraulic fracturing beginning in the 1940s. In 1947, Floyd Farris of Stanolind Oil and Gas started to study the relationship between oil and gas production output and the amount of pressurized treatment used on each well. This study led to the first experiment at the Hugoton gas field in Grant County, Kansas.[1187] Over the past few decades, billions have been spent with new advances and interest in fracking. According to the Environmental Protection Agency (EPA), of the approximately one million U.S. wells fractured between 1940 and 2014, one-third of those were fractured after 2000.[1188]

Hydraulic fracking requires extensive amounts of equipment. It needs high-pressure, high-volume fracking pumps; blenders for fracking fluids; and storage tanks for water, sand, chemicals, and wastewater.[1189] Water is the main component of nearly all hydraulic fracturing fluids, making up 90-97% of the fluid volume injected into a well. Typically freshwater is taken from available groundwater and/or surface water resources. The median volume of water used per well for fracking was approximately 1.5 million gallons (5.7 million liters).[1190]

Environmental issues of concern associated with fracking include the type and combination of chemicals injected into the rock, groundwater safety, the impacts fracking may have on the water cycle, including drinking water, and the possibility of fracking technology to generate earthquakes.

A 2016 report by the American Environmental Protection Agency (EPA) found scientific evidence that hydraulic fracturing activities can impact drinking water under some circumstances. For example, drinking water can be affected where:

- Fracking withdrawals water in times or areas of low water availability
- Spills during the handling of fracking fluids and chemicals, or large volumes or high concentrations of chemicals reach ground water resources
- Injection of hydraulic fracturing fluids into wells with inadequate mechanical integrity, allowing gases or liquids to move to ground water resources
- Injection of fracking fluids directly into groundwater resources
- Discharge of inadequately treated fracking wastewater to surface water
- Disposal of storage of fracking wastewater in unlined pits resulting in contamination of groundwater resources.[1191]

Hydraulic fracturing fluids are engineered to create and grow fractures in the targeted rock formation and to carry proppant through the oil and gas production well into the newly created fractures. Proppant is a solid material, typically sand, treated sand, man-made ceramic materials, and sintered bauxite (a type of rock composed of mainly aluminum-baring minerals)[1192] designed to keep an induced hydraulic fracture open during or following treatment. Proppant is added to a fracking fluid of varying composition, depending on the type of fracturing used.[1193]

Fracturing fluids are typically made up of base fluids, proppant, and additives.[1194] Different chemicals are used for fracking based on the rock type and other specifics of the fracking site. Acids are used to dissolve minerals to help fossil fuels flow more easily, biocides are used to eliminate bacteria, gelling agents help proppants into fractures, and corrosion inhibitors prevent steel parts of the well from being damaged by the fracking fluid. After evaluating available data sources, the EPA compiled a list of 1,606 chemicals associated with the fracking water cycle, including 1,084 chemicals reportedly used in hydraulic fracturing fluids and 599 chemicals detected in produced water.[1195] Chemicals harmful to human health and those that don't degrade are of more significant concern when used in the environment.

Evaluating potential hazards from chemicals needs to be done at a local and/or regional scale because fracking chemical use can vary from well to well, influenced by the geochemistry of hydraulically fractured rock formations, the local landscape, and soil surface permeability. This can affect whether and how chemicals enter drinking water resources, directly influencing how long people may be exposed to specific chemicals and at what concentrations. The EPA identified chronic toxicity values from the selected data sources for 98 of the 1,084 chemicals reported to have been used in hydraulic fracturing fluids between 2005 and 2013. Some substances that are of great concern and persist for years are radium, selenium, lead,[1196] formaldehyde, benzene, hydrogen sulfide, and diesel fuel.[1197] Potential human health hazards associated with chronic oral exposure to these chemicals include cancer, immune system effects, changes in body weight, changes in blood chemistry, cardiotoxicity, neurotoxicity, liver and kidney toxicity, and reproductive and developmental toxicity.[1198]

Water produced from hydraulically fractured oil and gas production wells is often considered a waste product that needs to be managed. This wastewater is generally collected through injection in Class II underground injection wells (commonly used to dispose of oil and gas wastewater). Class II wells are regulated under the Underground Injection Control Program of the Safe Drinking Water Act.[1199] Other uses for the wastewater are that after fluid has been used down a well, the wastewater (flow-back water) is returned to the surface to be re-used in other hydraulic fracturing operations or be stored in lined water evaporation ponds. This prevents wastewater from seeping into groundwater and allows leftover material to be tested and then removed by a licensed waste contractor for disposal at a waste facility.[1200]

Some wastewater treatment facilities treat fracking wastewater and then release it to surface water. The solid or liquid by-products can be sent to landfills or injected underground. Evaporation ponds and percolation pits can be used for fracking wastewater disposal. The evaporation ponds allow liquid waste to naturally evaporate back into the environment. Percolation pits allow wastewater to move into the ground, although most states have discontinued this practice. Class II wells are used to inject wastewater associated with oil and gas production underground. Fluids can be injected for disposal or enhance oil or gas production from nearby oil and gas wells.[1201]

Because aboveground disposal practices release treated, or under certain conditions, untreated wastewater directly to surface water or other land surfaces, treated or untreated wastewater can move through the soil to groundwater resources. This can impact drinking water resources. Impacts on drinking water resources from aboveground disposal have been documented. For example, early wastewater management practices in the Marcellus Shale region of Pennsylvania included wastewater treatment facilities that released treated wastewater to surface waters.

The wastewater treatment facilities could not adequately remove the high amounts of dissolved solids found in produced water from Marcellus Shale gas wells. As a result, the discharges contributed to elevated levels of total dissolved solids (particularly bromide) in the Monongahela River Basin. In the Allegheny River Basin, high bromide levels were linked to increases in the concentration of hazardous disinfection by-products in at least one downstream drinking water facility.[1202]

Scientific literature and data from the Pennsylvania Department of Environmental Protection suggest that other produced water constituents (e.g., barium, strontium, and radium) may have been introduced to surface waters by releasing inadequately treated fracking wastewater. These products can be caught up in sediments, and if disturbed during flooding or dredging, they can migrate, further polluting groundwater or surface water resources. The impact on humans from using these chemicals and combinations of chemicals to date is not known.[1203]

Air pollution from oil and natural gas production is a serious problem that threatens the health of nearby communities. Flaring (a controlled burn used for testing, safety, and waste-management purposes), venting (the direct release of gas into the atmosphere), leaking, combustion, and release of contaminants through production, processing, transmission, and distribution of oil and natural gas are significant sources of air pollution.[1204]

There has been an unprecedented increase in earthquakes in the U.S. mid-continent beginning in 2009. According to the USGS, many of these earthquakes have been documented as induced by wastewater injection. On examining the relationship between wastewater injection and the U.S. mid-continent seismicity, they found that the entire increase in earthquake rate is associated with fluid injection wells, with high-rate injection wells (>300,000 barrels per month) much more likely to be associated with earthquakes than the lower-rate wells.[1205] High-rate SWD (Salt Water Disposal) wells are nearly twice as likely as low-rate wells to be near an earthquake. The high-rate wells perturb the ambient reservoir pressure by a larger magnitude, thus increasing the likelihood that pressure changes will reach an optimally oriented, critically stressed fault.[1206]

There are many challenges when moving away from fracking to more renewable energy sources. Since the early 2000s, the U.S has increased oil and natural gas production to levels that ensure energy security for the country. Suppose a ban on high-volume fracking starts in 2021? In that case, the trajectory that has led to the United States becoming a major global liquefied natural gas (LNG) exporter will be reversed and once again it will be reliant on other countries as net importers of natural gas by 2025. This will result in diminished U.S. energy security due to increased reliance on Middle Eastern and Russian energy supplies.

The hydraulic fracturing ban may affect other renewable energy technology growth areas such as wind and solar, as natural gas liquids are an essential raw material input, particularly in plastics and metals production. Manufacturing costs for solar panels, wind turbines, and associated materials will rise with equipment not immune from cost increases.[1207]

A report published by the Biden administration has made the case that a failure to adopt clean energy technologies and reduce carbon emissions will contribute to rising economic costs and warns that the United States has fallen behind in its efforts to develop strategies to combat the effects of climate change. The administration made a promise to cut emissions in half by the end of the decade.[1208]

The report published by President Biden's Council of Economic Advisers shows that the rate of global carbon emissions is expected to stay roughly the same through to the next decade, conflicting with goals that Mr. Biden's administration has made to lower them over the next 10 years and reach a net-zero level by 2050.[1209]

Deadly Pollution

According to a 2017 study from The Lancet medical journal, pollution is one of the "great existential challenges" of the human-dominated era. Altogether pollution of the water, air, and soil are linked to an estimated 9 million deaths globally. This accounts for a staggering 16% or one in six deaths.[1210] Exposure to contaminated air, water, and soil kills more people than a high-sodium diet, obesity, alcohol, road accidents, or child and maternal malnutrition. Global pollution is also responsible for three times as many deaths as AIDS, tuberculosis, and malaria combined. And nearly fifteen times as many deaths as war and all forms of violence.

About 92% of the pollution-related deaths were found in poor or middle-income nations. In quickly industrializing countries, such as China, India, Pakistan, Madagascar, and Bangladesh, pollution was connected to as many as 1 in 4 deaths.[1211] The authors of a study published in 2017 noted that "pollution endangers the stability of the Earth's support systems and threatens the continuing survival of human societies." Prof Philip Landrigan, from the Icahn School of Medicine at Mount Sinai, who co-led the study, said,

We fear that with nine million deaths a year, we are pushing the envelope on the amount of pollution the Earth can carry. For example air pollution deaths in south-east Asia are on track to double by 2050.[1212]

The editor-in-chief of the Lancet, Dr. Richard Horton, and the executive editor, Dr. Pamela Das, noted, "No country is unaffected by pollution. Human activities, including industrialisation, urbanisation, and globalisation, are all drivers of pollution."[1213]

Poisoned Planet

Over the last 500 years, populations have spread worldwide, with millions of people benefiting from technological advances such as plumbing and electricity. Simultaneously, many parts of the once pristine world had been transformed from clean, full-of-life environments to places filled with waste and toxins. From the 19th into the 20th-century, societies in the Western world learned how to reverse some human-created environmental damage and better regulate pollution. These positive actions created relatively clean water and environments.

Yet, the lessons of the past are fleeting as the drive for more material goods has supplanted the desire to maintain a clean and sustainable world. Millions suffer and die to provide inexpensive merchandise for those that are wealthier. The poor people who manufacture these cheap products and the enormous quantities of pollution dumped into the water, air, and land are primarily invisible, ignored, or forgotten by those that benefit.

Decades of toxic waste from agriculture, industrialization, and urbanization have inundated many of the world's waterways. The once surging currents of many healthy rivers have slowed to a coagulated trickle, saturated with raw sewage and industrial runoff. The air has become unbreathable and deadly in many places. Discarded plastics and other garbage litter every corner of the world. We're already condemned to live with yesterday's plastic and other toxic pollution, and we are exacerbating the situation with each passing day of unchanged behavior.

Virtually no place remains unspoiled. As Pope Francis wrote in 2015,

> *The Earth, our home, is beginning to look more and more like an immense pile of filth. In many parts of the planet, the elderly lament that once beautiful landscapes are now covered with rubbish.*[1214]

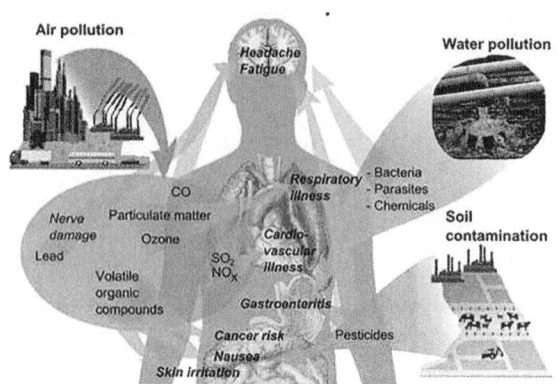

Health effects of pollution.

We all live and breathe on a finite planet, and it is becoming increasingly apparent that what happens in places that may be out of sight can and will affect us all. The massive and expanding problems of air, water, and soil pollution produced by mostly apathetic worldwide societies have significantly impacted the globe. This alarming trend calls for rethinking our current developmental pathway with higher environmental investments and water supply infrastructure. Clean air, water, and land are a matter of will and priority. *Will humanity shift in a sustainable direction or continue down the road towards an increasingly toxic and deadly planet?*

What you can do!

Our once pristine planet is awash in an ever-increasing quantity of toxic chemicals. Those chemicals are not only causing damage to the environment but also to human health. We all live on the same small planet, and each of us can make a difference in how we live our lives. Here are some simple steps that you can take at a personal level to make a difference.

Avoid toxic ingredients – Cleaning products, personal care products, cosmetics, and other merchandise can contain harmful ingredients to the environment and your health. Health and wellness are not merely about diet and exercise but also about limiting exposure to toxins. Become aware of these toxic products and take steps to remove them wherever possible. For instance, vinegar, baking soda, cornstarch, black tea, and lemon[1215] are cheap and non-toxic cleaning ingredients that can also save you money.

Choose responsibly manufactured products – Products from countries with little or no environmental standards are rapidly contaminating our planet. In addition, many overseas goods are often less expensive because workers are usually paid meager or slave wages. Plus, the toxins generated during their production can simply be dumped into the environment. Choose products that are responsibly produced and where the workers that make these products are fairly compensated.

Avoid plastics – Plastic pollution is everywhere. We are eating, drinking, and even breathing in plastics. It's not only the plastics themselves, but the toxins they bring along that can cause health problems that we are only just starting to be aware of. Unfortunately, the plastic pollution problem is only getting worse. Read the chapter "Plastic Oceans" to find out more.

Avoid pharmaceuticals – We live in a predominately a quick-fix, pharmaceutical slanted world. We don't often consider that we may not need many of these products. Instead, consider how your lifestyle could be contributing to the health issues you are experiencing. Take the time to investigate alternatives instead of using products that often have their own adverse effects. For instance, headaches and migraines might be related to dehydration,[1216] falling blood sugar (hypoglycemia),[1217] muscle tension,[1218] or caffeine withdrawal.[1219] These root causes can be corrected with lifestyle changes letting you avoid headaches in the first place.

Properly dispose of pharmaceuticals – If you choose or need to use a medication, don't flush them down the toilet, but dispose of them properly. Look for drug disposal locations near you.

Eat a primarily plant-based diet – Animal agriculture has a disastrous impact on our world. If you consume meat, then choose organic, grass-fed, hormone and antibiotic-free products from your local area. Read the chapter "Dead Zones" to find out more.

Choose organic – Pesticides and fertilizers are taking an increasingly devastating toll on our planet. Read the chapter "Exterminate" to find out more.

Plant some native vegetation – Find out which plants are native to your area and plant a bee/insect-friendly garden to promote the pollination of plants. Native plants generally use fewer resources as they are suited to the climate.

AIR POLLUTION CATCH-22

It is an enduring truth, which can never be altered that every infraction of the Law of nature must carry its punitive consequences with it. We can never get beyond that range of cause and effect.
– Thomas Troward

Civilization exists by geological consent, subject to change without notice.
– Will Durant

I had a dream, which was not all a dream.
The bright sun was extinguish'd, and the stars
Did wander darkling in the eternal space,
Rayless, and pathless, and the icy earth.
– Darkness by Lord Byron, 1816

Insanity is contagious.
– Joseph Heller, Catch-22

The year without a summer

On June 7th, heavy snow fell in New England with 18 to 20-inch drifts in Philadelphia. Frozen birds dropped dead in the streets of Montreal, where a foot of snow had accumulated. Frost, ice, and snow were common in June, killing almost all green plants, and on July 4th, people were still wearing heavy overcoats during the day.[1220] On the same day in the south, Savannah, Georgia, recorded a high temperature of 46°F (7.8°C.)[1221] Throughout New England on the morning of July 6th, there was ice as thick as window-glass, encrusting lakes and rivers as far south as Pennsylvania, causing crops to wither.[1222] Food was so scarce that people resorted to eating raccoons and pigeons, and a few people subsisted on porcupine and boiled nettles.[1223]

During the entire month of August, there were no sunny, warm days. Severe frosts produced ice that was half an inch thick in many places, freezing the growing corn. As a result, very little corn ripened in New England and scarcely any in the middle of the country.[1224] In September, the small amount of corn that had survived was entirely frozen, causing the ears of corn to rot. The stench from the decaying corn was so offensive that people avoided passing downwind of a cornfield. During the New England growing season, the frosts had killed almost all of the main crops.[1225]

Things weren't any better in Europe as the cold and wet summer led to famine and food riots, as crop yields across the British Isles and Western Europe plummeted by 75%.[1226] This resulted in one of the worst typhus epidemics in history. In England, refugee families traveled long distances looking for food, and in Ireland, there was widespread famine following crop failures.[1227] In Switzerland, the food crisis was so bad that people resorted to eating moss as well as cat,[1228] horse, and dog meat.[1229] Swiss women who couldn't feed their children committed infanticide rather than see them starve to death. Those mothers were later prosecuted and decapitated for their actions.[1230]

This abnormally cold summer and resultant tragic events happened over 200 years ago in the year 1816. It was known as the "Year Without a Summer" or as farmers referred to it as "Eighteen Hundred and Starve to Death." That year inspired Lord Byron to write his apocalyptic poem Darkness and Mary Shelley to pen her most famous novel, Frankenstein.[1231]

Mount Tambora

The previous year in Indonesia, an event occurred that directly impacted what was happening on the other side of the globe, from North America to Europe. In April 1815, Mount Tambora, a volcano in the northern part of the island of Sumbawa, erupted violently. The eruption was 100 times more energetic than the 1980 blast from Mount St. Helens, situated in Washington state, and 10 times more than the famous

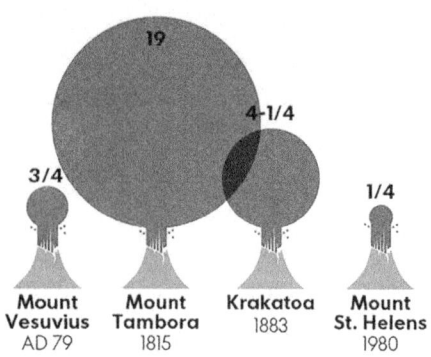

Violent Volcanoes - The eruption of Mount Tambora in 1815 was the biggest in recorded human history. The diagram shows the cubic miles of material ejected.

1883 eruption of Krakatau located in Indonesia.[1232] It ejected 100 cubic kilometers (24 cubic miles) of gas, rock, and ash into the air, with enough ash and pumice to cover 260 square kilometers to a depth of over 3.5 meters (100 square miles, 12 feet deep).[1233] It is the most massive eruption on record, roughly 20 times bigger than Mount Vesuvius in Italy, that destroyed the cities of Pompeii and Herculaneum.[1234] This eruption destroyed the city of Tambora and caused tsunamis over 4.5 meters (14 feet) high. These catastrophes and the resulting destruction of livestock and crops led to an estimated 71,000 to 121,000 deaths. [1235]

> *The eruption destroyed the top three thousand feet of the volcano, blasting it into the air in pieces, leaving behind only a large crater three miles wide and half a mile deep, as though the mountain had been struck by a meteor.*[1236]

The Tambora volcanic eruption spewed ash containing sulfur dioxide (SO_2) about 44 kilometers (27 miles) high into the stratosphere.[1237] It left many places within a 600 kilometer (370 miles) radius pitch black for one or two days,[1238] while Tambora's ash cloud expanded to cover a region nearly the size of the continental United States.[1239]

The material spread across the stratosphere of the globe, forming a planetary-wide molecular sunscreen, dimming the sun, and dropping surface temperatures, as well as wreaking havoc on weather patterns. Over the subsequent three years, the particles in the stratosphere increased the amount of solar radiation reflected back into space. This decreased the amount of energy reaching the lower atmosphere and the Earth's surface.[1240] The overall effect was that global temperatures were lowered by about 1.7°C (3°F).[1241] Although this temperature change may seem small, it was enough to bring about the widespread crop failures that resulted in famine and disease outbreaks in 1816.[1242]

The principal cause of this global cooling was aerosols injected high into the atmosphere by Tambora's eruption. Aerosols are tiny particles suspended in the atmosphere that scatter a portion of the incoming light from the sun back into space.[1243] In the case of volcanic eruptions, the primary aerosol is SO_2.[1244] In the atmosphere, the SO_2 combines with water vapor to form sulfuric acid (H_2SO_4), which then quickly condenses into tiny sulfate (SO_4) aerosols.[1245] Because these tiny particles are two hundred times finer than the width of human hair, they easily remain suspended in the upper atmosphere as an aerosol cloud.[1246]

Winds in the stratosphere spread these aerosols until they cover most of the globe. While volcanic ash will fall out of the lower atmosphere within a short period, these sulfate aerosols remained in the stratosphere for approximately two or three years while also affecting the climate.[1247] In the case of Mount Tambora, it's estimated that it expelled into the atmosphere 60 Tg (teragrams) or 60,000,000 metric tons (66,000,000 tons) of SO_2,[1248] equal to the weight of 180 Empire State Buildings or 40 million cars.[1249]

> One teragram (Tg) equals one trillion grams or one million metric tons or about the weight of three Empire State Buildings.

The June 1991 eruption of the Mount Pinatubo volcano in the Philippines left no doubt that sulfur dioxide [SO$_2$] from volcanoes is the primary way they affect climate. Pinatubo was the 20th century's second-largest eruption, behind only the 1912 eruption of Novarupta on the Alaska Peninsula in 1912... In the United States, the summer of 1992, was the third coldest and third wettest in 77 years.[1250]

Fossil fuels

The after-effects of the Tambora eruption are a dramatic illustration of how the smallest particles can alter the climate. Yet, volcanic eruptions are not the only source of these types of particles. Aerosols also come from human activities that originate from the burning of forests, coal, and oil.

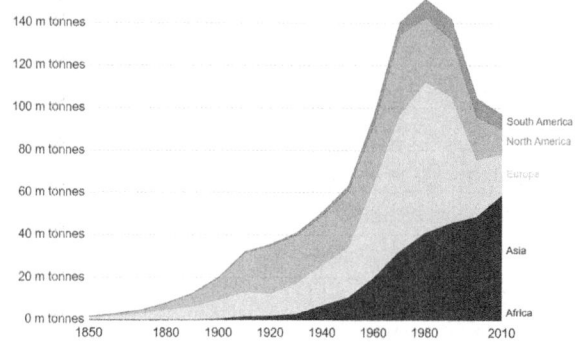

Annual sulfur dioxide (SO$_2$) emissions in million metric tons from 1850-2010.

The burning of fossil fuels refers to the burning of oil, natural gas, and coal to generate energy. This energy is used to generate electricity and to power transportation, such as cars, planes, boats, etc., and industrial processes. Burning fossil fuels results in the greenhouse gases that cause global warming and produce these tiny airborne particles that create a cloudy haze that sits over many industrial parts of the planet. The concentration of human-made sulfate aerosols has grown since the Industrial Revolution. In 1860, the amount of anthropogenic [human-caused] sulfur dioxide (SO$_2$) was roughly 4,000,000 metric tons, reaching 20,000,000 metric tons by 1900.[1251] By 1950, SO$_2$ had reached 63,000,000 metric tons, or about the amount thrown into the air by the Tambora eruption.[1252] By 1980 it had more than doubled to 151,000,000 metric tons (the equivalent of over 450 Empire State Buildings worth or over 100 million cars).

The troposphere is the lower layer of the atmosphere from the surface to about an altitude of 10 kilometers (6.2 miles). The stratosphere is from the end of the troposphere to about 50 kilometers (31 miles). Unlike the Tambora eruption that launched

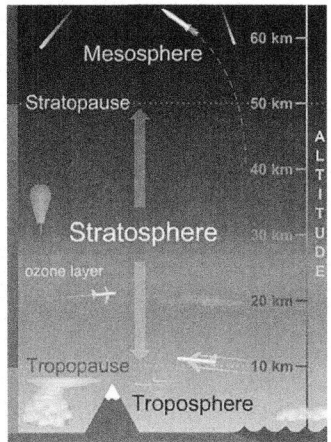

The troposphere and stratosphere.

SO_2 high into the stratosphere, the gases emitted by most volcanic eruptions and by human-made sources never leave the troposphere to make it that high up in the atmosphere.[1253] Because Tambora ejected material high into the stratosphere, it took 2-3 years to dissipate. The aerosols pollutants from vehicles, factories, power plants, and the like stay in the lower atmosphere, where they are believed to survive for only about 3-5 days.[1254] While these human-generated aerosols only last a short while in the atmosphere, they are continuously replenished by the waste products of a modern carbon-based, energy-hungry society.

A certain amount of natural background pollution is generated by the Earth's many volcanoes, and it sometimes spikes with larger eruptions. Every year volcanoes and other natural processes release

> Sulfur (S) has an atomic mass of about 32 and oxygen is about 16. Since there are two oxygen atoms sulfur-dioxide (SO_2) has about twice the weight at 64.

approximately 24 Tg of sulfur (42 Tg of SO_2) into the atmosphere.[1255] Since 2000, 17 volcanoes, including Nabro in Eritrea, Kasatochi in Alaska, and Merapi in Indonesia, have ejected vast amounts of sun-blocking sulfur.[1256] On top of this naturally generated pollution, there is a large amount of human-generated pollution that currently adds about double that amount. Just as aerosols produced by volcanoes reduce global temperatures, human-made aerosols have the same effect. The more aerosols created by burning fossil fuels, the less light reaches Earth's lower atmosphere and surface, lowering the temperature.

Scientists also believe particle pollution has an impact on the composition of clouds. Because aerosol particles form the core around which water droplets form within clouds,[1257] polluted clouds, sometimes referred to as "brown clouds,"[1258] contain more water droplets than they would otherwise. The more water droplets within a cloud, the more sunlight it reflects back into space.

Global dimming

One of the first people to detect this decrease in sunlight phenomenon was an English scientist named Gerry Stanhill. In the 1950s, Stanhill was making climate measurements in Israel, including the amount of solar radiation received, to determine the amount of water that crops would require. Later, when he repeated his studies in the 1980s, he discovered a sharp decline of 22% in sunlight.[1259] Another young German climatologist, Beate Gertrud Liepert, was making similar observations over the Bavarian Alps. Later, Liepert and Stanhill began to independently gather data from around the world.

> *All across the globe, the amount of sunlight reaching the earth's surface had seen a sharp decrease over the years. Between 1950 and 1990, there was a 9% decrease in Antarctica, a 10% decrease in the United States, a 30% decrease in Russia, and a 16% drop in the British Isles.*[1260]

A wide variety of data and several independent studies demonstrated and confirmed that there were substantial declines in the amount of the sun's energy reaching the Earth's surface. Stanhill named this phenomenon "global dimming." However, the apparent paradox was that the Earth seemed to be getting hotter despite a decline in the amount of sunlight reaching the surface, which should have made the world significantly cooler. Although aerosol pollution had a cooling effect, from 1960 to 2000, land surface temperatures had increased by around 0.8°C (1.4°F).[1261]

It turns out that our climate is being influenced by two competing forces that affect temperature.[1262] Increasing greenhouse gases trap heat from escaping the Earth's surface, making it warmer, while atmospheric aerosols keep some sunlight from reaching the surface, keeping it cooler. The cooling effect of aerosols has actually masked the full magnitude of greenhouse warming. According to Peter Cox, who is the Professor of Climate System Dynamics within Mathematics and Computer Science at the University of Exeter,

> *Climate change, to the current date, appears to have been a tug of war, really, between two manmade pollutants. On the one side, we've got greenhouse gases that are pulling the system towards a warmer state, on the other hand, we've got particles from pollution that are cooling it down. And there's a kind of tug of war going on between the two, in which the middle of the rope if you like, determines where the climate system is going in terms of warming or cooling.*[1263]

> A watt is a measurement of power or energy per unit of time.

Another way to look at this is by measuring the amount of energy that greenhouse gases and aerosols contribute to the climate. The extra energy trapped by human-caused greenhouse gases is enough to run a 100-watt lightbulb placed every six square meters over the entire surface of the globe. That is equivalent to an additional 2.6 to 3 watts for every square meter. On the other hand, global dimming has subtracted about 1.5 watts effectively, making the energy 1.1 to 1.5 watts for every square meter.[1264] While land surface temperatures have actually increased by around 0.8°C (1.4°F) from about 1960 to 2000, without the cooling effects of the aerosols, temperatures would have increased by 1.2°C (2.2°F). The full impact of the greenhouse effect has been concealed from us by global dimming.

> *In other words, about one-third of potential continental warming attributable to increased greenhouse gas concentrations has been masked by aerosol cooling during this time period.*[1265]

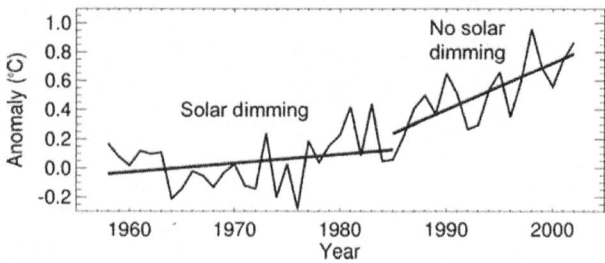

Temperature change over global land surfaces from 1958 to 2002 with respect to 1960. While the temperature rise during the period of solar dimming from the 1960s to the 1980s is moderate, temperature rise is more rapid in the last two decades.

After decades of an increasing solar dimming effect, it appears to have reversed direction after the mid-1980s, with increased sunlight reaching the Earth's surface.[1266] Scientists that observed this new phenomenon named it "global brightening." Some parts of the world, such as Europe and North America, have seen brightening. Other regions, such as China and India, have generally seen stable or further dimming.[1267]

Since the 1980s, North America and Europe have substantially reduced air pollution, which is a significant contributor to dimming. Scrubbers in power stations, catalytic converters in cars, and low sulfur fuels, although they do nothing to reduce greenhouse gases, have already led to a significant reduction in visible air pollution, accounting for the observed brightening.

However, China, Indonesia, and India are now producing pollution to levels that are partially replacing the dimming that was once generated by Western countries.[1268] Despite some countries' increases, overall, SO_2 levels had decreased by about 35% from their peak of 151,000,000 metric tons in 1980 to 97,000,000 in 2010. SO_2 levels now are close to what they were in 1960. As global pollution levels have declined, so has overall dimming.

A 2018 paper published in the Journal of Climate by the American Meteorological Society found that 23% of the warming in the Arctic by greenhouse gases was offset by cooling from aerosols.[1269] The problem is that once the air pollution is reduced, so is the dimming and thus cooling effect it has on the planet. This means that once pollution is significantly decreased, we could see temperatures increase by one or two degrees.[1270]

It means that if/when humans reduce our aerosol pollution, the warming in the Arctic and the ice loss, there will be worse. This puts us into a Faustian bargain. We want to reduce airborne pollution, like sulfur aerosols. But, if we do that, it makes the effects of greenhouse gas pollution worse.[1271]

The sea ice loss in the Arctic is already significant, and the losses could have been even faster without the aerosol dampening effect. As aerosol emissions decrease, "the recent trend of amplified Arctic warming will be further strengthened."[1272] As CO_2 levels continue to rise, it will increase the amount of heat, and as pollution particles decrease, this will also increase the amount of heat. As the world's industrial nations seem to finally be reining in particle pollution, ironically, that could be the catalyst that speeds up global warming. The decrease in worldwide dimming and the ensuing increase in global warming could result in a stark situation for us all. According to Professor Cox,

We're going to be in a situation, unless we act, where the cooling pollutant is dropping off while the warming pollutant is going up. CO_2 will be going up, and particles will be dropping off, and that means that we'll get an accelerated warming. We'll get a double whammy. We'll get reducing cooling and increased heating at the same time, and that's, that's a problem for us.[1273]

Distressingly, if all human-made aerosols were removed from the atmosphere, the increased solar energy would rapidly heat the lower atmosphere by a considerable amount.

...the globally averaged surface air temperature and amount of precipitation could increase in less than a decade by 0.8 K (Kelvin) [equals 0.8°C or 1.4°F] and 3%, respectively, if the entire amount of anthropogenic sulfate aerosols were removed from the atmosphere.[1274]

The effects of volcanic eruptions have led some scientists to propose artificially mimicking volcanic activity as a temporary "Band-Aid" to mitigate the warming effects of greenhouse gases.[1275] However, the SO_2 of most volcanic eruptions does not reach the higher stratosphere, where it could last years instead of days. To duplicate the impact of volcanoes such as Tambora, sulfur particles would need to be released high into the stratosphere.

> *Researchers have envisioned duplicating the phenomenon by launching jets equipped to fly to 70,000 feet [21 kilometers or 13 miles], the lower reaches of the stratosphere, where they would release a sulfur compound. The effort would bleach blue skies a lighter color and make sunsets more vivid while shielding Earth from some of the sun's rays.*[1276]

However, such a scheme would by no means provide a long-term resolution to our societal excesses, as there would be an ever-increasing price tag and health consequences. According to Piers Forster, Professor of Climate Change at the University of Leeds, "Volcanoes give us only a temporary respite from the relentless warming pressure of continued increases in carbon dioxide."[1277] Also, once the geoengineering strategy begins, it would be disastrous if quickly discontinued. Halting geoengineering would cause atmospheric aerosols to dissipate in just a few years. As a result, there would be a significant increase in global temperature as occurred just after the few cold years following the Tambora eruption. This "termination shock" would wreak havoc and doom many amphibians, mammals, corals, and land plants to local or global extinction because they could not adapt to the rapid temperature rise.[1278]

In 2016, 195 countries, including the United States, signed an agreement to keep global temperature rise to less than 2°C (3.6°F) above pre-industrial levels, as well as attempting to limit it further to less than 1.5°C (2.7°F). The year 2019 ranked as the second warmest year on record globally, with land/ocean surface combined at 0.95°C (1.71°F)[1279] edging closer to that agreed-upon limit. According to Professor Cox,

> *We need to reduce the particulates in the air for human and ecosystem health, but this could lead to an increase in global warming of 0.5 to 1 K [C]. Ironically, improvements in air could make the Paris climate targets unattainable.*[1280]

Catch-22

A temperature increase of half to one degree might seem insignificant. Still, as with the 1.7°C (3°F) temperature drop with the Tambora eruption, it can significantly impact the climate. We like to believe our modern selves are mostly separate from the environment in which we actually exist. In 1815, Tambora, a single volcano, erupted and formed a crater 6 kilometers (4 miles) in diameter.[1281] This area is only about 0.0000005% of the Earth's surface, and the 60 Tg of SO_2 thrown into the 5.1 billion Tg total Earth's atmosphere[1282] is about 0.00000117% of that atmosphere. This seemingly small amount relative to the Earth's size chilled the global climate and had devastating impacts on human agriculture, food supply, and societies.

We don't see the full impact of greenhouse gases' warming because these aerosols hide global warming's real power. Pollution is itself a considerable risk to humanity, but it seems to have been protecting us from perhaps an even greater one. As we reduce dimming through alternative clean energy, we could be accelerating a bigger problem with increased global temperatures. While warming greenhouse gases such as CO_2 are removed from the atmosphere on a scale of decades to centuries, aerosols' cooling effect has very short lifetimes of a few days or weeks as they are cleaned out of the atmosphere by rainfall.[1283] Cleaning up these pollutants will immediately affect warming the planet while the impact of the CO_2 already in the atmosphere will persist.

Scientists are beginning to understand that the entire greenhouse effect has been camouflaged by global dimming. This is the tricky, no-win catch-22. *If we solve the global dimming problem with cleaner energy technologies and reduce air pollution impacts, the world could get significantly warmer. On the other hand, the decrease in worldwide dimming and the ensuing increase in global warming could result in a grim situation for us all.*

What you can do!

The effects of global dimming from aerosol cooling have placed the planet in quite a quandary. We must do all we can to reduce CO_2 and other greenhouse gases emitted into the atmosphere to mitigate long-term warming. Read the chapter "Acid Seas" to find out more.

On an individual level, you can help reduce air pollution by carpooling, using public transport, walking, or riding a bike. Switch to cars that have greater fuel efficiency. Save power by switching off appliances when you are not using them. You can also help by becoming aware of products manufactured using fossil fuels and purchasing fewer of these products.

To compensate for the effects of reduced global dimming, some action may be required that can't be addressed at an individual level. It may require more drastic measures, such as releasing calcium carbonate into the upper atmosphere to reflect some of the sun's rays back into space.[1284] Whether this action is genuinely warranted or effective remains an open question.

Invention is the most important product of man's creative brain. The ultimate purpose is the complete mastery of mind over the material world, the harnessing of human nature to human needs.
– Nikola Tesla, My Inventions

For as long as anyone can remember, reliable, cheap electricity has been taken for granted in the United States.
– Alex Berenson

Electricity is something we take for granted as part of our everyday reality. We flip a switch, and what would have seemed like magic a century or so ago creates the miracle of light. Electricity powers our phones, computers, televisions, refrigerators, and most of our modern life. It is the lifeblood that pumps through our entire contemporary world. That lifeblood is made up of a stream of electrons delivered to our homes and businesses through the arteries of a massive electrical grid. For generations, this grid has almost always been there to loyally transport energy from power plants where it is generated to our electronic devices on demand.

The electric grid is a complex and vital system and one of the most impressive engineering feats of the human race. The grid is an electricity transmission system made up of a massive number of interconnected groups of power lines and associated equipment for moving high voltage electrical energy from the power supplying utilities to their customers. The electricity is generated from coal, natural gas, nuclear, hydroelectric, and other sources. There are 19,000 individual generators at about 7,000 power plants that make up the United States electrical grid, the largest machine in the world. The generated electrical power is distributed over 642,000 miles of high-voltage transmission lines and 6.3 million miles of distribution lines which could stretch over 14 times to the moon and back.[1285] Electric power grids are amazingly complicated and intricate systems, consisting of many millions of interdependent turbines, conductors, transmission lines, insulators, switches, and people.[1286] It is this secure and reliable delivery of electricity that is the cornerstone of modern society.

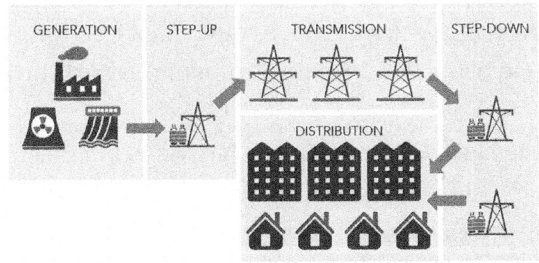

Centrally-generated power flows from generation plants through the bulk power system to industrial, commercial, and residential users.

313

The various technologies that supply electricity and the ever-changing demand are continually monitored and managed by grid operators to ensure everything runs smoothly. Because electricity is not stored, but instead is constantly being generated and used, it is the job of the grid operators to make sure that there's not too much or too little electricity on the grid at any given time. As power demand changes over the day, electric grid operators must match electric supply to this demand in real-time. If supply and demand are not correctly matched, power quality will suffer, causing issues such as flickering lights and brownouts.[1287]

From the 1950s to the 1980s, significant power outages averaged fewer than five per year. But as the grid has grown, the number of outages has increased. In 2007 there were 76, and in 2011 there were more than 300.[1288] In the summer of 2003, a power line came in contact with a tree limb, and this started a cascading chain reaction that eventually brought some 100 power plants down. This affected more than 55 million people in eight U.S. states and Ontario.[1289] It took at least six hours and as long as two days to return electric service to the affected areas. A 2013 study estimated the blackout caused approximately 90 excess deaths in New York City alone.[1290] Yet these disturbances are infrequent with this modern marvel essentially working day and night, keeping the lights on and our machines running.

Reliable electricity is paramount because, unlike the previous decades and centuries before electricity became commonplace, our modern societies are entirely dependent on this source of power for our complex high-tech systems. Moreover, humankind is increasingly more technology-reliant all the time. While this scientific and engineering miracle provides us with the luxuries of modern life which we have come to expect, there is an ominous menace that could bring this all to an abrupt and devastating halt.

Looking up at the sky, we see the sun. It provides the Earth with light and warmth that makes life possible. Like electricity, we don't often pay attention to our solar benefactor, but day after day, year after year, century after century, the bright yellow orb radiantly glows in a seemingly perfectly constant fashion. The sun is an average yellow star that accounts for over 99% of the mass of the solar system. It is an immense spinning ball of scorching gas and plasma made mostly of hydrogen and helium atoms. Nuclear reactions provide the source of the sun's energy which produces 100 billion one-megaton hydrogen bombs worth of energy every second.[1291] The sun's enormous size is what gives it its relative stability. However, the sun is not entirely as unwavering as it may first appear.

> *The surface of the sun writhes and dances. Far from the still, whitish-yellow disk it appears to be from the ground, the sun sports twisting, towering loops and swirling cyclones that reach into the solar upper atmosphere, the million-degree corona [outermost part of the atmosphere of a star]... the sun is a giant magnetic star, made of material that moves in concert with the laws of electromagnetism.*[1292]

An image of giant magnetic loops on the sun from NASA's Solar Dynamic Observatory (SDO) mission.

The Sun is made of positively charged ions and negatively charged electrons in a state of matter called plasma. Plasma is a super-heated ionized gas consisting of approximately equal numbers of these positively charged ions and electrons. The characteristics of plasma are significantly different enough from a gas that plasma is considered a distinct fourth state of matter. This vast sphere of plasma rotates more quickly at its equator than at its poles, and these different rotation rates can cause the sun's magnetic fields to become twisted and tangled. These tangled magnetic field lines can produce powerful localized magnetic fields resulting in incredibly powerful energy occurrences.[1293]

Solar flares are the most violent events on the surface of the sun. They occur when the energy stored up in the sun's magnetic field is suddenly released or converted from magnetic energy into heat and motion energy, potentially slamming into the Earth approximately 8 minutes after they occur. Solar flares are classified into 3 categories, with X-class flares being the highest category that can trigger worldwide radio blackouts and radiation storms in the upper atmosphere. These X class flares have been rated from 1 to 9. An X2 flare is twice as powerful as an X1, and an X9 is nine times as powerful as an X1.

> *There are flares more than 10 times the power of an X1, so X-class flares can go higher than 9. The most powerful flare measured with modern methods was in 2003, during the last solar maximum, and it was so powerful that it overloaded the sensors measuring it. The sensors cut out at X28.*[1294]

The Space Weather Prediction Center, part of the National Oceanic and Atmospheric Administration (NOAA), has also categorized these storms on a scale of G1 to G5.[1295] Although our planet is shielded by a vast, invisible magnetic field, large storms can penetrate it and wreak havoc. G4 (severe) and G5 (extreme) would have the most significant impact on electrical grid operations. While a G4 storm would cause some protective systems to perhaps mistakenly trip, a G5 storm could cause transformer damage and blackouts.

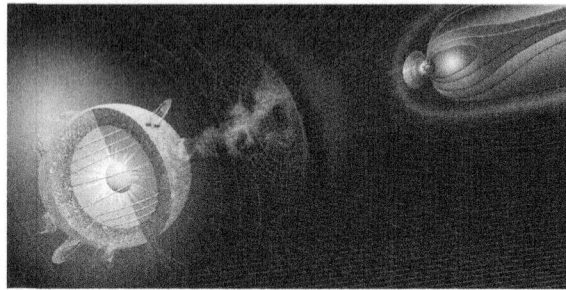

The arc of light heading towards the Earth is a coronal mass ejection that impacts Earth's magnetic field, causing magnetic storms.

A Coronal Mass Ejections (CME) is a discharge of materials from the sun's corona. A CME is a giant billion-ton bubble of plasma that escapes the sun's gravitational field and travels through space at about one million miles an hour, which can hit the Earth in about 2 to 4 days.[1296] When CMEs impact our planet, it temporarily deforms the Earth's magnetic field, inducing sizeable electrical ground currents in the Earth. This event is also known as a geomagnetic storm.[1297]

The largest known recorded geomagnetic storm occurred from August 27 to September 6, 1859, known as the Carrington Event. At its height, a deep crimson red aurora was described to be so bright that even at 1:00 in the morning, it was possible to read a newspaper without any other source of light.[1298] Modern analysis of the Carrington event indicates it was approximately an X42-X45 class solar flare[1299,1300], meaning it was over 40 times more powerful than an X1 class solar flare.

This solar event occurred just before electricity came into use some 20 years later with the creation of the light bulb by Thomas Edison, Lewis Latimer, and others.[1301] However, at that time, there was a relatively new technology being used called the telegraph. The telegraph worked by transmitting electrical signals over wires that were strung along with polls between stations. During the storm, a significant portion of the world's telegraph lines was severely affected.

> *Brilliant sparks were drawn from the telegraph wires... a spark of fire jumped from the forehead of a telegraph operator when his forehead touched a ground wire... a flame of fire burned through a dozen thicknesses of paper; the paper was set on fire and produced considerable smoke... some of the telegraph operators received severe shocks when they touched the telegraph wires.[1302]*

During a solar flare, a massive amount of energy is liberated. The plasma released during a CME connects with the Earth's magnetosphere causing it to temporarily warp the Earth's magnetic field, changing the direction of compass needles and inducing large electrical ground currents in the Earth. A geomagnetically-induced current (GIC) is caused by variations in Earth's magnetic field, producing currents in transmission lines that can damage transformers and other electrical equipment. While massive solar events are not frequent, they can trigger enormous failures across large parts of the electric grid when they occur.

> *When magnetic fields move in the vicinity of a conductor such as a wire, a geomagnetically induced current is produced in the conductor. This happens on a ground scale during geomagnetic storms (the same mechanism also influences telephone/telegraph lines) on all transmission lines. Power companies which operate long transmission lines (many kilometers in length) are thus subject to damage by this effect. The (nearly direct) currents induced in these lines from geomagnetic storms are harmful to electrical transmission equipment, especially generators and transformer-since they induce core saturation, constraining their performance (as well as tripping various safety devices), and causes coils and cores to heat up. This heat can disable or destroy them, even inducing a chain reaction that can blow transformers throughout the system.*[1303]

Sunspots are regions on the sun where the relative temperature is low and where the magnetic field is powerful. The number of sunspots has been shown to increase and decrease over time in an approximately 11-year cycle, although the cycle can be as short as 8 years and as long as 14 years.[1304] A comparison of the solar cycle to geomagnetic storms shows they can occur at any time during that cycle and therefore pose a near-continuous threat, not just when there are many sunspots as once thought.[1305] In the fall of 2017, powerful flares, including an X9.3 on September 6, occurred while the sun was entering a solar minimum which is supposed to be the quiet part of the sun's 11-year cycle.[1306]

A solar storm as big as the Carrington event hasn't struck the Earth since then, although there have been smaller ones. On March 13th, 1989, a severe geomagnetic storm caused the Hydro-Québec power grid to collapse as equipment protection relays tripped in a cascading sequence of events. Six million people were left without power for nine hours, with significant economic loss. The storm even caused auroras to be seen all the way to Texas.[1307]

In May 1921, an extreme solar storm hit the Earth that was approximately 10 times as strong as the 1989 storm.[1308] Also known as the Railroad Storm, auroras could be seen over Europe and the Eastern United States for several evenings. A 1921 New York Times article noted the event.

> *The sunspot which caused a brilliant aurora borealis on Saturday night and the worst electrical disturbance in memory on the telegraph systems was credited with an unprecedented thing at 7:04 o'clock yesterday morning, when the entire signal and switching system of the New York Central Railroad below 125th Street was put out of operation, followed by a fire in the control tower at Fifty-seventh Street and Park Avenue.*[1309]

Most of the East Coast was subject to a blackout in communication in the wake of the solar storm. A telegraph operator stated that his switchboard burst in flames, which caused the entire building to become engulfed by fire. The solar storm impacted telephone, telegraph, and cable traffic throughout Europe.[1310] This storm hit when the electrical grid and communications were in their infancy, so it did not significantly impact the newly forming technological world. The New York Times article also reported a large solar storm that hit the Earth on September 25, 1909.

> *Its effects were spread pretty well over the earth. It was observed as far south from the North Pole as Northern Italy, and as far north from the South Pole as Australia and South Africa.*[1311]

With technological advancements, the electrical grid is actually more vulnerable than when the 1989 solar storm caused the grid to fail, throwing millions into darkness. The most significant vulnerabilities are found within transformers and transmission lines because the higher the voltage rating of a network, the lower its resistance to geomagnetic induced currents. High-voltage networks have increased from 100-200 kV [Kilovolt] during the 1950s to today's 345-765-kV extra-high voltage threshold.[1312] The high voltage grid has expanded by almost a factor of 10 during this time as well as increasing in complexity.[1313] This has made today's sprawling, high-voltage power grids more susceptible to space weather impacts than ever before.

Much of the grid is also made up of decaying and outdated infrastructure. Many major components, including transmissions, distributors, generators, and transformers, are over 25 years old and undergo minor repairs to keep them functioning.[1314] In New England, 25% of electricity generation is more than four decades old and must be replaced or kept running with costly upgrades.[1315] A power plant built during the rapid expansion of the power sector after World War II is 50 years old or older and likely needs to be replaced. The decaying energy infrastructure of the United States was only given a D+ by the American Society of Civil Engineers and would cost almost $5 trillion to replace.[1316]

> *Some parts of the U.S. electric grid predate the turn of the 20th century. Most T&D [transmission and distribution] lines were constructed in the 1950s and 1960s with a 50-year life expectancy, and were not originally engineered to meet today's demand, nor severe weather events... the lower 48 states' power grid is at full capacity, with many lines operating well beyond their design... Often a single line cannot be taken out of service to perform maintenance as it will overload other interconnected lines in operation... As a result of aging infrastructure, severe weather events, and attacks and vandalism, in 2015 Americans experienced a reported 3,571 total outages...[1317]*

High voltage transformers are a vulnerable portion of the grid due to their size, the complexity of their manufacturing, and the lack of production in the United States.[1318] Replacements must be manufactured overseas with a 12-15 month lead-time to replace any damaged parts or update older components. Typically, only a handful of transformers of this size are purchased for U.S. locations on an annual basis. During an extreme solar event that damages large portions of the grid, the global demand for transformer replacements could vastly outstrip the world's capability to manufacture and supply replacements.

> *Transformers of this size class (600MVA+) [Mega Volt Amp] are large and expensive devices to replace. Operators of facilities such as these generally do not have spares readily available. In the best of cases, a large operator of a number of plants may have one or*

two un-energized spare transformers that could be re-located to one of many plants that they operate over a region. If no transformer is readily available, the delivery time on a newly manufactured unit typically runs to approximately 12 months or longer. Even with a readily available spare, the process of removing the old transformer, disassembling, shipping, assembling, and installing the spare transformer is a process with a timeline of a few weeks or longer.[1319]

Up to 15 months are required for the manufacturing and testing of the equipment. Then the equipment also needs to be transported to the site and put into service. Because of the size and weight of these transformers, they can only be transported by ocean vessels, taking several weeks. Once the transformer arrives, even under ideal circumstances, it may take a week or more to transport a transformer even a short distance using special trailers. Special heavy lift cranes are also needed to move these transformers. Once on site, the installation process can take several days. From all these steps, it is easy to see that restoration of the grid could easily take years if numerous transformers need to be replaced.

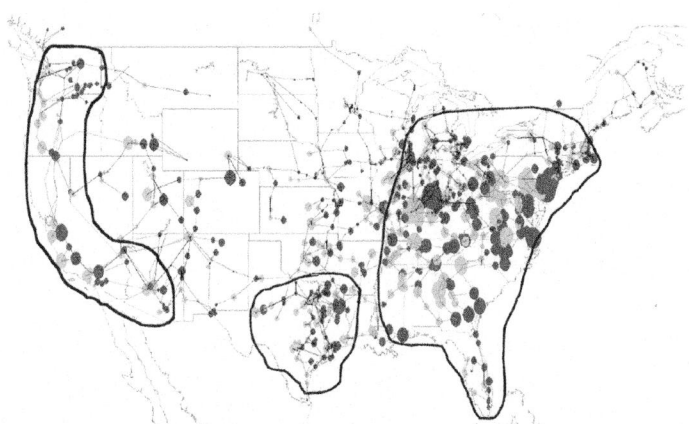

The possible results of an extreme geomagnetic storm. This United States map shows the outlined regions that are susceptible to system collapse due to the effects of the GIC disturbance.

Even with advanced warning, utilities are limited in their ability to quickly harden the grid in advance of a geomagnetic storm. Many researchers have predicted that if a geomagnetic storm at the level of the Carrington

Event were to occur, it could take up to ten years to recover and could cost between $1 trillion to $2 trillion in the first year alone. Physicist Ying Liu of China's State Key Laboratory of Space Weather stated that,

> *An extreme space weather storm – a solar superstorm – is a low-probability, high-consequence event that poses severe threats to critical infrastructures of the modern society. The cost of an extreme space weather event, if it hits Earth, could reach trillions of dollars with a potential recovery time of 4-10 years.*[1320]

While a severe solar storm is a low-probability event, it can potentially damage large portions of the power grid with a ripple effect on almost every aspect of modern society.

> *Impacts would be felt on interdependent infrastructures, with, for example, potable water distribution affected within several hours; perishable foods and medications lost in about 12-24 hours; and immediate or eventual loss of heating/air conditioning, sewage disposal, phone service, transportation, fuel resupply, and so on... the effects on these interdependent infrastructures could persist for multiple years, with a potential for significant societal impacts and with economic costs that could be measurable in the several-trillion dollars- per-year range.*[1321]

Jon Wellinghoff, who served as chairman of the Federal Energy Regulatory Commission from 2009 to 2013, noted,

> *Once your electricity is out, your gasoline is out, because you can't pump the gas anymore. All your transportation's out, all of your financial transactions are out, of course because there's no electronics.*[1322]

Because of the collapse of almost every aspect of modern life, Peter Vincent Pry, Executive Director of the Task Force on National and Homeland Security, dreadfully predicted that if the U.S. power grid were to go down from a solar storm, or a similar pulse generated from a nuclear explosion in the upper atmosphere, that approximately 90% of the population would die.

Natural EMP [Electromagnetic Pulse] from a geomagnetic super-storm, like the 1859 Carrington Event or 1921 Railroad Storm, and nuclear EMP attack from terrorists or rogue states, as practiced by North Korea during the nuclear crisis of 2013, are both existential threats that could kill 9 of 10 Americans through starvation, disease, and societal collapse.[1323]

United States Representative Trent Frank of Arizona testified before the House Homeland Security Committee's Subcommittee on Cybersecurity, Infrastructure Protection, and Security Technologies that explored the effects of an EMP. He declared that,

The thing that people don't realize is it's the length of the blackout that begins to make it dangerous. Everybody thinks, 'Oh, well we'll go outside and we'll build a campfire and we'll have a nice evening at home, we'll break out the candles, it'll be nice.' A couple of days like that is okay. A week like that might be okay. But you start looking at two or three months, you start looking at a very dangerous and unthinkable scenario in a society that is as dependent on electric supply as we are.[1324]

Even without an extreme solar storm, much of the electric transmission and distribution lines were constructed in the 1950s and 1960s, with a 50-year life expectancy nearing or exceeding this lifespan. Many of the high-voltage transmission lines in the lower 48 states' power grids are at maximum capacity. With aging equipment, capacity bottlenecks, increased demand, and increasing storm and climate impacts, Americans will likely experience longer and more frequent power interruptions.[1325] In addition, natural disasters, such as earthquakes, could take down a large portion of the national power structure. For example, a colossal magnitude earthquake on the west coast of the United States of over 9.0 on the Richter scale and possible ensuing tsunami[1326] could destroy California's entire energy system.[1327]

In July 2012, a rapid succession of coronal mass ejections hit where the Earth had been just nine days earlier.[1328] The strength of the storm was, in all respects, at least as strong as the 1859 Carrington event, according to Daniel Baker of the University of Colorado. If the timing was different, the event could have been catastrophic to the electrical grid as well as knocking out satellites. Baker noted that "If it had hit, we would still be picking up the pieces."[1329]

Some people believe these massive X25+ solar flares to be somewhat rare, with an estimate that they happen once every 30 to 150 years.[1330] However, physicist Pete Riley of Predictive Science estimated a 12% chance of being hit with an extreme solar event every ten years.[1331] Riley noted that he was surprised that his analysis showed that the odds were so high and observed that "it is a sobering figure." This means that there is just slightly over a 1 in 100 chance that we could possibly experience a civilization extinguishing event every year. Because of the limited amount of data on these storms, scientists can't really know how often these events actually hit the Earth. The Earth may experience a massive game-changing X event tomorrow or many years from now. But what is certain is that eventually, one will hit.

Geomagnetic storms can develop almost instantaneously over large geographic areas and produce significant collateral damage to critical systems. Many warnings from concerned scientists say that electromagnetic space storms from the sun may wipe out telephone lines and television signals, cause bank accounts to disappear, cripple aircraft navigation systems, and leave cities without power supplies for months or years. The devastation caused by a large geomagnetic storm can be extensive as old, and decaying power grids grow in vulnerability to disturbances from the space environment.

The general awareness of the geomagnetic storms and their inherently more threatening impacts are much less appreciated than more familiar hazards such as earthquakes and hurricanes. Whether we like it or not, all major critical infrastructure providers are interconnected and interdependent. If the grid goes down, everything further down in the supply chain goes down. Every single facet of modern human life would be brought to a grinding halt by domino effects that span well past the electric sector.

In an instant, society could evaporate, and it would be like being seized in a global time machine that transports the world back to how life was in the 1800s. The Carrington event occurred in America's horse-and-buggy era when a solar storm of this magnitude did not really matter. Today's society is completely unprepared to live in a 19th-century environment. Our forefathers' skills had been long forgotten, replaced by a primarily electronics-dependent or modern-world-centric education. Because of our complete reliance on technology and general lack of knowledge of how to live without it, we would be in a substantially worse situation than those who lived before our modern electrical era. *Because we know these extreme solar events occur reasonably frequently, it's not a question of if a massive solar flare will obliterate the backbone of the modern world, but when.*

What you can do!

The long-term loss of everyday electric power is one of the most devastating events that could happen to our modern civilization. The following are some basic steps that you can take to make a difference on an individual level.

Add solar or alternative power – For those who own their own homes, solar panels with battery storage offer some electricity without entirely relying on the power grid. Other possible sources of power include residential wind power, micro-hydropower that harnesses flowing water, geothermal heat pumps,[1332] biomass that generates power from organic waste,[1333] and more. Each type of home energy production is based on your own local environment, and all these approaches take personal research and effort to implement.[1334] Any power supplied without being attached to a grid that can run essentials, such as refrigeration, would be critical if there is a sudden loss of electrical power. Other ways to compensate for a loss of power, such as solar cookers, solar hot water heaters, and more, are also ways to become energy independent.

Get healthy – In the modern era, many people have become grossly unhealthy. The increased use of cars, trucks, and other transportation has resulted in a significant decrease in most physical activity. Also, the widespread use of televisions, computers, phones, and other electronics has resulted in Western societies becoming increasingly sedentary. Most people have 24/7 access to enormous quantities of high-calorie but nutritionally deficient junk foods. Many people have complete reliance on a medical system to take care of every health issue. Many of these health issues are related to a poor lifestyle, and by treating symptoms with medications, the underlying problems are not often addressed. These are some of the factors that have resulted in a global obesity epidemic.

As of 2018, 82.3% of American adults are overweight or heavier.[1335] Also, 18.5% of children and adolescents are obese.[1336] Those with obesity have a higher increased risk of overall mortality, high cholesterol, diabetes, heart disease, stroke, many types of cancer, depression, anxiety, body pain, and difficulty with physical functioning.[1337] In 2019, 1,408,647 Americans died from heart disease, cancer, and stroke, or on average 3,850 every day.[1338] Due to poor lifestyle choices, more than half of Americans take prescription drugs.[1339] Medications have serious consequences, with 128,000 Americans dying each year due to taking medications as prescribed.[1340]

Trying to survive during extreme events, such as the collapse of the electrical grid when you are not physically fit, will be exceptionally challenging, if not deadly. Therefore, losing weight and getting yourself in better health will not only benefit you in the present, but it will also be important later if a natural disaster strikes.

Stock up – Have long-term supplies of food, water, candles, and flashlights. Have a medical kit stocked with bandages and other items to allow you to take care of health problems that may arise. These are essential items to have on hand in the event of other disasters such as hurricanes, earthquakes, etc. Buy an emergency portable generator for your home so that you will be able to run necessary appliances for some time.

Learn self-reliance skills – Learn how to take care of a minor wound, make a meal from scratch, start and build a fire, sew a rip in clothing, make repairs to structures and machines, and other skills that might be useful without any electric power.

Plant a garden – Having a garden that can supply food to yourself and even share with your neighbors could be critical during any natural disaster. If you grow larger quantities of food, consider canning. Before the advent of refrigeration, it was a lifesaving skill to preserve food to feed families through the long winter months.

Become more mentally resilient – Today, with every form of electronic device and entertainment at our fingertips, we can't imagine a world without these things. Yet, for many centuries, people functioned without electricity, and while life was sometimes challenging, people survived. Spend time without all these modern conveniences, read books, participate in personal conversations, learn new skills, and engage in other electronic-free activities. Being free of the modern electronic world will make you more resilient, and you may find yourself more well-rounded and happier.

Be a good neighbor – In the modern era of social media, we may become associated with people worldwide, but at the cost of hardly knowing our actual local neighbors. Spending time to know your neighbors can foster trust and a feeling of kinship that can pay real dividends during times of need.

Lights Out

If we unbalance Nature, humankind will suffer. Furthermore, as people alive today, we must consider future generations: a clean environment is a human right like any other. It is, therefore, part of our responsibility towards others to ensure that the world we pass on is as healthy, if not healthier, than when we found it. This is not quite such a difficult proposition as it might sound. For although there is a limit to what we as individuals can do, there is no limit to what a universal response might achieve. It is up to us as individuals to do what we can, however little that may be. Just because switching off the light when leaving the room seems inconsequential, it does not mean that we should not do it.
– Dalai Lama

Making the world work for 100% of humanity in the shortest possible time through spontaneous cooperation without ecological offense or the disadvantage of anyone.
– Buckminster Fuller

The greatness of a man is not in how much wealth he acquires, but in his integrity and his ability to affect those around him positively.
– Bob Marley

Live simply so that others may simply live.
– Mahatma Gandhi

Choices

Throughout the day, we make numerous choices that impact us and the world we live in. When we consider everyday options, they are not often broader than our immediate convenience, what we perceive as opportune, or what others have or do. If our perspective is our own expediency, accessibility, user-friendliness, and fitting in with what we see as best, we will perceive something as positive – because for us in the moment, it is. But the long-term impact can be negative. Our personal viewpoint determines our choices, and in our consumer-oriented society, too often and for too long, many of those choices have had considerable widespread harmful impacts. Price, convenience, and social forces, all greatly influenced by product-promoting advertisers, have driven rampant consumerism.

Wealth in the form of money and material possessions is revered and used by most as the primary barometer of success. Excessive, opulent materialism is widely worshipped, and a large six-figure or more salary represents for many a milestone of having "made it." Those that make millions or billions of dollars attain the level of celebrity regardless of their integrity or how they acquired such enormous wealth. Adverse impacts of products, their manufacturing, and disposal are virtually meaningless in a society where money is often the only measure of success. This almost singular focus on the blind accumulation of material wealth has negatively impacted the environment, people, and health.

Life in our oceans is being decimated, tropical forests are vanishing, species are disappearing and becoming extinct, while the world becomes increasingly filled with plastic and other toxic waste. Uncontrolled human plundering of the world's limited natural resources from fish to minerals to endangered species is relentless and accelerating. The vibrant and living world of the past is heading toward a barren, lifeless planet filled mainly with people, livestock, and products. The natural world, as we have known it, has been and continues to perish in rapid slow motion.

Our health is impacted by toxins through the fabrication of products for mass consumption and by the products themselves. An obesity epidemic due to the overconsumption of junk foods and other poor lifestyle decisions is escalating across the planet, significantly affecting health and wellbeing. This has given rise to a massive medical system that is utilized to compensate essentially for a polluted environment and poor lifestyle choices. This medical system uses pharmaceuticals and other massively profit-generating interventions that largely ignore the root causes of the problems it is attempting to address. The system itself becomes part of the mass consumption problem as its ever-growing desire for more profits forces it to try and solve health issues with an often single-minded, limited ideology.

With the race to the bottom line, millions of less fortunate people on the planet live horribly difficult lives to supply the more privileged. Many people are in an almost never-ending struggle to keep a roof over their heads and feed themselves. Others are worse off working in horrifying conditions to provide the world's wealthier inhabitants with clothes, electronics, gold, timber, seafood, and other goods. Some live their lives as slaves to support a massively imbalanced system. Elites at the top of the economic pyramid have a material wealth equal to more than most of the combined wealth of a large portion of the rest of the people that live on the planet. Others who are not part of the elite crowd may even consider themselves "poor" because they compare themselves to millionaires and the media production of "success." Yet, compared to many others around the world, they live relatively opulent lives free of the struggles to survive.

This is the unhealthy, imbalanced world today. It is a world currently experiencing and escalating towards numerous ecological, health, and social cataclysms. It may seem overwhelming and that we are doomed to destroy our home, planet Earth, but it doesn't need to be that way. We all can make changes that make it a better world. The changes can start off small, grow, and build toward a brighter future. A future where humanity lives in concert with nature and each other. A future that is not based on acquiring enormous sums of money and things but on improving human health and potential through a spirit of cooperation, creativity, and peacefulness. It all begins with each of us taking small steps.

Small significant steps

Although each decision you make daily might seem inconsequential, those small changes grow and magnify when repeated throughout your life. It also sets the tone for how you live your life, which influences others to make similar decisions multiplying those changes further until they are large and significant.

- Avoid single-use plastics, such as plastic bottles, bags, straws, utensils, and takeout food containers.
- Purchase clothing made of natural instead of synthetic materials.
- Quit smoking, eliminating a source of toxic plastic cigarette butts.
- Avoid buying industrial seafood, sticking with only locally caught.
- Avoid farmed seafood.
- Avoid any industrial-raised foods, using locally grown whenever possible.
- Use organic foods and products whenever possible.
- Buy locally whenever possible.
- Avoid toxic sunscreens and similar products.
- Reduce food waste.
- Avoid environmentally destructive products and corporations.
- Eat lower on the food chain with a primarily plant-based, unprocessed foods diet.
- Avoid products derived from threatened species.
- Consider buying clothes from thrift shops.
- Properly dispose of used electronics.
- Avoid buying things that you don't really need.
- Breastfeed when possible instead of using plastic baby bottles.
- Use modern cloth diapers instead of disposable diapers.
- Consider purchasing things you need that were previously owned.
- Share things you no longer need.
- Reuse and recycle.

As you integrate these and other steps into your life, it reduces the negative impact you might have had on the world. In many cases, these choices also improve your health and wellbeing. Some of these actions you may already be doing or have thought about doing. Each step takes varying amounts of effort and a shift in how you live your life. Research how to approach these changes and determine how you can merge them into your lifestyle. As you proceed, you may think of other ideas to enact that make a difference in your life and its impact on the planet. Change is not a linear course but requires thinking and shifting as you determine how to advance. Perhaps the biggest step is the first one on the path.

Bigger steps

Other more extensive actions you can adopt will probably take more time. You may not be able to immediately take some steps because of your personal situation and others you may already be doing. Nevertheless, all these steps are a positive shift in living and how we interact with the world.

- Grow an organic garden.
- Set up a compost bin for food scraps.
- Plant native vegetation.
- Learn about preserving and canning food.
- Make your home as organic and natural as possible.
- Make your home more energy efficient.
- Become aware that advertising is often manipulating you to purchase things you don't need.
- Work on your health and avoid pharmaceuticals where possible.
- Learn how to fix things yourself.
- Become more conscious and avoid products that are environmentally or health destructive or involve worker abuse.

Shifting consciousness

All these steps to change how we interact with the environment are about becoming more conscious of the world and taking action. Each positive choice impacts not only the environment but will make you healthier and happier. Nevertheless, they do require you, in some measure, to disconnect from how you might have previously interacted with society. It's about becoming more independent and self-sufficient, as well as working locally with others, sharing, and bartering.

We've been taught that happiness comes from outside of ourselves. What and how much we own is supposed to define us. We have convinced ourselves that any joy experienced results from the things we own or the amusements that can be bought. Yet, the endless wealth chase and trying to compete with others to be "successful" doesn't really work. True happiness is an inside job. No matter how much we try, we will never find true lasting happiness outside of ourselves. No one can make us happy – it is always a choice. Our choice. Mass media bombards the public daily with fear, panic, and nonsense, which only serves to make us powerless and dumb us down. We are sold fear repeatedly. So-called social media may have allowed us to keep in touch with people far away, but it has also weakened authentic connections, often making us lonelier and more anxious.

There is an ever-growing number of have-nots with a handful of elites at the very top who have everything and think they can and should control others. There are giant multinational corporations that use the Earth's resources recklessly, creating many useless things for mass consumption. They make tremendous amounts of waste and poisons that cannot be easily extracted from the environment. While there may have been ignorance about the impact at one time, it is no longer the case. These poisons continue to be created and used because of greed for profit, control, and power. The uncaring destruction continues as corporations fashion extensive facades to attempt to convince the masses otherwise.

But we can disconnect from these essentially destructive behaviors and reconnect with what truly makes people happy. Simple and slow living, healthy eating, exercise, good friends and family, engaging hobbies, and meaningful work are antidotes to the modern unhappiness treadmill. The most significant thing each of us can do is to change how we look at the world. Being happy and content is initially easier said than done. Being happy just because, being at peace, and feeling calm is a learned art. Something to work on each day.

Every day, we are presented with a series of events in life in which there are ample opportunities to practice this exact thing. We all have a choice as to how we react to something. Do we go into fear, anger, anxiety, worry, depression, fury or, resignation? Do we take these events to heart and let them determine our lives and our states of mind? Do we mull over them, playing them over and over in our mind, creating fantastical ideas, or do we see them for what they are?

Be willing to take a look at the truth of what choices you make in your life. Realize you are the source of your own happiness. The capacity is within you and, with the recognition, you can refuse to allow the outside world to dictate how you feel. You are the creator, and you project meaning onto an event. You can recontextualize your life and realize you alone have a choice on how you think. Once you recognize you have that power, it brings about a state of inner peace and joy.

We are told that we need the latest smart gadget, pretty knickknack, new wardrobe item, or new trinket to make us happy, but does all this accumulated "stuff" really make us happy? Is it long-lasting happiness? How long before we restlessly move on to buying the next thing?

When we look deep inside of ourselves, at our fears and anxieties, at our traumas and life experiences, that's where we find the source of our unhappiness and distress. It certainly can be challenging work healing these. But with the proper support, these traumas can be addressed and helped. We can become happier, healthier, more at one with the environment, and live life with a shift in our awareness. Our own inner transformation is what ultimately impacts others and the world.

Transformation to a better world

Experiencing nature and the wonders of our environment is not only beneficial for our physical body, but it is reviving for our soul and energy too. Humanity has become so enamored by a fast-paced, busy digital world that we have lost the vital aspect of not only human connection but also our connection to the Earth. Environmental damage is affecting every single living organism on this planet. Because of disconnect and lack of awareness and education over the years, we have developed a perspective that the Earth is here to serve and provide resources for us. The entire planet is there only for humanity to consume and use, with little awareness of the Earth's limitations or that as a living thing it too can die.

We need to expand as a world community and see beyond the smoke and mirrors of constant programming and marketing that happiness is only derived from consuming. It is a matter of health and perspective. Humans can be happy with very little. Simple things, such as walking in nature for 20 minutes, have been shown to improve mental health. When basic needs are met, there is potential to grow from there. Basic fundamentals such as connection, having a meaningful vocation, learning to know yourself, who you are, and what you want out of life all lead to different versions of success.

Today, health really has become the new wealth, the tide is turning, and it is health and meaningful relationships and life purpose that can really be seen as the new "success." No longer will it be acceptable to overwork, hoard wealth, ignore those around you, take, consume and use for selfish purposes.

We are entering a time where, if we do not drastically alter the course of destruction and ignorance, the future will be devastating and bleak. Humanity is at a major decision crossroads. As major environmental problems continue to manifest, we can choose to ignore them and continue our destructive patterns of behavior. Or we can follow a path that returns us to a world of empowerment and connection. We all have to act together to restore and keep a life-sustaining environment. We need to speak for those who are not heard. We need to honor the inalienable rights of all people and all living beings to have a livable planet.

We have created this disaster over the last few hundred years step by greedy step. The bottomless need for things is depleting the world of its natural bounty and filling it with toxic waste. So what kind of world are we leaving future generations? If we don't change direction, we will be judged harshly by the future inhabitants of this world. We have the potential, technology, intelligence, capability, and hopefully motivation to really turn this around. There is no reason why humanity can't live a contented, flourishing life while living in harmony with the Earth. We need to turn outward, with love and partnership, work together so that we can all benefit. We need to care for, empower, and build a future that honors life and the planet and honors the potential of human imagination.

The hour is late. The clock is ticking. The choice is each of ours. Like it or not, everything is rapidly changing. The result could be the most amazing transformation in history. Or the most horrible future you can imagine. Be active or abdicate – the future is in your hands. There is no one coming to save you. It's up to each of us. We have to pay attention to what's going on. We have to realize all the disasters unfolding on the planet are our problems. We have no other home, and only we can deal with them because we allowed it to happen. We let them happen because we got distracted. We traded conveniences for our future – 24/7 fast food, a massive medical system to take care of every little sneeze, media feeding us junk news and information.

Never before in history have we needed unity more than we do now. We can still save the world. We can stop the destruction of our planet and start to work on reversing some of the damage. We can find a way out of this mess, but it's going to take all of us working globally. It's going to take a shift in consciousness.

It's freedom. It's personal responsibility. We have many freedoms but what we need to focus on is personal responsibility. We each have an individual choice. We have the freedom to choose to be part of the problem or be part of the solution. There is no one way out of all these problems. We can all take different paths and find different answers, but unless we are moving towards the overall good of the world, then we are in for a devastating future.

Nothing is impossible. The only thing that is impossible is what you believe is impossible. We can create a future that is bright and beautiful if we choose to do so. There is nothing more powerful and capable of transforming the world than human will and imagination. There is nothing that we can't achieve if we care about each other and work together. There is hope for a wonderful future for all of us here on mother Earth. We can recognize our true power. We can make it a better world. It all depends on each of us.

Turning Back

Glossary

Acute Radiation Syndrome: An acute illness caused by irradiation of the entire body (or most of the body) by a high dose of penetrating radiation in a very short period of time.

AIDS: A disease in which there is a severe loss of the body's cellular immunity, greatly lowering the resistance to infection and malignancy.

Aigrette: Long white back feathers of egrets that were once in great demand to adorn women's hats.

Algae: Organisms of aquatic or moist habitats that use photosynthesis to live and range in size from single cells to large seaweeds such as kelp.

Aerobic: Relating to, involving, or requiring free oxygen.

Amazon Rainforest: Also known as the "Lungs of the World," it is the largest and most biodiverse tropical rainforest in the world, occupying the drainage basin of the Amazon River and its tributaries in northern South America.

Annihilation Trawling: Indiscriminate trawling where a specific fish species is not targeted, but instead, all sea life is caught and usually turned into fishmeal, fish oil, chicken feed, or surimi.

Anoxia: The absence of oxygen.

Anthropogenic: Of, relating to, or resulting from, the influence of human beings on nature.

Aquaculture: Also known as fish farming or shellfish farming.

Arc of Deforestation: A heavily deforested crescent-shaped area along the eastern and southern edges of the Amazon rainforest.

BHRRC: Business and Human Rights Resource Center

Bioaccumulation: The gradual accumulation over time of a substance and especially a contaminant, such as a pesticide or heavy metal, in a living organism.

Biochemical Oxygen Demand (BOD): The amount of dissolved oxygen needed by aerobic biological organisms to break down organic material present in a given water sample at a certain temperature over a specific time period.

Biodegradation: Biodegradation is the breakdown of an object into environmentally natural components such as water, carbon dioxide, etc., by the action of naturally available microorganisms under normal environmental conditions.

Biodiversity: Variety of life.

Biofuel: A fuel produced from renewable resources, especially plant biomass, vegetable oils, and treated municipal and industrial wastes.

Biomagnification: The increasing buildup of toxic substances within organisms at each stage of the food chain.

Biota: The plant and animal life of a particular region.

Biotic pump theory: Through the action of the enormous amount of water that evaporates from the trees, forests create the mechanism in which the rainforest "sucks in" moist air from the ocean.

Bison: The American bison, also commonly known as the American buffalo.

Black carbon: Soot-like byproduct of the use of coal, burning of agricultural waste, or forest fires.

BMI (Body Mass Index): A measure of body fat that is the ratio of the weight of the body in kilograms to the square of its height in meters.

BPA (Bisphenol-A): BPA is a chemical produced for use primarily in the production of polycarbonate plastics and epoxy resins.

Bromide: A compound of bromine with another element or group, especially a salt containing the anion Br– or an organic compound with bromine bonded to an alkyl radical.

Bryozoans: A family of small filter-feeding invertebrates that live as colonies.

Bunker fuel: A primary fuel oil used aboard ships, also known as marine heavy fuel.

Bycatch: Fish and other animals such as dolphins, whales, sea turtles, and seabirds that become hooked or entangled in fishing gear. They are discarded creatures that fishermen do not want, cannot sell, or are not allowed to keep.

$C_6H_{12}O_6$: Chemical formula for glucose.

$CaCO_3$: Chemical formula for calcium carbonate.

Cardiotoxicity: A condition when there is damage to the heart muscle.

CCLME (California Current Large Marine Ecosystem): An oceanic ecosystem in the eastern North Pacific Ocean.

CDC: Centers for Disease Control and Prevention

Cerrado: Brazilian savanna.

Cerebrovascular: The blood vessels of the brain.

CH_4: Chemical formula for methane.

Cholera: An infectious and often fatal bacterial disease of the small intestine, typically contracted from infected water supplies and causing severe vomiting and diarrhea.

Climate change: A significant and long-lasting change in the Earth's climate and weather patterns.

CME (Coronal Mass Ejections): A giant billion-ton bubble of plasma that escapes the sun's gravitational field and travels through space at about one million miles an hour.

CO_2: Chemical formula for carbon dioxide.

CoECRS: ARC Centre of Excellence for Coral Reef Studies

Coral bleaching: When corals experience stressful conditions, such as high temperatures, it causes them to expel the symbiotic algae, which are the source of their brilliant colors. Once the algae are gone, only the bone-white skeletons of the polyps remain, which gives the coral a bleached look.

Coral reef: An underwater ecosystem characterized by reef-building corals.

Coral polyp: Tiny animals that build up coral reefs.

CoTS: Coral-eating Crown-of-Thorns Starfish

DDE: A breakdown product from the insecticide DDT.

DDT: Short for dichlorodiphenyltrichloroethane, it is an insecticide that is poisonous to humans and animals as well. DDT has remained active in the environment for many years and has been banned in the United States for most uses since 1972.

Dead Zone: Extremely-low oxygen ocean areas.

Debt Bondage: A practice in which employers give high-interest loans to workers who then labor at low wages to pay off the debt.

Decontamination: The neutralization or removal of dangerous substances, radioactivity, or germs from an area, object, or person.

Developmental Toxicity: Any structural or functional alteration, reversible or irreversible, which interferes with homeostasis, normal growth, differentiation, development, or behavior, and which is caused by environmental insult.

DFO: Department of Fisheries and Oceans

Dioxin: A group of highly toxic chemically-related compounds that are persistent environmental pollutants. Dioxins are found throughout the world in the environment, and they accumulate in the food chain, mainly in the fatty tissue of animals.

DO (Dissolved Oxygen): The amount of gaseous oxygen dissolved in water.

DRC: Democratic Republic of Congo

Dysentery: Infection of the intestines resulting in severe diarrhea with the presence of blood and mucus in the feces.

Ecosystem: A geographic area where plants, animals, and other organisms exist together to form a bubble of life.

EISA: The Energy Independence and Security Act is a required renewable fuel standard set by the United States Congress requiring that at least 136 billion liters of biofuels be used by 2022.

El Niño: An irregularly recurring flow of unusually warm surface waters from the Pacific Ocean toward and along the western coast of South America that prevents upwelling of nutrient-rich cold deep water and disrupts typical regional and global weather patterns.

Endangered Species Act: A United States federal law passed in 1973 that obligates federal and state governments to protect all species threatened with extinction that fall within the borders of the United States and its outlying territories.

EPA: Environmental Protection Agency

Ethanol: An alcohol fuel that's distilled from plant materials, such as corn and sugar.

Ethylene: A widely used chemical that is used to make polyethylene, an extensively used plastic.

EU: European Union

Eutrophication: Excessive richness of nutrients in a lake or other body of water, frequently due to runoff from the land, which causes a dense growth of plant life and death of animal life from lack of oxygen.

E-waste: Electronic Waste

Excreta: Animal waste as urine, feces, or sweat.

FAO: United Nations Food and Agriculture Organization

FDA: United States Food and Drug Administration

Flaring: The process of burning unwanted natural gas released in oil extraction.

Fracking: The process of injecting liquid at high pressure into subterranean rocks, boreholes, etc., to force open existing fissures and extract oil or gas.

Fractures: The cracking or breaking of a hard object or material.

Fracturing Fluids: A chemical mixture used in drilling operations to increase the quantity of hydrocarbons that can be extracted.

Ganges River: The Ganges is a transboundary river of Asia, which flows through India and Bangladesh.

GBR (Great Barrier Reef): The most extensive coral reef in the world, which sits off the northeast coast of Australia.

GCRMN: Global Coral Reef Monitoring Network

GDP (Gross Domestic Product): The total monetary or market value of all the finished goods and services produced within a country's borders in a specific period.

Geochemistry: The study of the chemical composition of the earth and its rocks and minerals.

Geomagnetic storm: Solar activity that results in a temporary disturbance of the Earth's magnetic field.

GHG (Green House Gas): Various gaseous compounds, such as carbon dioxide or methane that absorb infrared radiation, trapping heat in the atmosphere.

Ghost fishing: Knotted masses of lost but relatively intact nets and fishing lines that can retain the ability to trap fish and other species for long periods.

GIC (Geomagnetically-Induced Current): Usually triggered by coronal mass ejections (CMEs), electric currents in the atmosphere experience significant variations, manifesting in the Earth's magnetic field. These variations induce currents in railroad tracks, underground pipelines, and power grids, and in extreme cases, they can cause blackouts.

Gigaton: A billion tons.

Glacier: Slow-moving rivers of ice that are pushed along by accumulating snow and ice at higher altitudes.

Globalization: The process by which businesses or other organizations develop international influence or start operating on an international scale.

Global Carbon Emissions: The release of carbon into the atmosphere, primarily from the combustion of fossil fuels.

Global Brightening: An increase in the amount of sunlight reaching the surface of the earth caused by a decrease in light-absorbing compounds in the Earth's atmosphere.

Global Dimming: A decrease in the amount of sunlight reaching the earth's surface caused by an increase in light-absorbing compounds in the Earth's atmosphere.

Global Warming: A gradual increase in the overall temperature of the earth's atmosphere generally attributed to the greenhouse effect caused by increased levels of carbon dioxide, CFCs, and other pollutants.

GMO: Genetically Modified Organism

Great Dying: From 1492 to 1650, the widespread death of the Native American people due to widespread epidemics, along with warfare, famine, and slavery.

Great Pacific Garbage Patch: A popular name for concentrations of marine debris in the North Pacific Ocean.

Green House Effect: A warming of Earth's surface and troposphere (the lowest layer of the atmosphere) caused by the presence of water vapor, carbon dioxide, methane, and certain other gases in the air.

Green House Gases: A gas that absorbs and emits radiant energy within the thermal infrared range, causing the greenhouse effect.

Green Revolution: A significant increase in the production of food grains, such as wheat and rice which occurred between 1950 and the late 1960s.

H_2O: Chemical formula for water.

H_2SO_4: Chemical formula for sulfuric acid.

Heavy Metals: A metal of relatively high density or of high relative atomic weight.

High-density polyethylene (HDPE): Used to make plastic storage containers for milk and detergents.

High-grading: The discarding of a smaller-sized fish previously captured and retained in favor of larger, more profitable sized fish.

HKH (Hindu Kush-Himalaya): The Hindu Kush Himalaya is arguably the world's most important "water tower," being the source of ten of Asia's largest rivers and the largest volume of ice and snow outside of the Arctic and Antarctica.

Hydraulic Fracturing: See Fracking.

Hypoxia: Low oxygen.

IJM: International Justice Mission

IMO: International Maritime Organization

Industrialization: The development of industries in a country or region on a wide scale.

Isotopes: Each of two or more forms of the same element that contain equal numbers of protons but different numbers of neutrons in their nuclei, and hence differ in relative atomic mass but not in chemical properties; in particular, a radioactive form of an element.

IUCN: International Union for Conservation of Nature

IWC: International Whaling Commission

Jambiya: Ceremonial daggers that use rhino horns to make their ornately carved handles.

Influenza: A highly contagious viral infection of the respiratory passages causing fever, severe aching, and catarrh, often occurring in epidemics.

kV: Kilovolt

Little Ice Age: The Little Ice Age (LIA) was a period of regional cooling, particularly pronounced in the North Atlantic region that occurred after the Medieval Warm Period.

London Fog: Toxic and deadly smog that covered the city of London.

Longline Fishing: A commercial fishing technique that uses a long line with baited hooks attached at intervals.

Macroalgae: Seaweed.

Malaria: - An intermittent and remittent fever caused by a protozoan parasite that invades the red blood cells and is transmitted by mosquitoes in many tropical and subtropical regions.

Mangrove: Mangroves are a group of trees and shrubs that live in the coastal intertidal zone.

Manna: Divinely supplied spiritual nourishment.

Marine Snow: A shower of organic material falling from upper waters to the deep ocean.

Measles: An infectious viral disease causing fever and a red rash, typically occurring in childhood.

Mesopelagic: The mesopelagic is an ocean zone at a depth of 200 to 1000 meters (660 to 3,300 feet).

Microbeads: A tiny sphere of plastic, such as polyethylene or polypropylene. Microbeads are used primarily as exfoliants in facial scrubs and body washes and add texture to other personal care products, such as toothpaste and lip balms.

Microplastic: Plastic pieces shorter than 5 millimeters (just under two-tenths of an inch) or about a pencil eraser's size.

Milliner: A women's hat maker.

Mollusk: An invertebrate of a large phylum that includes snails, slugs, mussels, and octopuses.

Monoculture: The cultivation of a single crop in a given area.

N_2O: Chemical formula for nitrous oxide.

Nanometer: One billionth of a meter.

Nanoplastic: Plastic particles less than 100 nanometers.

NASA: National Aeronautics and Space Administration

NEFSC: NOAA's Northeast Fisheries Science Center

Nephritis: Inflammation of the kidneys.

Net Zero Emissions: Refers to achieving an overall balance between greenhouse gas emissions produced and greenhouse gas emissions taken out of the atmosphere.

Neurotoxicity: Occurs when the exposure to natural or manmade toxic substances (neurotoxicants) alters the normal activity of the nervous system

Nitrate: A chemical compound that includes nitrogen and oxygen and is used as fertilizers in agriculture.

Nitrogen: The chemical element of atomic number 7, a colorless, odorless unreactive gas that forms about 78 percent of the earth's atmosphere.

NOAA: National Oceanic and Atmospheric Administration

NRDC: Natural Resources Defense Council

NSAIDs: Nonsteroidal Anti-inflammatory Drugs

Nuclear Power: Electric or motive power generated by a nuclear reactor.

Nuclear Waste: Radioactive waste material, for example, from the use or reprocessing of nuclear fuel.

Nurdles: Tiny pellets of plastic resin that manufacturers use to create plastic packaging and products.

Ocean Acidification: A result of too much carbon dioxide reacting with seawater to form carbonic acid.

Ocean Dead Zone: A low or reduced oxygen concentration area.

Ocean Deoxygenation: A process where the ocean loses a substantial amount of oxygen in response to global warming.

Ocean Garbage Patch: Large ocean areas where litter, fishing gear, and other debris collect.

OEA-2 (Oceanic Anoxic Event-2): An extinction event that suffocated about 27% of marine invertebrates in Earth's oceans.

OMZ (Oxygen Minimum Zone): Ocean water that is permanently oxygen-deprived.

OSU: Oregon State University

Oxybenzone: A sunscreen agent that protects primarily from UVB rays.

Ozone: A gas made up of three oxygen atoms (O_3) that occurs naturally in trace amounts in the stratosphere, protecting life on Earth from the sun's ultraviolet (UV) radiation.

PCB (Polychlorinated Biphenyl): PCBs were once widely deployed as dielectric and coolant fluids in electrical apparatus, carbonless copy paper, and heat transfer fluids.

PDBE (Polybrominated Diphenyl Ether): A class of chemicals added to certain manufactured products as a flame retardant. Not produced in the United States since the 1970s, it was used in furniture foam padding; wire insulation; rugs, draperies, upholstery; and plastic cabinets for televisions, personal computers, and small appliances.

Pesticide: A substance used for destroying insects or other organisms harmful to cultivated plants or to animals.

pH: A value expressing the acidity or alkalinity of a solution on a logarithmic scale on which 7 is neutral, lower values are more acid, and higher values are more alkaline.

Photosynthesis: The process by which green plants and some other organisms use sunlight to synthesize food from carbon dioxide and water.

Phosphorus: The chemical element of atomic number 15, a poisonous, combustible non-metal.

Phthalates: A group of chemicals used to make plastics more durable.

Phytoplankton: Microscopic marine single-cell plants.

Plankton: Microscopic organisms drifting or floating in the sea or freshwater.

Plasma: Super-heated ionized gas consisting of approximately equal numbers of positively charged ions and electrons.

Plastic: A synthetic material made from a wide range of organic polymers such as polyethylene, PVC, nylon, etc., that can be molded into shape while soft and then set into a rigid or slightly elastic form.

Plasticizer: A substance added to a material to produce or promote plasticity and flexibility and reduce brittleness.

$PM_{2.5}$: Atmospheric particulate matter (PM) with a smaller than 2.5 micrometers diameter.

Polyamide (PA): A plastic that has high stability and stiffness, which is optimal for the production of toothbrushes, electric plugs, etc.

Polyester (PES): A generalized term for any fabric or textile made using polyester yarns or fibers.

Polyethylene (PE): A light, usually thin, soft plastic, often used in making bags and other products.

Polyethylene Terephthalate (PET): A form of polyester that is used in clothing fabric. It is extruded or molded into plastic bottles and containers for packaging foods and beverages, personal care products, and many other consumer products.

Polypropylene (PP): A type of plastic used especially for ropes, fabrics, and molded objects such as drinking straws.

Polystyrene (PS): Polystyrene is a versatile plastic used to make a wide variety of consumer products. Polystyrene also is made into a foam material, called expanded polystyrene, aka Styrofoam.

Polyvinyl chloride (PVC): A high-strength thermoplastic material widely used in applications, such as pipes, medical devices, wire, and cable insulation.

PPM: Parts Per Million

Proppant: A solid material, typically sand, treated sand, or manmade ceramic materials, designed to keep an induced hydraulic fracture open during or following a fracturing treatment.

PTB (Permo-Triassic Boundary): Also known as the Great Dying, happened 251 million years ago, over the span of 60,000 years. During this mass extinction, 96% of all species were lost.

Quintal: An antiquated unit of weight equivalent to 100 kilograms or 220 pounds.

Radiation: The emission of energy as electromagnetic waves or as moving subatomic particles, especially high-energy particles which cause ionization.

Radioactive: Emitting or relating to the emission of ionizing radiation or particles.

Radioactive decay: The process by which an unstable atomic nucleus loses energy by radiation.

Radioactive waste: Radioactive waste is a type of hazardous waste that contains radioactive material.

Radionuclides: A radioactive nuclide

RCP (Representative Concentration Pathway): A greenhouse gas concentration (not emissions) trajectory adopted by the Intergovernmental Panel on Climate Change.

Reactive Nitrogen: A variety of nitrogen compounds that support growth directly or indirectly.

Renewable Energy: Energy from a source that is not depleted when used, such as wind or solar power.

Salmonid: A fish of the salmon family.

Salt Water Disposal Well (SWD): A disposal site for water produced as a result of the oil and gas extraction process.

SCS: South China Sea

Scurvy: A disease caused by a deficiency of vitamin C.

Seamount: Mountains that rise from the ocean seafloor and do not quite reach the ocean's surface.

Smallpox: An acute contagious viral disease with fever and pustules that usually leave permanent scars.

Solvent: Able to dissolve other substances.

SO_2: Chemical formula for sulfur dioxide.

SO_4: Sulfate aerosols.

Solar flare: A tremendous explosion on the sun that happens when energy stored in "twisted" magnetic fields (usually above sunspots) is suddenly released. Solar flares are classified into 3 categories, with X-class flares being the highest category that can trigger worldwide radio blackouts and radiation storms in the upper atmosphere. These X class flares have been rated from 1 to 9, although they can go higher than 9.

Solar storm: Another name for a geomagnetic storm.

SSRI: Selective Serotonin Reuptake Inhibitor

SST: Sea Surface Temperature

Still-hunting: When hunting buffalo or other animals, a hunter would secure a position far from his quarry from which he could safely shoot and kill it.

Stratosphere: A layer of the atmosphere from the end of the troposphere to about 50 kilometers (31 miles).

Surimi: A paste of inexpensive fish, shaped, colored, and flavored in imitation of lobster meat, crabmeat, etc.

Symbiosis: A mutually beneficial relationship.

Taxa: Group of organisms.

Tibetan Plateau: The largest high-altitude landmass on Earth, sometimes referred to as the "Third Pole."

Tg: Teragram

Transuranic waste: Radioactive waste that contains manmade elements heavier than uranium on the periodic table.

Transformer: A device that transfers electric energy from one alternating-current circuit to one or more other circuits, either increasing (stepping up) or reducing (stepping down) the voltage.

Trawling: A method of fishing that involves dragging a net through the water behind a boat.

Tributyltin (TBT): A biocide used to prevent the growth of bacteria.

Troposphere: The lower layer of the atmosphere from the surface to about an altitude of 10 kilometers (6.2 miles).

Tsunami: A long, high sea wave caused by an earthquake or other disturbance.

Tuberculosis: An infectious bacterial disease characterized by the growth of nodules (tubercles) in the tissues, especially the lungs.

Typhoid: An infectious bacterial fever with an eruption of red spots on the chest and abdomen and severe intestinal irritation.

UFP: Ultra-Fine Particle

UNEP: UN Environment Programme

Uranium: The chemical element of atomic number 92, a dense grey radioactive metal used as a fuel in nuclear reactors.

Urbanization: The process of making an area more urban.

USDA: United States Department of Agriculture

VOC: Volatile Organic Compound

Water Column: A conceptual column of water from the surface of a sea, river, or lake to the bottom sediment.

Watt: A measurement of power or energy per unit of time.

WWF: World Wide Fund for nature

Zooxanthellae: Microscopic marine algae.

References

1   "As many as six billion Earth-like planets in our galaxy, according to new estimates," Science Daily, June 16, 2020, https://www.sciencedaily.com/releases/2020/06/200616100831.htm
2   Brooke, Lindsay, "A universe of 2 trillion galaxies," Phys.org, January 16, 2017, https://phys.org/news/2017-01-universe-trillion-galaxies.html
3   "The Nearest Star," NASA, https://imagine.gsfc.nasa.gov/features/cosmic/nearest_star_info.html
4   "NASA Probe Sees Solar Wind Decline," NASA, https://voyager.jpl.nasa.gov/news/details.php?article_id=20
5   Torbet, Georgina, "It will take 20,000 years for our earliest probes to reach Alpha Centauri," Digital Trends, December 29, 2019, https://www.digitaltrends.com/cool-tech/spacecraft-time-to-travel-nearest-stars
6   Bonsor, Kevin, "How Terraforming Mars Will Work," How Stuff Works, https://science.howstuffworks.com/terraforming2.htm
7   Lamb, Robert, "What is it about Earth that makes it just right for life?" How Stuff Works, https://science.howstuffworks.com/life/evolution/earth-just-right-for-life.htm
8   Elizabeth Claire Alberts, "Humanity's 'ecological Ponzi scheme' sets up bleak future, scientists warn," Mongabay, January 14, 2021, https://news.mongabay.com/2021/01/humanitys-ecological-ponzi-scheme-sets-up-bleak-future-scientists-warn
9   Peter G. Ryan, et al., "Monitoring the abundance of plastic debris in the marine environment," Philosophical Transaction of the Royal Society B, 2009, vol. 364, p. 2008, doi:10.1098/rstb.2008.0207
10 "What We Know About: Plastic Marine Debris," NOAA (National Oceanic and Atmospheric Administration) Marine Debris Program, http://marinedebris.noaa.gov/info/plastic.html
11 Jenna R. Jambeck, et al., "Plastic waste inputs from land into the ocean," Science Magazine, February 13, 2015, vol. 347, issue 6223, p. 768.
12 Laura Parker, "Planet or Plastic?" National Geographic, May 15, 2018, https://www.nationalgeographic.com/magazine/2018/06/plastic-planet-waste-pollution-trash-crisis
13 "Global plastic production by feedstock type 2018," Statista, January 27, 2021, https://www.statista.com/statistics/1114289/global-cumulative-plastic-production-by-type
14 Deborah Sullivan Brennan, "Study: 9 billion tons of plastic produced since '40s, and most is still out there," The San Diego Union-Tribute, July 24, 2017, http://www.sandiegouniontribune.com/communities/north-county/sd-no-plastic-study-20170724-story.html
15 Laura Parker "Planet or Plastic?" National Geographic, May 15, 2018, https://www.nationalgeographic.com/magazine/2018/06/plastic-planet-waste-pollution-trash-crisis
16 Jonathan Amos, "Earth is becoming 'Planet Plastic'," BBC, July 19, 2017, http://www.bbc.com/news/science-environment-40654915
17 Empire State Building Fact Sheet, http://www.esbnyc.com/sites/default/files/esb_fact_sheet_4_9_14_4.pdf
18 Jenna R. Jambeck, et al., "Plastic waste inputs from land into the ocean," Sci-

ence Magazine, February 13, 2015, vol. 347, issue 6223, p. 770.

19   http://www.marinemammalcenter.org/education/marine-mammal-information/cetaceans/blue-whale.html

20   Stefan Ogbac, "20 of the lightest cars sold in the U.S.," Motortrend, October 12, 2015, https://www.motortrend.com/news/20-of-the-lightest-cars-sold-in-the-u-s

21   Emily Holden, "US produces far more waste and recycles far less of it than other developed countries," The Guardian, July 3, 2019, https://www.theguardian.com/us-news/2019/jul/02/us-plastic-waste-recycling

22   "1 to 2 million tons of US plastic trash go astray, study finds," The New York Post, October 31, 2020, https://nypost.com/2020/10/31/1-to-2-million-tons-of-us-plastic-trash-go-astray-study-finds

23   "1 to 2 million tons of US plastic trash go astray, study finds," The New York Post, October 31, 2020, https://nypost.com/2020/10/31/1-to-2-million-tons-of-us-plastic-trash-go-astray-study-finds

24   Olivia Rosane, "U.S. Leads the World in Plastic Pollution, New Study Finds," EcoWatch, November 3, 2020, https://www.ecowatch.com/amp/world-plastic-pollution-leader-us-2648607158

25   "Most ocean plastic comes from Asian rivers: study," Japan Times, June 9, 2017, http://www.japantimes.co.jp/news/2017/06/09/world/science-health-world/ocean-plastic-comes-asian-rivers-study

26   "Tracking Trash, 25 Years of action for the Ocean, 2011 Report," Ocean Conservancy, 2011, http://act.oceanconservancy.org/pdf/Marine_Debris_2011_Report_OC.pdf

27   Elizabeth A Smith and Thomas E Novotny, "Whose butt is it? tobacco industry research about smokers and cigarette butt waste," Tobacco Control, 2011, pp. i2-i9, doi:10.1136/tc.2010.040105

28   Giuliano Bonanomi, et al., "Cigarette Butt Decomposition and Associated Chemical Changes Assessed by C CPMAS NMR," PLOS ONE, January 27, 2015, pp. 1-16, DOI:10.1371/journal.pone.0117393

29   Cindy Mullet, "Cigarette butts can take years to decompose," Sidney Herald, April 30, 2010, https://www.sidneyherald.com/archives/cigarette-butts-can-take-years-to-decompose/article_5fa4aee9-f6b9-5871-aae7-ec662449192c.html

30   "Why cigarette butts threaten to stub out marine life," The Guardian, June 9, 2015, https://www.theguardian.com/sustainable-business/2015/jun/09/why-cigarette-butts-threaten-to-stub-out-marine-life

31   Thomas E. Novotny, et al., "Cigarettes Butts and the Case for an Environmental Policy on Hazardous Cigarette Waste," International Journal of Environmental Research and Public Health, 2009, pp. 1691-1705, doi:10.3390/ijerph6051691

32   "Risks Associated with Smoking Cigarettes with Low Machine-Measured Yields of Tar and Nicotine," Public Health Service, National Institutes of Health, and the National Cancer Institute, 2001.

33   "Science for Environment Policy, In-depth Reports, Plastic Waste: Ecological and Human Health Impacts," November 2011, p. 4., http://ec.europa.eu/environment/integration/research/newsalert/pdf/IR1_en.pdf

34   Graeme Macfadyen, Tim Huntington and Rod Cappell, "Abandoned, lost or otherwise discarded fishing gear," United Nations Environment Programme Food and Agriculture Organization of the United Nations.

35   Peter G. Ryan, et al., "Monitoring the abundance of plastic debris in the marine environment," Philosophical Transaction of the Royal Society B, 2009, vol. 364, p. 1999, doi:10.1098/rstb.2008.0207

36   Jenna R. Jambeck, et al., "Plastic waste inputs from land into the ocean," Science Magazine, February 13, 2015, vol. 347, issue 6223, p. 770.
37   The IUCN (International Union for the Conservation of Nature) Red List of Threatened Species - http://www.iucnredlist.org/details/full/2477/0
38   Daniel Hoornweg, Perinaz Bhada-Tata and Chris Kennedy, "Waste production must peak this century," Nature, October 31, 2013, vol. 502, p. 615.
39   Emma Newburger, "Enormous amount of plastic will fill oceans, land by 2040 even with immediate global action, report says," CNBC, July 23, 2020, https://www.cnbc.com/2020/07/23/enormous-amount-of-plastic-will-fill-oceans-and-land-by-2040-report.html
40   Kevin Loria, "What's Gone Wrong With Plastic Recycling," Consumer Reports, April 30, 2020, https://www.consumerreports.org/recycling/whats-gone-wrong-with-plastic-recycling
41   Sandra Laville and Matthew Taylor, "A million bottles a minute: world's plastic binge 'as dangerous as climate change'", The Guardian, June 28, 2017, https://www.theguardian.com/environment/2017/jun/28/a-million-a-minute-worlds-plastic-bottle-binge-as-dangerous-as-climate-change
42   Ben Kentish, "Coca-Cola 'increases production of plastic bottles by a billion'," The Independent, October 4, 2017, http://www.independent.co.uk/news/business/news/coca-cola-bottles-plastic-increase-coke-total-number-world-yearly-a7981561.html
43   Seth Borenstein, "Research: Amount of straws, plastic pollution huge," Allied News, April 24, 2018, http://www.alliednews.com/news/research-amount-of-straws-plastic-pollution-huge/article_636f2f32-684a-50f0-bfb8-f54ba1063759.html
44   Plastic Bag Consumption Facts, Conserving Now, https://conservingnow.com/plastic-bag-consumption-facts
45   "Hamburg becomes the first city to reject single-use coffee pods," CBC Radio, February 2016, http://www.cbc.ca/radio/asithappens/as-it-happens-monday-edition-1.3458647/hamburg-becomes-the-first-city-to-reject-single-use-coffee-pods-1.3458649
46   Laura Parker, "Planet or Plastic?" National Geographic, May 15, 2018, https://www.nationalgeographic.com/magazine/2018/06/plastic-planet-waste-pollution-trash-crisis
47   Lorenzo Kyle Subido, "Filipinos Throw Out Over 163 Million Sachets and 93 Million Plastic Bags Each Day," Esquire, March 8, 2019, https://www.esquiremag.ph/life/health-and-fitness/philippines-plastic-pollution-statistics-a00288-20190308
48   Rowan Carter and Murray R. Gregory, "Bryozoan encrusted plastic from the continental slope: eastern South Island, New Zealand," New Zealand Natural Sciences, 2005, p. 52.
49   Dennis Jefferson, et al., "Physical interventions to interrupt or reduce the spread of respiratory of viruses (Review)," The Cochrane Collaboration, November 20, 2020, DOI: 10.1002/14651858.CD006207.pub5
50   Jingyi Xiao, et al., "Nonpharmaceutical Measures for Pandemic Influenza in Nonhealthcare Settings—Personal Protective and Environmental Measures," Emerging Infectious Diseases, May 2020, vol. 26, no. 5, https://dx.doi.org/10.3201/eid2605.190994
51   Henning Bundgaard, DMSc, et al., "Effectiveness of Adding a Mask Recommendation to Other Public Health Measures to Prevent SARS-CoV-2 Infection in Danish Mask Wearers," Annals of Internal Medicine, November 18, 2020, https://doi.org/10.7326/M20-6817

52 Dr. Teale Phelps Bondaroff and Sam Cooke, "Masks on the Beach: The Impact of COVID-19 on Marine Plastic Pollution," OceansAsia, December 2020, https://oceansasia.org/covid-19-facemasks

53 Riyaz Patel, "Over 1.5 billion face masks now believed to be polluting world's oceans," The South African, December 29, 2020, https://www.thesouthafrican.com/lifestyle/environment/over-1-5-billion-face-masks-now-believed-to-be-polluting-worlds-oceans

54 David Ford, "COVID-19 Has Worsened the Ocean Plastic Pollution Problem," Scientific American, August 17, 2020, https://www.scientificamerican.com/article/covid-19-has-worsened-the-ocean-plastic-pollution-problem

55 Monique Beals, "25K tons of pandemic-related plastic waste now in the ocean: research," The Hill, November 8, 2021, https://thehill.com/policy/energy-environment/580645-26k-tons-of-pandemic-related-plastic-waste-now-in-the-ocean

56 "COVID-19 Facemasks & Marine Plastic Pollution," OceansAsia, https://oceansasia.org/covid-19-facemasks

57 "Metro scientist fears PPE is 'nail in the coffin' for plastic pollution," WSB-TV Atlanta, February 18, 2021, https://www.wsbtv.com/news/local/atlanta/metro-scientist-fears-ppe-is-nail-coffin-plastic-pollution/IK2CR2OYZRH7ZHRVXTEGCHF5WU

58 "Metro scientist fears PPE is 'nail in the coffin' for plastic pollution," WSB-TV Atlanta, February 18, 2021, https://www.wsbtv.com/news/local/atlanta/metro-scientist-fears-ppe-is-nail-coffin-plastic-pollution/IK2CR2OYZRH7ZHRVXTEGCHF5WU

59 Marcus Eriksen, et al., "Plastic Pollution in the World's Oceans: More than 5 Trillion Plastic Pieces Weighing over 250,000 Tons Afloat at Sea," PLOS ONE, December 10, 2014, pp. 1-15, DOI:10.1371/journal.pone.0111913

60 Daniel Cressey, "Bottles, bags, ropes and toothbrushes: the struggle to track ocean plastics," Nature, August 17, 2016, vol. 536, no. 7616

61 C.G. Avio, et al., "Plastics and microplastics in the oceans: From emerging pollutants to emerged threat," Marine Environmental Research, 2016, p. 2, http://dx.doi.org/10.1016/j.marenvres.2016.05.012

62 Julia Lurie, "Your Toothpaste May Be Loaded With Tiny Plastic Beads That Never Go Away," Mother Jones, May 28, 2015, http://www.motherjones.com/environment/2015/05/microbeads-exfoliators-plastic-face-scrub-toothpaste

63 John H.Tibbetts, "The Global Plastic Breakdown How Microplastics Are Shredding Ocean Health," Costal Heritage, 2014, vol. 28, no. 3.

64 Hazel Pfeifer, "The UK now has one of world's toughest microbead bans," CNN, January 9, 2018, https://www.cnn.com/2018/01/09/health/microbead-ban-uk-intl/index.html

65 Chelsea M. Rochman, et al., "Scientific Evidence Supports a Ban on Microbeads," Environmental Science and Technology, September 2015, DOI: 10.1021/acs.est.5b03909

66 "People may be breathing in microplastics, health expert warns," The Guardian, May 9, 2016, https://www.theguardian.com/environment/2016/may/09/people-may-be-breathing-in-microplastics-health-expert-warns

67 A. M. Mahon, et al., "Microplastics in Sewage Sludge: Effects of Treatment," Environmental Science and Technology, 2017, pp. 810–818, DOI: 10.1021/acs.est.6b04048

68 Marzia Sesini, "The garbage patch in the oceans: The problem and possible solutions," August 2011, Earth Institute, Columbia University

69 Laura Parker, "The Great Pacific Garbage Patch Isn't What You Think it Is," National Geographic, March 22, 2018, https://news.nationalgeographic.com/2018/03/

great-pacific-garbage-patch-plastics-environment

70   "Full scale of plastic in the world's oceans revealed for first time," The Guardian, December 2014, https://www.theguardian.com/environment/2014/dec/10/full-scale-plastic-worlds-oceans-revealed-first-time-pollution

71   Robinson Meyer, "The Arctic Ocean Is Clogging With Billions of Plastic Bits," The Atlantic, April 20, 2017, https://www.theatlantic.com/science/archive/2017/04/the-arctic-ocean-is-filling-with-billions-of-plastic-bits/523713

72   Laura Parker, "Planet or Plastic?" National Geographic, May 15, 2018, https://www.nationalgeographic.com/magazine/2018/06/plastic-planet-waste-pollution-trash-crisis

73   "Single-use plastic has reached the world's deepest ocean trench," UN Environment, April 20, 2018, https://www.unenvironment.org/news-and-stories/story/single-use-plastic-has-reached-worlds-deepest-ocean-trench

74   M. L. Taylor, et al., "Plastic microfibre ingestion by deep-sea organisms," Nature Scientific Reports, 2016, p. 1, DOI: 10.1038/srep33997

75   Lianne Kolirin, "Highest ever concentration of microplastics found on sea floor," CNN, April 30, 2020, https://amp.cnn.com/cnn/2020/04/30/world/microplastics-seafloor-concentration-scn-scli-intl/index.html

76   M. L. Taylor, et al., "Plastic microfibre ingestion by deep-sea organisms," Nature Scientific Reports, 2016, p. 1, DOI: 10.1038/srep33997

77   C. Anela Choy, Jeffrey C. Drazen, "Plastic for dinner? Observations of frequent debris ingestion by pelagic predatory fishes from the central North Pacific," Marine Ecology Progress Series, 2013, vol. 485, pp. 155-163, doi: 10.3354/meps10342

78   Lisbeth Van Cauwenberghe and Colin R. Janssen, "Microplastics in bivalves cultured for human consumption," Environmental Pollution, 2014, vol. 193, pp. 65-70, http://dx.doi.org/10.1016/j.envpol.2014.06.010

79   Kirsti Marohn, "Microplastics could pose big treatment challenges," MPR (Minnesota Public Radio) News, August 15, 2017, https://www.mprnews.org/story/2017/08/15/microplastics-could-pose-big-treatment-challenges

80   Damian Carrington, "Microplastic pollution in oceans is far worse than feared, say scientists," The Guardian, March 12, 2018, https://www.theguardian.com/environment/2018/mar/12/microplastic-pollution-in-oceans-is-far-greater-than-thought-say-scientists

81   Damian Carrington, "Microplastic pollution revealed 'absolutely everywhere' by new research," The Guardian, March 6, 2019, https://amp.theguardian.com/environment/2019/mar/07/microplastic-pollution-revealed-absolutely-everywhere-by-new-research

82   "High levels of microplastics found in Northwest Atlantic fish," Phys.org, February 16, 2018, https://phys.org/news/2018-02-high-microplastics-northwest-atlantic-fish.html

83   Karen Duis and Anga Coors, "Microplastics in the aquatic and terrestrial environment: sources (with a specific focus on personal care products), fate and effects," Environmental Sciences Europe, January 6, 2016, doi: 10.1186/s12302-015-0069-y

84   Matthew Cole and Tamara S. Galloway, "Ingestion of Nanoplastics and Microplastics by Pacific Oyster Larvae," Environmental Science & Technology, 2015, pp. 14625–14632, DOI: 10.1021/acs.est.5b04099

85   M. L. Taylor, et al., "Plastic microfibre ingestion by deep-sea organisms," Nature Scientific Reports, 2016, p. 1, DOI: 10.1038/srep33997

86   John H. Tibbetts, "The Global Plastic Breakdown How Microplastics Are Shredding Ocean Health," Costal Heritage, 2014, vol. 28, no. 3.

87   "Most ocean plastic comes from Asian rivers: study," Japan Times, June 9, 2017, http://www.japantimes.co.jp/news/2017/06/09/world/science-health-world/ocean-plastic-comes-asian-rivers-study

88   Brandon Specktor, "65 Pounds of Plastic Trash Tore This Whale Apart from the Inside," Live Science, April 10, 2018, https://www.livescience.com/62266-dead-sperm-whale-plastic-bags.html

89   Matthew Robinson, "Dead whale found with 40 kilograms of plastic bags in its stomach," CNN, March 18, 2019, https://edition.cnn.com/2019/03/18/asia/dead-whale-philippines-40kg-plastic-stomach-intl-scli/index.html

90   Gianluca Mezzofiore, "Pregnant whale washed up in Italian tourist spot had 22 kilograms of plastic in its stomach," CNN, April 1, 2019, https://edition.cnn.com/2019/04/01/europe/sperm-whale-plastic-stomach-italy-scli-intl/index.html

91   Elaina Zachos, "How This Whale Got Nearly 20 Pounds of Plastic in Its Stomach," National Geographic, June 4, 2018, https://news.nationalgeographic.com/2018/06/whale-dead-plastic-bags-thailand-animals

92   Anthee Carassava, "Sperm whales killed in Mediterranean by supermarket bags," The Times, May 23, 2018, https://www.thetimes.co.uk/article/mediterranean-sperm-whales-were-killed-by-plastic-waste-say-scientists-07wkjd2m5

93   Laura Parker, "Nearly Every Seabird on Earth Is Eating Plastic," National Geographic, September 2, 2015, https://news.nationalgeographic.com/2015/09/15092-plastic-seabirds-albatross-australia

94   "UN's mission to keep plastics out of oceans and marine life," UN News Centre, April 27, 2017, http://www.un.org/apps/news/story.asp?NewsID=56638

95   Matt McGrath, "Fish eat plastic like teens eat fast food, researchers say," BBC News, June 2, 2016, http://www.bbc.com/news/science-environment-36435288

96   "Microplastics killing fish before they reach reproductive age, study finds," The Guardian, June 2016, https://www.theguardian.com/environment/2016/jun/02/microplastics-killing-fish-before-they-reach-reproductive-age-study-finds

97   "Taste, not appearance, drives corals to eat plastics," Science Daily, October 24, 2017, https://www.sciencedaily.com/releases/2017/10/171024141710.htm

98   Yvaine Ye, "Mosquitoes are eating plastic and spreading it to new food chains," NewScientist, September 19, 2018, https://www.newscientist.com/article/2180055-mosquitoes-are-eating-plastic-and-spreading-it-to-new-food-chains

99   Katherine J. Wu, "Mosquitoes Are Passing Microplastics Up the Food Chain," The Smithsonian, September 21, 2018, https://www.smithsonianmag.com/smart-news/mosquitoes-are-passing-microplastics-up-food-chain-180970373

100   "UN's mission to keep plastics out of oceans and marine life," UN News Centre, April 27, 2017, http://www.un.org/apps/news/story.asp?NewsID=56638

101   Jan Wesner Childs, "Discarded Fishing Gear is a Major Source of Ocean Pollution, Greenpeace Says," The Weather Channel, November 9, 2019, https://weather.com/news/news/2019-11-09-fishing-gear-pollutes-oceans-greenpeace

102   Laura Parker, "The Great Pacific Garbage Patch Isn't What You Think it Is," National Geographic, March 22, 2018, https://news.nationalgeographic.com/2018/03/great-pacific-garbage-patch-plastics-environment

103   Murray R. Gregory, "Environmental implications of plastic debris in marine settings—entanglement, ingestion, smothering, hangers-on, hitch-hiking and alien invasions," Philosophical Transactions of the Royal Society B, 2009, vol. 364, p. 2014, doi:10.1098/rstb.2008.0265

104   Gwynn Guilford, "Derelict fishing nets have turned the bottom of the sea into a death trap," Quartz, August 13, 2014, https://qz.com/247942/derelict-fishing-nets-

have-turned-the-bottom-of-the-sea-into-a-death-trap-2

105   "Plastic Plague in Our Oceans," Science In Society, Spring 2015, issue 65, p. 7.
106   "2-Hydroxy-4-methoxybenzophenone," PubChem, https://pubchem.ncbi.nlm.nih.gov/source/hsdb/4503
107   "Perils of plastics? Survey of risks to human health and the environment," Science News, March 20, 2010, https://www.sciencedaily.com/releases/2010/03/100319115631.htm
108   Amy Westervelt, "Phthalates are everywhere, and the health risks are worrying. How bad are they really?" The Guardian, February 10, 2015, https://www.theguardian.com/lifeandstyle/2015/feb/10/phthalates-plastics-chemicals-research-analysis
109   Sarah Gibbens, "Chemicals From Plastics, Cosmetics Found in Wild Dolphins," National Geographic, September 7, 2018, https://www.nationalgeographic.com/animals/2018/09/news-dolphins-plastic-chemical-traces-found
110   C.G. Avio, et al., "Plastics and microplastics in the oceans: From emerging pollutants to emerged threat,"Marine Environmental Research, 2016, p. 5, http://dx.doi.org/10.1016/j.marenvres.2016.05.012
111   Paul Rincon, "Banned chemicals persist in deep ocean," BBC News, February 13, 2017, http://www.bbc.com/news/science-environment-38957549
112   "What We Know About: Plastic Marine Debris," NOAA (National Oceanic and Atmospheric Administration) Marine Debris Program, http://marinedebris.noaa.gov/info/plastic.html
113   Watershed Discipleship: Reinhabiting Bioregional Faith and Practice, 2016
114   Victoria A. Sleight, et al., "Assessment of microplastic-sorbed contaminant bioavailability through analysis of biomarker gene expression in larval zebrafish," Marine Pollution Bulletin, 2016, http://dx.doi.org/10.1016/j.marpolbul.2016.12.055
115   Neo Chai Chin, "Toxic bacteria found on small pieces of plastic trash from S'pore beaches," Today, February 11, 2019, https://www.todayonline.com/singapore/toxic-bacteria-found-small-pieces-plastic-trash-singapores-beaches
116   Chelsea M. Rochman, et al., "Anthropogenic debris in seafood: Plastic debris and fibers from textiles in fish and bivalves sold for human consumption," Nature Scientific Reports, 2015, pp. 1-2, DOI: 10.1038/srep14340
117   Josh Gabbatiss, "All UK mussels contain plastic and other contaminants, study finds," The Independent, June 8, 2018, https://www.independent.co.uk/environment/mussels-plastic-microplastic-pollution-shellfish-seafood-oceans-uk-a8388486.html
118   Lisbeth Van Cauwenberghe and Colin R. Janssen, "Microplastics in bivalves cultured for human consumption," Environmental Pollution, 2014, vol. 193, pp. 65-70, http://dx.doi.org/10.1016/j.envpol.2014.06.010
119   Steph Yin, "Your seafood might contain tiny plastic particles - Is 'Trash to Table' the next dining trend?" Popular Science, September 24, 2015, http://www.popsci.com/your-seafood-might-contain-tiny-plastic-particles
120   Ali Karami, et al., "The presence of microplastics in commercial salts from different countries," Scientific Reports, April 2017, DOI: 10.1038/srep46173
121   Laura Parker, "Microplastics found in 90 percent of table salt," National Geographic, October 17, 2018, https://relay.nationalgeographic.com/proxy/distribution/public/amp/environment/2018/10/microplastics-found-90-percent-table-salt-sea-salt
122   Michael Allen, "There's Probably Plastic in Your Sea Salt," Hakai Magazine, May 8, 2017, https://www.hakaimagazine.com/article-short/theres-probably-plastic-your-sea-salt

123   Damian Carrington, "Microplastics revealed in the placentas of unborn babies," The Guardian, December 22, 2020, https://www.theguardian.com/environment/2020/dec/22/microplastics-revealed-in-placentas-unborn-babies

124   Damian Carrington, "Microplastics revealed in the placentas of unborn babies," The Guardian, December 22, 2020, https://www.theguardian.com/environment/2020/dec/22/microplastics-revealed-in-placentas-unborn-babies

125   Jessica Glenza, "Sea salt around the world is contaminated by plastic, studies show," The Guardian, September 8, 2017, https://www.theguardian.com/environment/2017/sep/08/sea-salt-around-world-contaminated-by-plastic-studies

126   Rob Picheta, "Microplastics found in human stools, research finds," CNN, October 23, 2018, https://www.cnn.com/2018/10/23/health/microplastics-human-stool-pollution-intl/index.html

127   Kevin Lui, "Plastic Fibers Are Found in 83% of the World's Tap Water, a New Study Reveals," Time Magazine, September 6, 2017, http://time.com/4928759/plastic-fiber-tap-water-study

128   Damian Carrington, "Plastic fibres found in tap water around the world, study reveals," The Guardian, September 5, 2017, https://www.theguardian.com/environment/2017/sep/06/plastic-fibres-found-tap-water-around-world-study-reveals

129   Pam Wright, "World's Leading Brands of Bottled Water Loaded With Microplastics, Study Says," Weather.com, March 16, 2018

130   Gerd Liebezeit and Elisabeth Liebezeit, "Origin of Synthetic Particles in Honeys," June 2015, Polish Journal of Food and Nutrition Sciences, vol. 65, no. 2, pp. 143–147, DOI: https://doi.org/10.1515/pjfns-2015-0025

131   Gerd Liebezeit and Elisabeth Liebezeit, "Synthetic particles as contaminants in German beers," Food Additives & Contaminants: Part A, 2014, vol. 31, no. 9, pp. 1574-1578, http://dx.doi.org/10.1080/19440049.2014.945099

132   Karl Mathiesen, "Wet wipes found on British beaches up more than 50% in 2014," The Guardian, March 19, 2015, https://www.theguardian.com/environment/2015/mar/19/dont-flush-wet-wipes-toilet-conservationists

133   Matt Kessler, "Are Wet Wipes Wrecking the World's Sewers?" The Atlantic, October 14, 2016, https://www.theatlantic.com/science/archive/2016/10/are-wet-wipes-wrecking-the-worlds-sewers/504098

134   "Facts and Figures about Materials, Waste and Recycling: Disposable Diapers," EPA (Environmental Protection Agency), https://www.epa.gov/facts-and-figures-about-materials-waste-and-recycling/nondurable-goods-product-specific-data#DisposableDiapers

135   "The Harmful Effects of Disposable Diapers," Unsustainable Magazine, https://www.unsustainablemagazine.com/the-effects-of-disposable-diapers-on-the-environment-and-human-health

136   Mfon Emmanuel Ntekpe, et al., "Disposable Diapers: Impact of Disposal Methods on Public Health and the Environment," American Journal of Medicine and Public Health, November 30, 2020.

137   Mfon Emmanuel Ntekpe, et al., "Disposable Diapers: Impact of Disposal Methods on Public Health and the Environment," American Journal of Medicine and Public Health, November 30, 2020.

138   CJ Park, R. Barakat, A. Ulanov, et al. "Sanitary pads and diapers contain higher phthalate contents than those in common commercial plastic products," Reproductive Toxicology, March 2019, pp.114-121, doi:10.1016/j.reprotox.2019.01.005

139   Helen Briggs, "Plastic from tyres 'major source' of ocean pollution," BBC News, February 22,2017, http://www.bbc.com/news/science-environment-39042655

140   Simon Hann, et al., "Investigating options for reducing releases in the aquatic environment of microplastics emitted by (but not intentionally added in) products," ICF, February 23, 2018, p. 12.

141   Victoria Bell, "Microplastics from tyres and roads which end up in the ocean make up 89% of ultra fine particles found in the air around busy motorways," Daily Mail, November 13, 2018, https://www.dailymail.co.uk/sciencetech/article-6384713/Particles-brake-systems-tires-roads-produce-microplastics-end-ocean.html

142   Maanvi Singh, "'Everywhere we looked': trillions of microplastics found in San Francisco bay," The Guardian, October 4, 2019, https://www.theguardian.com/environment/2019/oct/04/san-francisco-microplastics-study-bay

143   Eric Roston, "There's a Tiny Plastic Enemy Threatening the Planet's Oceans," Bloomberg, January 11, 2019, https://www.bloomberg.com/news/articles/2019-01-11/there-s-a-tiny-plastic-enemy-threatening-the-earth-s-oceans

144   "Industry is leaking huge amounts of microplastics, Swedish study shows," Science Daily, February 20, 2018, https://www.sciencedaily.com/releases/2018/02/180220124546.htm

145   Alan Williams, "Washing clothes releases thousands of microplastic particles into environment, study shows," Plymouth University News, September 27, 2016, https://www.plymouth.ac.uk/news/washing-clothes-releases-thousands-of-microplastic-particles-into-environment-study-shows

146   Katharine Gammon, "Groundbreaking study finds 13.3 quadrillion plastic fibers in California's environment," The Guardian, October 15, 2020, https://www.theguardian.com/us-news/2020/oct/16/plastic-waste-microfibers-california-study

147   Leah Messinger, "How your clothes are poisoning our oceans and food supply," The Guardian, June 20, 2016, https://www.theguardian.com/environment/2016/jun/20/microfibers-plastic-pollution-oceans-patagonia-synthetic-clothes-microbeads

148   Takeshi Noda, "Native Morphology of Influenza Virions," Frontiers in Microbiology, 2011.

149   Albert A. Koelmans, Ellen Besseling and Won J. Shim, Marine Anthropogenic Litter, 2015, pp. 325-340.

150   Damian Carrington, "Microplastic pollution in oceans is far worse than feared, say scientists," The Guardian, March 12, 2018, https://www.theguardian.com/environment/2018/mar/12/microplastic-pollution-in-oceans-is-far-greater-than-thought-say-scientists

151   Tommy Cedervall, et al., "Food Chain Transport of Nanoparticles Affects Behaviour and Fat Metabolism in Fish," PLOS ONE, February 2012, vol. 7, issue 2.

152   "Brain damage in fish from plastic nanoparticles in water," Science News, September 25, 2017, https://www.sciencedaily.com/releases/2017/09/170925104730.htm

153   "Brain damage in fish from plastic nanoparticles in water," Science News, September 25, 2017, https://www.sciencedaily.com/releases/2017/09/170925104730.htm

154   "Airborne particles from 3D printers could be as harmful to your health as cigarette smoke," Daily Mail, July 26, 2013, http://www.dailymail.co.uk/news/article-2378687/Airborne-particles-3D-printers-harmful-health-cigarette-smoke.html

155   Tom Rayner, "Plastic waste on beaches underestimated by 80% - study," Sky News, April 25, 2017, http://news.sky.com/story/up-to-400-more-plastic-in-oceans-than-thought-study-10849911

156   Kevin Loria, "What's Gone Wrong With Plastic Recycling," Consumer Reports, April 30, 2020, https://www.consumerreports.org/recycling/whats-gone-wrong-with-plastic-recycling

157   "Plastic's carbon footprint," Science Daily, April 15, 2019, https://www.sciencedaily.com/releases/2019/04/190415144004.htm

158   "Degrading plastics revealed as source of greenhouse gases," Science Daily, August 1, 2018, https://www.sciencedaily.com/releases/2018/08/180801182009.htm

159   "Degrading plastics revealed as source of greenhouse gases," Science Daily, August 1, 2018, https://www.sciencedaily.com/releases/2018/08/180801182009.htm

160   Jan Dell, 157,000 Shipping Containers of U.S. Plastic Waste Exported to Countries with Poor Waste Management in 2018," plasticpollutioncoalition.org, March 6 2019,

161   Olympic-size swimming pool, Wikipedia, https://en.wikipedia.org/wiki/Olympic-size_swimming_pool

162   Shashank Bengali, "How mountains of U.S. plastic waste ended up in Malaysia, broken down by workers for $10 a day," Los Angeles, December 29, 2018, https://www.latimes.com/world/asia/la-fg-malaysia-plastic-2018-story.html

163   "1 to 2 million tons of US plastic trash go astray, study finds," The New York Post, October 31, 2020, https://nypost.com/2020/10/31/1-to-2-million-tons-of-us-plastic-trash-go-astray-study-finds

164   Amy Westervelt, "Can Recycling Be Bad for the Environment?" Forbes, April 25, 2012, https://www.forbes.com/sites/amywestervelt/2012/04/25/can-recycling-be-bad-for-the-environment/?sh=42ab4d533bec

165   Amy Westervelt, "Can Recycling Be Bad for the Environment?" Forbes, April 25, 2012, https://www.forbes.com/sites/amywestervelt/2012/04/25/can-recycling-be-bad-for-the-environment/?sh=42ab4d533bec

166   Tiffany Duong, "The Recycling Industry in America Is Broken," EcoWatch, April 20, 2021, https://www.ecowatch.com/us-recycling-industry-2652630035.html

167   Kevin Loria, "What's Gone Wrong With Plastic Recycling," Consumer Reports, April 30, 2020, https://www.consumerreports.org/recycling/whats-gone-wrong-with-plastic-recycling

168   Tiffany Duong, "The Recycling Industry in America Is Broken," EcoWatch, April 20, 2021, https://www.ecowatch.com/us-recycling-industry-2652630035.html

169   Tiffany Duong, "The Recycling Industry in America Is Broken," EcoWatch, April 20, 2021, https://www.ecowatch.com/us-recycling-industry-2652630035.html

170   Kevin Loria, "What's Gone Wrong With Plastic Recycling," Consumer Reports, April 30, 2020, https://www.consumerreports.org/recycling/whats-gone-wrong-with-plastic-recycling

171   "Conservationists say sea garbage is hindering the search for missing Malaysia Airlines plane MH370," Courier Mail, April 2 2014, http://www.couriermail.com.au/news/conservationists-say-sea-garbage-is-hindering-the-search-for-missing-malaysia-airlines-plane-mh370/news-story/21e328976305c1392ee8a8d1f4243ea1

172   Laura Parker, "Nearly Every Seabird on Earth Is Eating Plastic," National Geographic, September 2, 2015, https://news.nationalgeographic.com/2015/09/15092-plastic-seabirds-albatross-australia

173   Harrington, Rebecca, "By 2050, the oceans could have more plastic than fish," Business Insider, January 26, 2017, https://www.businessinsider.com/plastic-in-ocean-outweighs-fish-evidence-report-2017-1

174   "The New Plastics Economy- Rethinking the future of plastics," Ellen MacArthur Foundation, 2017, https://www.ellenmacarthurfoundation.org/publications/the-new-plastics-economy-rethinking-the-future-of-plastics
175   "Perils of plastics? Survey of risks to human health and the environment," Science News, March 20, 2010, https://www.sciencedaily.com/releases/2010/03/100319115631.htm
176   Roger Harrabin, "Ocean plastic a 'planetary crisis' - UN," BBC News, December 5, 2017, http://www.bbc.com/news/science-environment-42225915
177   Colin Fernandez, "Pledge to stop planetary crisis: United Nations chief demands halt to the 'ocean armageddon'... and even Donald Trump could sign up," The Daily Mail, December 5, 2017, http://www.dailymail.co.uk/news/article-5148405/UN-gets-Daily-Mails-campaign-plastic-waste.html
178   Mihai Florin-Constantin, "One Global Map but Different Worlds: Worldwide Survey of Human Access to Basic Utilities," Human Ecology, June 2017, vol. 45, issue 3, pp. 425-429, doi:10.1007/s10745-017-9904-7
179   Katherine J. Wu, "Ocean Life Eats Tons of Plastic—Here's Why That Matters," National Geographic, August 16, 2017, https://news.nationalgeographic.com/2017/08/ocean-life-eats-plastic-larvaceans-anchovy-environment
180   P.H. Gleick and H.S. Coole, "Energy implications of bottled water," Environmental Research Letters, January 2009, doi:10.1088/1748-9326/4/1/014009
181   "Sugary Drinks," Harvard School of Public Health, https://www.hsph.harvard.edu/nutritionsource/healthy-drinks/sugary-drinks
182   "The Throwaway Generation: 25 Billion Styrofoam Cups a Year," Environmental Magazine, https://emagazine.com/the-throwaway-generation-25-billion-styrofoam-cups-a-year
183   "Glass straws? Straw straws? Here are some eco-friendly alternatives to plastic," USA Today, https://www.usatoday.com/story/money/nation-now/2018/05/23/sustainable-alternatives-plastic-straws-recyling/632993002
184   Nancy Trent, "Ending Take Out Waste," Wholefood Magazine, January 34, 2011, https://wholefoodsmagazine.com/blog/ending-take-out-waste
185   "The problem with plastic bags," Center for Biological Diversity, https://www.biologicaldiversity.org/programs/population_and_sustainability/sustainability/plastic_bag_facts.html
186   Amanda Ogle, "4 Reasons to Ditch Your Shampoo Bottle for a Bar," National Geographic, August 14, 2018, https://www.nationalgeographic.com/travel/lists/travel-gear/shampoo-bar-sustainable-toiletries-plastic-free
187   Kirsten Brodde, "What are microfibers and why are our clothes polluting the oceans?" Greenpeace, March 2, 2017, https://www.greenpeace.org/international/story/6956/what-are-microfibers-and-why-are-our-clothes-polluting-the-oceans
188   Jane Dalton, "Seafood giants 'let thousands of whales, dolphins and seals die in agony each year from discarded fishing equipment'," The Independent, March 8, 2018, https://www.independent.co.uk/news/world/seafood-firms-discarded-lost-fishing-equipment-thousands-whales-dolphins-seals-die-plastic-pollution-a8244181.htmlv
189   "Breast-feeding vs. formula-feeding: What's best?" Mayo Clinic, https://www.mayoclinic.org/healthy-lifestyle/infant-and-toddler-health/in-depth/breast-feeding/art-20047898
190   Thomas E. Novotny, et al., "Cigarettes Butts and the Case for an Environmental Policy on Hazardous Cigarette Waste," International Journal of Environmental Research and Public Health, 2009, pp. 1691-1705, doi:10.3390/ijerph6051691

191   Farley Mowat, Sea of Slaughter, 1984, Key Porter Books Limited, Toronto Canada, p. 208.
192   Steve Nicholls, Paradise Found: Nature in America at the Time of Discovery, 2009, The University of Chicago Press, Chicago & London, p. 39.
193   Farley Mowat, Sea of Slaughter, 1984, Key Porter Books Limited, Toronto Canada, p. 153.
194   Baron De Lahontan, New Voyages to North-America, Volume I, 1905, Chicago, A. C. McClurg & Co., p. 27.
195   The Penny Cyclopaedia of The Society for the Diffusion of Useful Knowledge, 1838, p. 288.
196   A Dictionary of Mechanical Science, Arts, Manufactures, and Miscellaneous Knowledge, 1827, London, p. 33.
197   The Nordic Fisheries in the New Consumer Era: Challenges Ahead for Nordic Fisheries Sector, April 1998, p. 54.
198   David Able, "Something new in the chill, salt air: Hope," The Boston Globe, August 6, 2016, https://www.bostonglobe.com/metro/2016/08/06/after-years-decline-cod-and-community-rebound-newfoundland/oNxKF14RpE47yc6500Ay6O/story.html
199   Tim Hirsch, "Cod's warning from Newfoundland," BBC News, December 16, 2002, http://news.bbc.co.uk/2/hi/science/nature/2580733.stm
200   Steve Nicholls, Paradise Found: Nature in America at the Time of Discovery, 2009, The University of Chicago Press, Chicago & London, p. 39.
201   Rosa Garcia-Orellan, Terranova: The Spanish Cod Fishery on the Grand Banks of Newfoundland in the Twentieth Century, 2010, p. 115.
202   Ashley Strub, and Daniel Pauly, "Atlantic cod: past and present," The Sea Around Us Project Newsletter, March/April 2011, p. 1.
203   Steve Nicholls, Paradise Found: Nature in America at the Time of Discovery, 2009, The University of Chicago Press, Chicago & London, p. 39.
204   Gib Brogan, "A Knockout Blow for American Fish Stocks," The New York Times, July 7, 2015, https://www.nytimes.com/2015/07/07/opinion/a-knockout-blow-for-american-fish-stocks.html
205   Ashley Strub, and Daniel Pauly, "Atlantic cod: past and present," The Sea Around Us Project Newsletter, March/April 2011, p. 3.
206   Steve Nicholls, Paradise Found: Nature in America at the Time of Discovery, 2009, The University of Chicago Press, Chicago & London, p. 13.
207   Farley Mowat, Sea of Slaughter, 1984, Stackpole books, Mechanicsburg, PA, p. 167.
208   Steve Nicholls, Paradise Found: Nature in America at the Time of Discovery, 2009, The University of Chicago Press, Chicago & London, p. 16.
209   Mowat, Farley, Sea of Slaughter, 1984, Stackpole books, Mechanicsburg, PA, p. 168.
210   Patrick Whittle, "Conservationists: Imperiled Atlantic Salmon Decline Worsens," US News and World Report, June 18, 2017, https://www.usnews.com/news/news/articles/2017-06-18/conservationists-imperiled-atlantic-salmon-decline-worsens
211   Monte Burke, "Endangered Atlantic Salmon Are Facing A New And Potentially Devastating Threat," Forbes, June 14, 2013, https://www.forbes.com/sites/monteburke/2013/06/14/endangered-atlantic-salmon-are-facing-a-new-and-potentially-devastating-threat
212   Laurel A. Col and Christopher M. Legault, The 2008 Assessment of Atlantic

Halibut in the Gulf of Maine-Georges Bank Region, April 2009, Northeast Fisheries Science Center, 166 Water Street, Woods Hole, MA 02543

213 "The State of World Fisheries and Aquaculture 2016," Food and Agriculture Organization of the United Nations (FAO), p. 43, www.fao.org

214 Amitav Ghosh and Aaron Savio Lobo, "Bay of Bengal: depleted fish stocks and huge dead zone signal tipping point," January 31 2017, https://www.theguardian.com/environment/2017/jan/31/bay-bengal-depleted-fish-stocks-pollution-climate-change-migration

215 Inam Ahmed and Sohel Parvex, "Fish stock set to crash," The Daily Star, August 7, 2010, http://www.thedailystar.net/news-detail-149731

216 Abul Kalam Azad and Charlotte Pamment, "Bangladesh overfishing: Almost all species pushed to brink," BBC, April 16 2020, https://www.bbc.com/news/world-asia-52227735

217 Amitav Ghosh and Aaron Savio Lobo, "Bay of Bengal: depleted fish stocks and huge dead zone signal tipping point," January 31, 2017, https://www.theguardian.com/environment/2017/jan/31/bay-bengal-depleted-fish-stocks-pollution-climate-change-migration

218 James H. Tidwell, and Geoff L. Allan, "Fish as food: aquaculture's contribution - Ecological and economic impacts and contributions of fish farming and capture fisheries," European Molecular Biology Organization Reports, 2001, vol. 2, no. 11, pp. 958-962.

219 Andrew Jacbos, "China's Appetite Pushes Fisheries to the Brink," New York Times, April 30, 2017, https://www.nytimes.com/2017/04/30/world/asia/chinas-appetite-pushes-fisheries-to-the-brink.html

220 Andrew Jacbos, "China's Appetite Pushes Fisheries to the Brink," New York Times, April 30, 2017, https://www.nytimes.com/2017/04/30/world/asia/chinas-appetite-pushes-fisheries-to-the-brink.html

221 Nicholas K. Dulvy, et al., "Extinction risk and conservation of the world's sharks and rays," eLife Sciences, January 2014, DOI: 10.7554/eLife.00590

222 Erik Vance, "The Push to Stop the Killing of Sharks for Their Fins," National Geographic, June 2016, http://www.nationalgeographic.com/magazine/2016/07/shark-fin-soup-campaign-illegal

223 Rhian Deutrom, "Aussie shark population in staggering decline," news.com.au, December 13, 2018, https://www.news.com.au/technology/science/animals/aussie-shark-population-is-staggering-decline/news-story/49e910c828b6e2b735d1c68e-6b2c956e

224 "Sharks at risk of extinction from overfishing, say scientists," The Guardian, March 2, 2013, https://www.theguardian.com/environment/2013/mar/02/sharks-risk-extinction-overfishing-scientists

225 "100 million sharks killed each year, say scientists," The Guardian, March 1, 2013, https://www.theguardian.com/environment/2013/mar/01/100-million-sharks-killed-each-year

226 "Oceanic shark and ray populations have collapsed by 70 percent over 50 years," National Geographic, January 27, 2021, https://www.nationalgeographic.com/animals/2021/01/oceanic-sharks-and-rays-collapsed-by-seventy-percent

227 "The State of World Fisheries and Aquaculture 2016," Food and Agriculture Organization of the United Nations (FAO), p. 38, www.fao.org

228 James Wilt, "Global Fish Stocks Are in Even Worse Shape Than We Thought," The Vice, December 15, 2016, https://www.vice.com/en_ca/article/global-fish-stocks-are-in-even-worse-shape-than-we-thought

229   Christopher Pala, "Official statistics understate global fish catch, new estimate concludes," Science Magazine, January 19, 2016, https://www.sciencemag.org/news/2016/01/official-statistics-understate-global-fish-catch-new-estimate-concludes
230   Keiran Kelleher, Discards in the world's marine fisheries - An update, Food and Agriculture Organization of the United Nations, Rome, 2005
231   Ecological Effects of Fishing - NOAA's National Ocean Service
232   Jamail Dahr, "Global Fisheries Are Collapsing—What Happens When There Are No Fish Left?" Alternet, April 17, 2016, http://www.alternet.org/environment/global-fisheries-are-collapsing-what-happens-when-there-are-no-fish-left
233   Dirk Zeller, et al., "Global marine fisheries discards: A synthesis of reconstructed data," Fish and Fisheries, 2017, DOI: 10.1111/faf.12233
234   "Ten million tons of fish wasted every year despite declining fish stocks," Science News, June 26, 2017, https://www.sciencedaily.com/releases/2017/06/170626131743.htm
235   Rebecca Kessler, "'Annihilation trawling': Q&A with marine biologist Amanda Vincent," Mongabay, March 27, 2018, https://news.mongabay.com/2018/03/annihilation-trawling-qa-with-marine-biologist-amanda-vincent/
236   "The State of World Fisheries and Aquaculture 2020," Food and Agriculture Organization of the United Nations (FAO), p. 47, www.fao.org
237   Neslen Arthur, "Global fish production approaching sustainable limit, UN warns," The Guardian, July 7, 2016, https://www.theguardian.com/environment/2016/jul/07/global-fish-production-approaching-sustainable-limit-un-warns
238   Ransom A. Myers and Boris Worm, "Rapid worldwide depletion of predatory fish communities," Nature, May 15, 2003, vol. 423
239   Jamail Dahr, "Global Fisheries Are Collapsing—What Happens When There Are No Fish Left?" Alternet, April 17, 2016, http://www.alternet.org/environment/global-fisheries-are-collapsing-what-happens-when-there-are-no-fish-left
240   Zak Smith, et al., "Net Loss: The Killing of Marine Mammals in Foreign Fisheries," NRDC (Natural Resources Defense Council) Report, January 2014
241   "Industrial fisheries are starving seabirds all around the world," Phys.org, December 6, 2018, https://phys.org/news/2018-12-industrial-fisheries-starving-seabirds-world.amp
242   Orea R. J. Anderson, et al., "Global seabird bycatch in longline fisheries," Endangered Species Research, vol. 11, 2011, pp. 91-106, DOI: https://doi.org/10.3354/esr00347
243   Karen McVeigh, "Industrial fishing ushers the albatross closer to extinction, say researchers," The Guardian, January 31, 2019, https://www.theguardian.com/environment/2019/jan/31/industrial-fishing-ushers-albatross-closer-to-extinction-say-researchers
244   Robin McKie, "The oceans' last chance: 'It has taken years of negotiations to set this up,'" The Guardian, August 5, 2018, https://www.theguardian.com/environment/2018/aug/05/last-chance-save-oceans-fishing-un-biodiversity-conference
245   "The State of World Fisheries and Aquaculture 2020," Food and Agriculture Organization of the United Nations (FAO), p. 7, www.fao.org
246   Jamail Dahr, "Global Fisheries Are Collapsing—What Happens When There Are No Fish Left?" Alternet, April 17, 2016, http://www.alternet.org/environment/global-fisheries-are-collapsing-what-happens-when-there-are-no-fish-left
247   Morato Gomes, Telmo Alexandre Fernandes, and Pauly, Daniel, Seamounts: Biodiversity and Fisheries. Fisheries Centre Research Reports, 2004 vol. 12, no. 5, p. 61.

248   Ransom A. Myers and Boris Worm, "Rapid worldwide depletion of predatory fish communities," Nature, May 15, 2003, vol. 423, p. 282.
249   Murray R. Gregory, "Environmental implications of plastic debris in marine settings—entanglement, ingestion, smothering, hangers-on, hitch-hiking and alien invasions," Philosophical Transactions of the Royal Society B, 2009, vol. 364, p. 2015, doi:10.1098/rstb.2008.0265
250   Morato Gomes, Telmo Alexandre Fernandes, and Pauly, Daniel, Seamounts: Biodiversity and Fisheries. Fisheries Centre Research Reports, 2004 vol. 12, no. 5, pp. 63-64
251   Andrew Jacbos, "China's Appetite Pushes Fisheries to the Brink," New York Times, April 30, 2017, https://www.nytimes.com/2017/04/30/world/asia/chinas-appetite-pushes-fisheries-to-the-brink.html
252   Richard Black, "'Only 50 years left' for sea fish," BBC News, November 2, 2006, http://news.bbc.co.uk/2/hi/science/nature/6108414.stm
253   "Grinding Nemo - a film about fish meal," YouTube, https://www.youtube.com/watch?v=MqW8V4Qjl1I
254   Taylor Mcneil, "Overfishing and modern-day slavery," Phys.org, October 12, 2018, https://phys.org/news/2018-10-overfishing-modern-day-slavery.html
255   Zoe Tabary, "Trafficking, debt bondage rampant in Thai fishing industry, study finds," Reuters, September 21, 2017, https://www.reuters.com/article/us-trafficking-thailand-fishing-idUSKCN1BW2XP
256   Zoe Tabary, "Trafficking, debt bondage rampant in Thai fishing industry, study finds," Reuters, September 21, 2017, https://www.reuters.com/article/us-trafficking-thailand-fishing-idUSKCN1BW2XP
257   Zoe Tabary, "Trafficking, debt bondage rampant in Thai fishing industry, study finds," Reuters, September 21, 2017, https://www.reuters.com/article/us-trafficking-thailand-fishing-idUSKCN1BW2XP
258   Desmond Ng, "Thailand's seafood slavery: Why the abuse of fishermen just won't go away," Channel News Asia, June 13, 2020, https://www.channelnewsasia.com/news/cnainsider/thailand-seafood-slavery-why-abuse-fishermen-will-not-go-away-12831948
259   Lin Taylor, "Canned tuna brands found failing to combat slavery in supply chains," Reuters, June 3, 2019, https://www.reuters.com/article/us-global-labour-fish-idUSKCN1T41AD
260   Lin Taylor, "Canned tuna brands found failing to combat slavery in supply chains," Reuters, June 3, 2019, https://www.reuters.com/article/us-global-labour-fish-idUSKCN1T41AD
261   "Was Your Seafood Caught By Slaves? AP Uncovers Unsavory Trade," NPR, March 27, 2015, https://www.npr.org/sections/thesalt/2015/03/27/395589154/was-your-seafood-caught-by-slaves-ap-uncovers-unsavory-trade
262   "Five Reasons Slaves Still Catch Your Seafood," NBC News, March 10, 2014, https://www.nbcnews.com/news/world/five-reasons-slaves-still-catch-your-seafood-n48796
263   Robin McDowell, "AP Investigation: Slaves may have caught the fish you bought," Associated Press, March 25, 2015, https://www.ap.org/explore/seafood-from-slaves/ap-investigation-slaves-may-have-caught-the-fish-you-bought.html
264   Katrina Nakamura, et al., "Seeing slavery in seafood supply chains," Science Advances, July 25, 2018, vol. 4, no. 7, DOI: 10.1126/sciadv.1701833
265   Marianne Lavelle, "Collapse of New England's iconic cod tied to climate change," Science Magazine, October 29, 2015, http://www.sciencemag.org/

news/2015/10/collapse-new-england-s-iconic-cod-tied-climate-change
266   Jamail Dahr, "Global Fisheries Are Collapsing—What Happens When There Are No Fish Left?" Alternet, April 17, 2016, https://www.alternet.org/2016/04/global-fisheries-are-collapsing-what-happens-when-there-are-no-fish-left
267   Colin Woodard, "Big changes are occurring in one of the fastest-warming spots on Earth," Portland Press Herald, October 25, 2015, https://www.pressherald.com/2015/10/25/climate-change-imperils-gulf-maine-people-plants-species-rely
268   "Summer in the Gulf of Maine Getting Longer and Stronger," The University of Maine, September 13, 2017, https://umaine.edu/marine/2017/09/13/summer-gulf-maine-getting-longer-stronger
269   Colin Woodard, "Big changes are occurring in one of the fastest-warming spots on Earth," Portland Press Herald, October 25, 2015, https://www.pressherald.com/2015/10/25/climate-change-imperils-gulf-maine-people-plants-species-rely
270   "Summer in the Gulf of Maine Getting Longer and Stronger," The University of Maine, September 13, 2017, https://umaine.edu/marine/2017/09/13/summer-gulf-maine-getting-longer-stronger
271   Andrew Pershing, Ph.D., "2020 Gulf of Maine Warming Update," Gulf of Maine Research Institute, August 13, 2020, https://gmri.org/stories/2020-gulf-maine-warming-update
272   Andrew Pershing, Ph.D., "2020 Gulf of Maine Warming Update," Gulf of Maine Research Institute, August 13, 2020, https://gmri.org/stories/2020-gulf-maine-warming-update
273   "North Atlantic Fish Populations Shifting as Ocean Temperatures Warm," November 2, 2009, http://www.nefsc.noaa.gov/press_release/2009/SciSpot/SS0916
274   William W. L. Cheung, Reg Watson and Daniel Pauly, "Signature of ocean warming in global fisheries catch," Nature, May 16, 2003, vol. 497, no. 365, doi:10.1038/nature12156
275   "'Fish thermometer' reveals long-standing, global impact of climate change," Science News, May 15, 2013, https://www.sciencedaily.com/releases/2013/05/130515131552.htm
276   Water pollution from agriculture: a global review - Executive summary, FAO & IWMI, 2017.
277   Neslen Arthur, "Global fish production approaching sustainable limit, UN warns," The Guardian, July 7, 2016, https://www.theguardian.com/environment/2016/jul/07/global-fish-production-approaching-sustainable-limit-un-warns
278   Samantha Andrews, "Controlling the uncontrollable? Sea lice in salmon aquaculture," sustainablefoodtrust.org, October 27, 2016, https://sustainablefoodtrust.org/articles/sea-lice-salmon-aquaculture
279   O. Torrissen, et al., "Salmon lice – impact on wild salmonids and salmon aquaculture," Journal of Fish Diseases, 2013, vol. 36, pp. 171-194, doi:10.1111/jfd.12061
280   "Water pollution from agriculture: a global review," Food and Agriculture Organization of the United Nations and the International Water Management Institute, 2017, p. 3.
281   Water pollution from agriculture: a global review - Executive summary, FAO & IWMI, 2017.
282   Mark J. Costello, "How sea lice from salmon farms may cause wild salmonid declines in Europe and North America and be a threat to fishes elsewhere." Proceedings for the Royal Society B, July 8, 2009, pp. 3385–3394, doi:10.1098/rspb.2009.0771
283   Jeff Matthews, "Seals And Sea Lions Pay The Price For B.C. Salmon Farming," April 12 2017, Huffington Post CA, http://www.huffingtonpost.ca/jeff-matthews/salm-

on-farms-bc_b_9656554.html

284   Jennifer Welsh, "The Disgusting Truth About Fish And Shrimp From Asian Farms," Business Insider, October 23, 2012, http://www.businessinsider.com/disgusting-truths-about-asian-aquaculture-2012-10

285   David Barboza, "In China, Farming Fish in Toxic Waters," New York Times, December 15, 2007, http://www.nytimes.com/2007/12/15/world/asia/15fish.html

286   Irene Luo, "Are You Eating Tainted Seafood From China?" The Epoch Times, July 6, 2015, https://www.theepochtimes.com/why-you-should-beware-of-seafood-from-china_1418351.html

287   Rona Rui, "Heavily Polluted Yangtze River Endangers 400 Million Chinese," Epoch Times, September 2, 2010, https://www.theepochtimes.com/heavily-polluted-yangtze-river-endangers-400-million-chinese_1508892.html

288   Ariana Eunjung Cha, "Farmed in China's Foul Waters, Imported Fish Treated With Drugs," Washington Post, July 6, 2007, http://www.washingtonpost.com/wp-dyn/content/article/2007/07/05/AR2007070502240.html

289   David Barboza, "In China, Farming Fish in Toxic Waters," New York Times, December 15, 2007, http://www.nytimes.com/2007/12/15/world/asia/15fish.html

290   Irene Luo,"Are You Eating Tainted Seafood From China?" The Epoch Times, July 6, 2015, https://www.theepochtimes.com/why-you-should-beware-of-seafood-from-china_1418351.html

291   David Barboza, "In China, Farming Fish in Toxic Waters," New York Times, December 15, 2007, http://www.nytimes.com/2007/12/15/world/asia/15fish.html

292   Suspicious Shrimp - The Health Risks of Industrialized Shrimp Production, Food & Water Watch, June, 2008

293   Emily Main, "3 Disgusting Facts About Shrimp. Do you know what you're really eating?" Prevention, June 8, 2012

294   Suspicious Shrimp - The Health Risks of Industrialized Shrimp Production, Food & Water Watch, June 2008

295   Water pollution from agriculture: a global review - Executive summary, FAO & IWMI, 2017.

296   "ENVIRONMENT-LATAM: Shrimp Industry Devastating Mangrove Forests," Inter Press Service News Agency, June 9 2005

297   "Grinding Nemo - a film about fish meal," YouTube, https://www.youtube.com/watch?v=MqW8V4Qjl1I

298   "Grinding Nemo - a film about fish meal," YouTube, https://www.youtube.com/watch?v=MqW8V4Qjl1I

299   "Effects of Aquaculture on World Fish Supplies," Issues in Ecology, Winter 2001, no. 8.

300   Living Planet Report - 2018: Aiming Higher, World Wildlife Fund, 2018

301   Boris Worm, et al., "Impacts of Biodiversity Loss on Ocean Ecosystem Services," Science, November 3, 2016, pp. 787-790.

302   Richard Black, "'Only 50 years left' for sea fish," BBC News, November 2, 2006, http://news.bbc.co.uk/2/hi/science/nature/6108414.stm

303   Faizan Hashmi , "World Ocean Might Be Depleted Of Seafood By 2030, Faster Than Initially Expected - Parley," Urdu Point, October 22, 2020, https://www.urdupoint.com/en/world/world-ocean-might-be-depleted-of-seafood-by-2-1064546.html

304   William Carmichael McIntosh, The resources of the sea, 1899, C. J. Clay and Sons, London, p. 235.

305   Jason Holland, "Rising incomes, increased urbanization to underpin seafood consumption growth," SeafoodSource, May 3, 2019, https://www.seafoodsource.com/

news/supply-trade/rising-incomes-increased-urbanization-to-underpin-seafood-consumption-growth

306   Jason Holland, "Global per capita fish consumption rises above 20 kilograms a year," Food and Agriculture Organization (FAO) of the United Nations, July 7, 2016, http://www.fao.org/news/story/en/item/421871/icode

307   "15 Best Protein Alternatives to Meat Besides Tofu," EcoWatch, October 6, 2015, https://www.ecowatch.com/15-best-protein-alternatives-to-meat-besides-tofu-1882106660.html

308   Rachael Link, MS, RD, "The 7 Best Plant Sources of Omega-3 Fatty Acids," HealthLine, July 17, 2017, https://www.healthline.com/nutrition/7-plant-sources-of-omega-3s

309   "Bureau of Mines exhibit on Government safety-first train," Yearbook of the Bureau of Mines, 1916, Washington, pp. 164-165.

310   "Use of Canaries in the Mine-Rescue Work," Coal Mine Management, August 1923, vol. 2, no. 8, p. 60.

311   Mark D. Spalding, Corinna Ravilious, and Edmund P. Green, World Atlas of Coral Reefs, 2001, p. 9.

312   Bijal P. Trivedi, "Scientists Check Coral Reef Health From Above," National Geographic Today, October 10, 2001.

313   Coral ecosystems, National Oceanic and Atmospheric Administration (NOAA), http://www.noaa.gov/resource-collections/coral-ecosystems

314   John W. Wells, Chapter 20 - Coral Reefs, The Geological Society of America Memoir 67, December 30, 1957, vol. 1, 1957, p. 609.

315   Michel J Kaiser, Marine Ecology: Processes, Systems, and Impacts, p. 313.

316   Facts about the Great Barrier Reef, Australian Government Great Barrier Reef Marine Park Authority, http://www.gbrmpa.gov.au/about-the-reef/facts-about-the-great-barrier-reef

317   Mark D. Spalding, Corinna Ravilious, and Edmund P. Green, World Atlas of Coral Reefs, 2001, p. 15.

318   Status of Coral Reefs of the World: 2008, Global Coral Reef Monitoring Network (GCRMN), p. 6.

319   John M. Pandolfi and Jeremy B.C. Jackson, "Ecological persistence interrupted in Caribbean coral reefs," Ecology Letters, 2006, vol. 9, pp. 818-826, doi: 10.1111/j.1461-0248.2006.00933.x

320   "Corals facing 'biggest impact in history'", Phys.org, June 19, 2006, https://phys.org/news/2006-06-corals-biggest-impact-history.html

321   John M. Pandolfi and Jeremy B.C. Jackson, "Ecological persistence interrupted in Caribbean coral reefs," Ecology Letters, 2006, vol. 9, pp. 818-826, doi: 10.1111/j.1461-0248.2006.00933.x

322   John M. Pandolfi, et al., "Global Trajectories of the Long-Term Decline of Coral Reef Ecosystems," Science, vol. 301, August 15 2003, pp. 955-958, DOI: 10.1126/science.1085706

323   John M. Pandolfi and Jeremy B.C. Jackson, "Ecological persistence interrupted in Caribbean coral reefs," Ecology Letters, 2006, vol. 9, pp. 818-826, doi: 10.1111/j.1461-0248.2006.00933.x

324   Loren McClenachan, Grace O'Connor, Benjamin P. Neal, John M. Pandolfi and Jeremy B. C. Jackson, "Ghost reefs: Nautical charts document large spatial scale of coral reef loss over 240 years," Science Advances, September 6, 2017

325   Rebecca L. Vega Thurber, et al., "Chronic nutrient enrichment increases prevalence and severity of coral disease and bleaching," Global Change Biology, 2014, vol. 20,

pp. 544-554, doi: 10.1111/gcb.12450

326   KeFu YU, "Coral reefs in the South China Sea: Their response to and records on past environmental changes," Science China Earth Sciences, August 2012, vol. 55, no. 8, pp. 1217–1229, doi: 10.1007/s11430-012-4449-5

327   Mark D. Spalding, Corinna Ravilious, and Edmund P. Green, World Atlas of Coral Reefs, 2001, p. 11.

328   "Large study shows pollution impact on coral reefs – and offers solution," Oregon State University News and Research Communications, November 26, 2013, http://oregonstate.edu/ua/ncs/archives/2013/nov/large-study-shows-pollution-impact-coral-reefs-%E2%80%93-and-offers-solution

329   Rebecca L. Vega Thurber, et al., "Chronic nutrient enrichment increases prevalence and severity of coral disease and bleaching," Global Change Biology, 2014, vol. 20, pp. 544-554, doi: 10.1111/gcb.12450

330   Toby A. Gardner, et al., "Long-Term Region-Wide Declines in Caribbean Corals," Science, vol. 301, August 15, 2003, pp. 958-960, DOI: 10.1126/science.1086050

331   Jeremy Jackson, et al., "Tropical Americas Coral Reef Resilience Workshop," Global Coral Reef Monitoring Network (GCRMN), April-May 2012, p. 8.

332   Jeremy Jackson, et al., "Tropical Americas Coral Reef Resilience Workshop," Global Coral Reef Monitoring Network (GCRMN), April-May 2012, pp. 10, 19.

333   Loren McClenachan, et al., "Ghost reefs: Nautical charts document large spatial scale of coral reef loss over 240 years," Science Advances, September 6, 2017

334   Alex Larson, "Video: Over Half A Million Corals Destroyed By Port Of Miami Dredging," May 31, 2019, Sea Voice News, http://seavoicenews.com/2019/05/31/video-over-half-a-million-corals-destroyed-by-port-of-miami-dredging

335   Patrick Beach, "Murky Waters - What is killing the Gulf of Mexico's majestic coral reefs?" December 2016, Texas Monthly, http://www.texasmonthly.com/articles/gulf-of-mexico-coral-reefs

336   Patrick Beach, "Murky Waters - What is killing the Gulf of Mexico's majestic coral reefs?" December 2016, Texas Monthly, http://www.texasmonthly.com/articles/gulf-of-mexico-coral-reefs

337   Status and Trends of Caribbean Coral Reefs 1970-2012, Global Coral Reef Monitoring Network (GCRMN), 2014

338   Jessica Aldred, "Caribbean coral reefs 'will be lost within 20 years' without protection, " The Guardian, July 2, 2014, https://www.theguardian.com/environment/2014/jul/02/caribbean-coral-reef-lost-fishing-pollution-report

339   Mark D. Spalding, Corinna Ravilious, and Edmund P. Green, World Atlas of Coral Reefs, 2001, p. 48.

340   Amitav Ghosh and Aaron Savio Lobo, "Bay of Bengal: depleted fish stocks and huge dead zone signal tipping point," January 31, 2017, https://www.theguardian.com/environment/2017/jan/31/bay-bengal-depleted-fish-stocks-pollution-climate-change-migration

341   Linda Markovina and Dr. Rhett Bennett, "The first large-scale survey of east Africa's coral reef fish," Africa Geographic Magazine, January 16, 2015

342   "Coral-eating starfish threaten Great Barrier Reef," New Scientist, November 2, 2020, https://www.newscientist.com/article/2258714-coral-eating-starfish-threaten-great-barrier-reef

343   Glenn De'ath, et al., "The 27–year decline of coral cover on the Great Barrier Reef and its causes," PNAS, October 30, 2012, vol. 109, no. 44, pp. 17995-17999, www.pnas.org/cgi/doi/10.1073/pnas.1208909109

344   S. Uthicke, et al., "Outbreak of coral-eating Crown-of-Thorns creates continu-

ous cloud of larvae over 320 km of the Great Barrier Reef," Scientific Reports, November 2015, DOI: 10.1038/srep16885

345   Jon Brodie and Jane Waterhouse, "A critical review of environmental management of the 'not so Great' Barrier Reef," Estuarine, Costal and Shelf Science, 2012, pp. 1-22.

346   Terry P. Hughes, Hui Wang and Matthew A. L. Young, "The wicked problem of China's disappearing coral reefs," Conservation Biology, November 9 2012, DOI: 10.1111/j.1523-1739.2012.01957.x

347   "China's Corals Have Declined By 80 Percent, Study," Asian Scientist, January 14, 2013, http://www.asianscientist.com/2013/01/in-the-lab/china-corals-declined-by-80-percent-2013

348   KeFu YU, "Coral reefs in the South China Sea: Their response to and records on past environmental changes," Science China Earth Sciences, August 2012, vol. 55, no. 8, p. 1219, doi: 10.1007/s11430-012-4449-5

349   "Panamax," Wikipedia, https://en.wikipedia.org/wiki/Panamax

350   Ross Cunning, et al., "Extensive coral mortality and critical habitat loss following dredging and their association with remotely-sensed sediment plumes," May 24, 2019, Marine Pollution Bulletin, pp. 185-199, https://doi.org/10.1016/j.marpolbul.2019.05.027

351   "How Many Cubic Yards Does a Dump Truck Hold?" Reference.com, https://www.reference.com/vehicles/many-cubic-yards-dump-truck-hold-600f9ed053d85267

352   Doyle Rice, "Coral reefs have another enemy: Dead zones," USA Today, March 20. 2017, http://www.usatoday.com/story/tech/sciencefair/2017/03/20/coral-reefs-have-another-enemy-dead-zones/99417660

353   "Coral bleaching and ocean acidification are two climate-related impacts to coral reefs," Florida Keys National Marine Sanctuary, https://floridakeys.noaa.gov/corals/climatethreat.html

354   Status of Coral Reefs of the World: 2008, Global Coral Reef Monitoring Network (GCRMN), p. 14.

355   Leonard Jones Chauka, "Tanzania: Coral Reefs Off Tanzania's Coast Are Being Destroyed, Most Beyond Repair," All Africa, January 15, 2017, http://allafrica.com/stories/201701200570.html

356   KeFu YU, "Coral reefs in the South China Sea: Their response to and records on past environmental changes," Science China Earth Sciences, August 2012, vol. 55, no. 8, pp. 1217–1229, doi: 10.1007/s11430-012-4449-5

357   John M. Pandolfi and Jeremy B.C. Jackson, "Ecological persistence interrupted in Caribbean coral reefs," Ecology Letters, vol. 9, pp. 818-826, doi: 10.1111/j.1461-0248.2006.00933.x

358   Dominique Mosbergen "Japan's Largest Coral Reef Is Dying," Huffington Post, January 25, 2017, https://www.huffpost.com/entry/japan-coral-reef-bleaching_n_58871616e4b070d8cad5292f

359   "99 Percent Of Japan's Biggest Coral Reef Is In A Very Bad Way," IFL Science, May 18. 2018, http://www.iflscience.com/environment/99-percent-of-japans-biggest-coral-reef-is-in-a-very-bad-way

360   Helen Briggs, "'Devastating' coral loss in South China Sea - scientists," March 24. 2017, http://www.bbc.com/news/science-environment-39365690

361   Chris Mooney, " 'An enormous loss': 900 miles of the Great Barrier Reef have bleached severely since 2016, " The Washington Post, April 9. 2017, https://www.washingtonpost.com/news/energy-environment/wp/2017/04/09/for-the-second-year-in-a-row-severe-coral-bleaching-has-struck-the-great-barrier-reef

362   Chris Mooney, " 'An enormous loss': 900 miles of the Great Barrier Reef have bleached severely since 2016, " The Washington Post, April 9. 2017, https://www.washingtonpost.com/news/energy-environment/wp/2017/04/09/for-the-second-year-in-a-row-severe-coral-bleaching-has-struck-the-great-barrier-reef

363   Terry P. Hughes, et al., "Global warming transforms coral reef assemblages," Nature, April 18. 2018, https://doi.org/10.1038/s41586-018-0041-2

364   Terry P. Hughes, et al., "Global warming transforms coral reef assemblages," Nature, April 18. 2018, https://doi.org/10.1038/s41586-018-0041-2

365   Chris Mooney, "Global warming has changed the Great Barrier Reef 'forever,' scientists say," Washington Post, April 18. 2018, https://www.washingtonpost.com/news/energy-environment/wp/2018/04/18/global-warming-has-changed-the-great-barrier-reef-forever-scientists-say

366   Trevor Nace, "The Great Barrier Reef Is In Its Final 'Terminal Stage'," Forbes, April 15. 2017, https://www.forbes.com/sites/trevornace/2017/04/15/the-great-barrier-reef-is-its-final-terminal-stage

367   Chris Mooney, "Global warming has changed the Great Barrier Reef 'forever,' scientists say," Washington Post, April 18. 2018, https://www.washingtonpost.com/news/energy-environment/wp/2018/04/18/global-warming-has-changed-the-great-barrier-reef-forever-scientists-say

368   "Outlook for Great Barrier Reef downgraded to 'very poor'," Times of Malta, August 30, 2019, https://timesofmalta.com/articles/view/outlook-for-great-barrier-reef-downgraded-to-very-poor.731890

369   Damian Carrington, "Billions of pieces of plastic on coral reefs send disease soaring, research reveals," The Guardian, January 25 2018, https://www.theguardian.com/environment/2018/jan/25/billions-of-pieces-of-plastic-on-coral-reefs-send-disease-soaring-research-reveals

370   Oxybenzone, Haereticus Environmental Laboratory, http://www.haereticus-lab.org/oxybenzone-2

371   Michelle Broder Van Dyke, "Your Sunscreen Could Be Killing The Ocean's Coral Reefs," BuzzFeed, October 20. 2015, https://www.buzzfeed.com/mbvd/sunscreen-ingredient-is-killing-coral-reefs-study-finds

372   Downs, et al., "Toxicopathological Effects of the Sunscreen UV Filter, Oxybenzone (Benzophenone-3), on Coral Planulae and Cultured Primary Cells and Its Environmental Contamination in Hawaii and the U.S. Virgin Islands," Archives of Environmental Contamination and Toxicology, February 2016, vol. 70, issue 2, pp. 265-288.

373   Downs, et al., "Toxicopathological Effects of the Sunscreen UV Filter, Oxybenzone (Benzophenone-3), on Coral Planulae and Cultured Primary Cells and Its Environmental Contamination in Hawaii and the U.S. Virgin Islands," Archives of Environmental Contamination and Toxicology, February 2016, vol. 70, issue 2, pp. 265-288.

374   Elizabeth Wood, "Impact of sunscreens on coral reefs," International Coral Reef Initiative (ICRI), February, 2018, p. 3.

375   Elaina Zachos, and Eric Rosen, "What sunscreens are best for you and the planet?" National Geographic, May 31, 2019, https://www.nationalgeographic.com/travel/features/sunscreen-destroying-coral-reefs-alternatives-travel-spd

376   Dobrina Zhekova and Rebecca Carhart, "How to Know If Your Sunscreen Is Killing Coral Reefs — and the Brands to Try Instead," Travel and Leisure, July 9, 2020, https://www.travelandleisure.com/style/beauty/reef-safe-sunscreen

377   "Marine pollutants," MarineSafe.org, http://www.marinesafe.org/scienc-and-data/marine-pollutants-identified-by-science

378   "Researchers Don't Know What's Killing Florida's Coral Reef," Futurism, May

16. 2018, https://futurism.com/white-syndrome-coral-reef-florida

379   Greg Allen, "Battered By Bleaching, Florida's Coral Reefs Now Face Mysterious Disease," NPR, May 15. 2018, https://www.npr.org/2018/05/15/611258056/battered-by-bleaching-floridas-coral-reefs-now-face-mysterious-disease

380   Status of Coral Reefs of the World: 2008, Global Coral Reef Monitoring Network (GCRMN), p. 5.

381   Glenn De'ath, et al., "The 27-year decline of coral cover on the Great Barrier Reef and its causes," PNAS, October 30, 2012, vol. 109, no. 44, p. 17998, www.pnas.org/cgi/doi/10.1073/pnas.1208909109

382   Elena Becatoros, "More than 90 percent of world's coral reefs will die by 2050," The Independent, March 13, 2017, http://www.independent.co.uk/environment/environment-90-percent-coral-reefs-die-2050-climate-change-bleaching-pollution-a7626911.html

383   Renee Cho, "Losing Our Coral Reefs," Colombia University, June 13, 2011, http://blogs.ei.columbia.edu/2011/06/13/losing-our-coral-reefs

384   Dennis Brady, "Why the death of coral reefs could be devastating for millions of humans," The Washington Post, November 9, 2016, https://www.washingtonpost.com/news/energy-environment/wp/2016/11/09/why-the-death-of-coral-reefs-could-be-devastating-for-millions-of-humans

385   Roger Bradbury, "A World Without Coral Reefs," The New York Times, July 13, 2012, http://www.nytimes.com/2012/07/14/opinion/a-world-without-coral-reefs.html

386   J.M. Pandolfi, et al., "Are U.S. Coral Reefs on the Slippery Slope to Slime?" Science, March 18, 2005, vol. 307, pp. 1725-1726.

387   World Trade Report 2013, World Trade Organization, p. 55.

388   Elaina Zachos and Eric Rosen, "What sunscreens are best for you and the planet?" National Geographic, May 31, 2019, https://www.nationalgeographic.com/travel/features/sunscreen-destroying-coral-reefs-alternatives-travel-spd

389   Elizabeth Wood, "Impact of sunscreens on coral reefs," International Coral Reef Initiative (ICRI), February, 2018, p. 16.

390   Anne Marie Uwitonze, BDT, MS; Mohammed S. Razzaque, MBBS, PhD, "Role of Magnesium in Vitamin D Activation and Function," J Am Osteopath Assoc., 2018, pp. 181–189, doi:10.7556/jaoa.2018.037

391   Zaher Fanari, et al., "Vitamin D deficiency plays an important role in cardiac disease and affects patient outcome: Still a myth or a fact that needs exploration?" J Saudi Heart Assoc, February 2015, pp. 264–271, doi: 10.1016/j.jsha.2015.02.003

392   Hasan Kweder and Housam Eidi, "Vitamin D deficiency in elderly: Risk factors and drugs impact on vitamin D status," Avicenna J Med., October-December 2018, pp. 139–146, doi: 10.4103/ajm.AJM_20_18

393   "'Dead Zone' Causing a Wave of Death Off Oregon Coast," Oregon State University News and Research Communications, August 9, 2006, http://oregonstate.edu/ua/ncs/archives/2006/aug/%E2%80%98dead-zone%E2%80%99-causing-wave-death-oregon-coast

394   "'Dead Zone' Causing a Wave of Death Off Oregon Coast," Oregon State University News and Research Communications, August 9, 2006, http://oregonstate.edu/ua/ncs/archives/2006/aug/%E2%80%98dead-zone%E2%80%99-causing-wave-death-oregon-coast

395   Kenneth R. Weiss, "Dead zones off Oregon and Washington likely tied to global warming, study says," Los Angeles Times, February 15, 2008, http://www.latimes.com/local/la-me-deadzone15feb15-story.html

396   F. Chan, et al., "Persistent spatial structuring of coastal ocean acidification in the California Current System," Scientific Reports, May 2017, doi:10.1038/s41598-017-02777-y
397   Keely Chalmers, "Dead zone off Oregon Coast is one of the worst in a decade," KGW-TV News Portland Oregon, October 10, 2017, http://www.kgw.com/news/local/central-coast/dead-zone-off-oregon-coast-is-one-of-the-worst-in-a-decade/482392480
398   Carolyn Raffensperger, "What The Ocean Dead Zones Tell Us," The Environmental Forum, September/October 2004, p. 14.
399   "Gulf of Mexico 'Dead Zone' larger than usual this year," USA Today, August 4, 2015, http://www.usatoday.com/story/weather/2015/08/04/dead-zone-gulf--mexico/31123029
400   Elizabeth Chuck, "Why This Year's 'Dead Zone' in Gulf of Mexico Is Bigger Than Ever," NBC News, August 5, 2017, http://www.nbcnews.com/science/environment/why-year-s-dead-zone-gulf-mexico-bigger-ever-n789636
401   Robert J. Diaz and Rutger Rosenberg, "Spreading Dead Zones and Consequences for Marine Ecosystems," Science Magazine, August 15, 2008, p. 928.
402   David Biello, "Oceanic Dead Zones Continue to Spread," Scientific American, August 15, 2008, https://www.scientificamerican.com/article/oceanic-dead-zones-spread
403   S. S. Rabotyagov, et al., "The Economics of Dead Zones: Causes, Impacts, Policy Challenges, and a Model of the Gulf of Mexico Hypoxic Zone," Review of Environmental Economics and Policy, January 4, 2014, pp. 58-79, https://doi.org/10.1093/reep/ret024
404   Damian Carrington, "Oceans suffocating as huge dead zones quadruple since 1950, scientists warn," The Guardian, January 4 2018, https://www.theguardian.com/environment/2018/jan/04/oceans-suffocating-dead-zones-oxygen-starved
405   Fred Pearce, "Can the World Find Solutions to the Nitrogen Pollution Crisis?" Yale Environment 360, February 6, 2018, https://e360.yale.edu/features/can-the-world-find-solutions-to-the-nitrogen-pollution-crisis
406   Robert J. Diaz and Rutger Rosenberg, "Spreading Dead Zones and Consequences for Marine Ecosystems," Science Magazine, August 15, 2008, p. 926.
407   Lucy Ngatia, et al., "Nitrogen and Phosphorus Eutrophication in Marine Ecosystems," IntechOpen, April 28, 2018, DOI: 10.5772/intechopen.81869
408   S. S. Rabotyagov, et al., "The Economics of Dead Zones: Causes, Impacts, Policy Challenges, and a Model of the Gulf of Mexico Hypoxic Zone," Review of Environmental Economics and Policy, January 4, 2014, pp. 58-79, https://doi.org/10.1093/reep/ret024
409   Dead Zones: Hypoxia in the Gulf of Mexico Fact Sheet, National Oceanic and Atmospheric Administration, 2009
410   "Gulf of Mexico Dead Zone—The Last 150 Years," U.S. Department of the Interior/U.S. Geological Survey, March 2006
411   Oliver Milman, "Meat industry blamed for largest-ever 'dead zone' in Gulf of Mexico," The Guardian, August 1, 2017, https://www.theguardian.com/environment/2017/aug/01/meat-industry-dead-zone-gulf-of-mexico-environment-pollution
412   Catey Hill, "This chart proves Americans love their meat," Market Watch, December 1, 2016, http://www.marketwatch.com/story/this-chart-proves-americans-love-their-meat-2016-08-15
413   Flavius Badau, "U.S. Beef and Pork Consumption Projected To Rebound," United States Department of Agriculture, September 6, 106, https://www.ers.usda.gov/

amber-waves/2016/september/us-beef-and-pork-consumption-projected-to-rebound
414   Mystery Meat II: The Industry Behind the Quiet Destruction of the American Heartland, Mighty, 2017.
415   Alastair Bland, "Is the Livestock Industry Destroying the Planet?" Smithsonian Magazine, August 1, 2012, http://www.smithsonianmag.com/travel/is-the-livestock-industry-destroying-the-planet-11308007
416   "Corn and Other Feedgrains - Feedgrains Sector at a Glance," United States Department of Agriculture – Economic Research Service, December 21, 2018, https://www.ers.usda.gov/topics/crops/corn-and-other-feedgrains/feedgrains-sector-at-a-glance
417   "Soybeans & Oil Crops - Related Data & Statistics," United States Department of Agriculture – Economic Research Service, October 6, 2017,https://www.ers.usda.gov/topics/crops/soybeans-oil-crops/related-data-statistics
418   USDA Coexistence Fact Sheets Soybeans, U.S. Department of Agriculture, February 2015, https://www.usda.gov/sites/default/files/documents/coexistence-soybeans-factsheet.pdf
419   USDA Coexistence Fact Sheets Corn, U.S. Department of Agriculture, February 2015, https://www.usda.gov/sites/default/files/documents/coexistence-corn-factsheet.pdf
420   Emerson Urry, "Tyson Foods Accused of Dumping More Poison into Waterways Than Exxon, Dow and Koch," Truthout, February 12, 2016, http://www.truth-out.org/news/item/34810-tyson-foods-accused-of-dumping-more-poison-to-waterways-than-exxon-dow-and-koch
421   Felicity Lawrence, "The global food crisis: ABCD of food – how the multinationals dominate trade," The Guardian, June 2, 2011, https://www.theguardian.com/global-development/poverty-matters/2011/jun/02/abcd-food-giants-dominate-trade
422   Shana Gallagher, "Tyson Foods Linked to Largest Toxic Dead Zone in U.S. History," EcoWatch, October 31, 2017, https://www.ecowatch.com/tyson-foods-toxic-dead-zone-2504305201.html
423   David Biello, "Oceanic Dead Zones Continue to Spread," Scientific American, August 15, 2008, https://www.scientificamerican.com/article/oceanic-dead-zones-spread
424   "Nitrogen Washing Off Midwest Farms Cause Billions in Annual Damage to Gulf of Mexico Fisheries and Marine Habitat, New Study Finds," Union of Concerned Scientists, June 1, 2020, https://www.ucsusa.org/about/news/nitrogen-farms-cause-24-billion-gulf-dead-zone-damage
425   Ben Potter, "Will China's Hunger for U.S. Soybeans Last?" AG Web, March 31, 2015, https://www.agweb.com/article/will-chinas-hunger-for-us-soybeans-last-naa-ben-potter
426   James Hansen, et al., "U.S. Agricultural Exports to China Increased Rapidly Making China the Number One Market," Choices, http://www.choicesmagazine.org/choices-magazine/theme-articles/us-commodity-markets-respond-to-changes-in-chinas-ag-policies/us-agricultural-exports-to-china-increased-rapidly-making-china-the-number-one-market
427   Kelly April Tyrrell, "Plowing prairies for grains: Biofuel crops replace grasslands nationwide," University of Wisconsin-Madison News, April 2, 2015, http://news.wisc.edu/plowing-prairies-for-grains-biofuel-crops-replace-grasslands-nationwide
428   USDA Coexistence Fact Sheets Corn, U.S. Department of Agriculture, February 2015, https://www.usda.gov/sites/default/files/documents/coexistence-corn-factsheet.pdf

429	Elizabeth Royte, "The Simple River-Cleaning Tactics That Big Farms Ignore," National Geographic, December 7, 2017
430	"New Evidence Shows Fertile Soil Gone From Midwestern Farms," NPR, February 24, 2021, https://www.npr.org/2021/02/24/967376880/new-evidence-shows-fertile-soil-gone-from-midwestern-farms
431	"New Evidence Shows Fertile Soil Gone From Midwestern Farms," NPR, February 24, 2021, https://www.npr.org/2021/02/24/967376880/new-evidence-shows-fertile-soil-gone-from-midwestern-farms
432	Peng Xiu, "Future changes in coastal upwelling ecosystems with global warming: The case of the California Current System," Scientific Reports, February 12, 2018
433	Kenneth R. Weiss, "Dead zones off Oregon and Washington likely tied to global warming, study says," Los Angeles Times, February 15, 2008, http://www.latimes.com/local/la-me-deadzone15feb15-story.html
434	Tony Barboza, "Coastal winds intensifying with climate change, study says," Los Angeles Times, July 3, 2014, http://www.latimes.com/science/sciencenow/la-sci-sn-coastal-upwelling-winds-climate-change-20140701-story.html
435	Virginia Gewin, "Dead in the Water," Nature, vol. 466, August 12, 2010
436	Stendardo and N. Gruber, "Oxygen trends over five decades in the North Atlantic," Journal of Geophysical Research, vol. 117, 2012, doi:10.1029/2012JC007909
437	Lothar Stramma, et al., "Ocean oxygen minima expansions and their biological impacts," Deep-Sea Research I, 2010, doi:10.1016/j.dsr.2010.01.005
438	Stendardo and N. Gruber, "Oxygen trends over five decades in the North Atlantic," Journal of Geophysical Research, vol. 117, 2012, doi:10.1029/2012JC007909
439	Lothar Stramma, et al., "Deoxygenation in the oxygen minimum zone of the eastern tropical North Atlantic," Geophysical Research Letters, 2009, vol. 36, doi:10.1029/2009GL039593
440	William F. Gilly, et al., "Oceanographic and Biological Effects of Shoaling of the Oxygen Minimum Zone," Annual Review of Marine Science, September 17, 2012, pp. 393-420, doi: 10.1146/annurev-marine-120710-100849
441	Craig Welch, "Oceans Are Losing Oxygen—and Becoming More Hostile to Life," National Geographic, March 13, 2015, http://news.nationalgeographic.com/2015/03/150313-oceans-marine-life-climate-change-acidification-oxygen-fish
442	E.Y. Kwon, et al., "The North Pacific Oxygen Uptake Rates over the Past Half Century," American Meteorological Society, 2016, pp. 61-76, DOI: 10.1175/JCLI-D-14-00157.1
443	Matthew C. Long, Curtis Deutsch, and Taka Ito, "Finding forced trends in oceanic oxygen," Global Biogeochemical Cycles, February 29, 2016, doi:10.1002/2015GB005310.
444	Ursula L. Kaly, "Review of Land-based sources of pollution to the coastal and marine environments in the BOBLME Region," March 1, 2004
445	Amitav Ghosh and Aaron Savio Lobo, "Bay of Bengal: depleted fish stocks and huge dead zone signal tipping point," January 31, 2017, https://www.theguardian.com/environment/2017/jan/31/bay-bengal-depleted-fish-stocks-pollution-climate-change-migration
446	L. A. Bristow, et al., "N2 production rates limited by nitrite availability in the Bay of Bengal oxygen minimum zone," Nature Geoscience, December 2016, DOI: 10.1038/ngeo2847
447	Prasun Sonwalkar, "Run-off from fertilisers has made Bay of Bengal reach 'tipping point', say experts," Hindustan Times, July 19, 2017, https://www.hindustantimes.com/world-news/run-off-from-fertilisers-has-made-bay-of-bengal-reaching-tipping-

point-say-experts/story-p1X5KfanCISCvPLykFcemO.html

448   "Growing 'dead zone' confirmed by underwater robots in the Gulf of Oman," Science News, April 27, 2018, https://www.sciencedaily.com/releases/2018/04/180427113251.htm

449   Jeremy Berke, "A 'dead zone' in the ocean - bigger than whole of KZN - has just been discovered," Business Insider South Africa, May 2018, https://www.businessinsider.co.za/dead-zone-in-the-arabian-sea-gulf-of-oman-is-biggest-in-world-2018-5

450   Jeremy Berke, "A 'dead zone' in the ocean - bigger than whole of KZN - has just been discovered," Business Insider South Africa, May 2018, https://www.businessinsider.co.za/dead-zone-in-the-arabian-sea-gulf-of-oman-is-biggest-in-world-2018-5

451   Fred Pearce, "Can the World Find Solutions to the Nitrogen Pollution Crisis?" Yale Environment 360, February 6, 2018, https://e360.yale.edu/features/can-the-world-find-solutions-to-the-nitrogen-pollution-crisis

452   Fred Pearce, "Can the World Find Solutions to the Nitrogen Pollution Crisis?" Yale Environment 360, February 6, 2018, https://e360.yale.edu/features/can-the-world-find-solutions-to-the-nitrogen-pollution-crisis

453   E. P. Chassignet, J. W. Jones, V. Misra, & J. Obeysekera, (Eds.). (2017). Florida's climate: Changes, variations, & impacts. Gainesville, FL: Florida Climate Institute. https://doi.org/10.17125/fci2017

454   Mallory Pickett, "Toxic 'red tide' algae bloom is killing Florida wildlife and menacing tourism," The Guardian, August 14, 2018, https://www.theguardian.com/us-news/2018/aug/13/florida-gulf-coast-red-tide-toxic-algae-bloom-killing-florida-wildlife

455   Bill Weir, "Scientists search for 'smoking gun' in the dead zone of Florida's red tide," CNN, August 8, 2018, https://www.cnn.com/2018/08/08/us/florida-red-tide-weir/index.html

456   Bill Weir, "Scientists search for 'smoking gun' in the dead zone of Florida's red tide," CNN, August 8, 2018, https://www.cnn.com/2018/08/08/us/florida-red-tide-weir/index.html

457   L. Stramma, A. Oschlies, and S. Schmidtko, "Anticorrelated observed and modeled trends in dissolved oceanic oxygen over the last 50 years," Biogeosciences Discussions, 2012, pp. 4594-4626, doi:10.5194/bgd-9-4595-2012

458   Matt Smith, "Huge 'Dead Zones' Could Appear in the World's Oceans by 2030 Because of Climate Change," Vice News, April 29, 2016, https://news.vice.com/article/huge-dead-zones-could-appear-in-the-worlds-oceans-by-2030-because-of-climate-change

459   " 'Dead zones' found in Atlantic open waters: Moving west, could lead to mass fish kills," Science News, April 15, 2015, https://www.sciencedaily.com/releases/2015/04/150430091825.htm

460   Sarah Zielinksi, "Ocean Dead Zones Are Getting Worse Globally Due to Climate Change - Warmer waters and other factors will cause nearly all areas of low oxygen to grow by the end of the century," Smithsonian Magazine, November 10, 2014

461   Virginia Gewin, "Dead in the Water," Nature, vol. 466, August 12, 2010

462   " 'Dead zones' found in Atlantic open waters: Moving west, could lead to mass fish kills," Science News, April 15, 2015, https://www.sciencedaily.com/releases/2015/04/150430091825.htm

463   Lothar Stramma, Sunke Schmidtko, Lisa A. Levin, Gregory C. Johnson, "Ocean oxygen minima expansions and their biological impacts," Deep-Sea Research I, 2010, doi:10.1016/j.dsr.2010.01.005

464   Damian Carrington, "Oceans suffocating as huge dead zones quadruple since

1950, scientists warn," The Guardian, January 4, 2018, https://www.theguardian.com/environment/2018/jan/04/oceans-suffocating-dead-zones-oxygen-starved

465   Empire State Building Fact Sheet, http://www.esbnyc.com/sites/default/files/esb_fact_sheet_4_9_14_4.pdf

466   Elizabeth Royte, "The Simple River-Cleaning Tactics That Big Farms Ignore,"" National Geographic, December 7, 2017

467   Robert J. Diaz and Rutger Rosenberg, "Spreading Dead Zones and Consequences for Marine Ecosystems," Science Magazine, August 15, 2008, p. 928.

468   The Times-Picayune Editorial Board, "Gulf of Mexico dead zone is going in the wrong direction: Editorial," New Orleans, Louisiana News, August 6, 2017, http://www.nola.com/opinions/index.ssf/2017/08/gulf_of_mexico_dead_zone.html

469   Tim Lucas, "Unprecedented levels of nitrogen could pose risks to Earth's environment," phys.org, September 6, 2017, https://phys.org/news/2017-09-unprecedented-nitrogen-pose-earth-environment.html

470   Elizabeth Royte, "The Simple River-Cleaning Tactics That Big Farms Ignore," National Geographic, December 7, 2017

471   Mohammad Alshawaf, Ellen Douglas, and Karen Ricciardi, "Estimating Nitrogen Load Resulting from Biofuel Mandates," International Journal of Environmental Research and Public Health, May 2016, doi: 10.3390/ijerph13050478

472   Nikos Alexandratos and Jelle Bruinsma, "World Agriculture Towards 2030/2050," Agricultural Development Economics Division, Food and Agriculture Organization of the United Nations, June 2012, www.fao.org/economic/esa

473   Fred Pearce, "Can the World Find Solutions to the Nitrogen Pollution Crisis?" Yale Environment 360, February 6, 2018, https://e360.yale.edu/features/can-the-world-find-solutions-to-the-nitrogen-pollution-crisis

474   E. Sinha, A. M. Michalak, V. Balaji, "Eutrophication will increase during the 21st century as a result of precipitation changes," Science, July 28, 2017, pp. 405-408.

475   Fred Pearce, "Can the World Find Solutions to the Nitrogen Pollution Crisis?" Yale Environment 360, February 6, 2018, https://e360.yale.edu/features/can-the-world-find-solutions-to-the-nitrogen-pollution-crisis

476   Lothar Stramma, et al., "Ocean oxygen minima expansions and their biological impacts," Deep-Sea Research I, 2010, doi:10.1016/j.dsr.2010.01.005

477   Carl Zimmer, "A Looming Oxygen Crisis and Its Impact on World's Oceans," August 5, 2010, Yale Environment 360, http://e360.yale.edu/features/a_looming_oxygen_crisis_and_its_impact_on_worlds_oceans

478   Karl Mathiesen, "Are jellyfish going to take over the oceans?" The Guardian, August 21, 2015, https://www.theguardian.com/environment/2015/aug/21/are-jellyfish-going-to-take-over-oceans

479   Chadlin M. Ostrander, Jeremy D. Owens, and Sune G. Nielsen, "Constraining the rate of oceanic deoxygenation
leading up to a Cretaceous Oceanic Anoxic Event (OAE-2: ~94 Ma)," Science Advances, August 9, 2017, vol. 3, no. 9, DOI: 10.1126/sciadv.1701020

480   "Submarine eruption bled Earth's oceans of oxygen," New Scientist, July 16, 2008

481   Ian Johnston, "Climate change, sewage and fertilisers could trigger mass extinction of life in oceans, scientists warn," The Independent, August 10, 2017, http://www.independent.co.uk/environment/climate-change-global-warming-sewage-fertilisers-mass-extinction-ocean-life-trigger-scientists-warn-a7884861.html

482   Damian Carrington, "Oceans suffocating as huge dead zones quadruple since 1950, scientists warn," The Guardian, January 4, 2018, https://www.theguardian.com/

environment/2018/jan/04/oceans-suffocating-dead-zones-oxygen-starved
483   Damian Carrington, "Oceans suffocating as huge dead zones quadruple since 1950, scientists warn," The Guardian, January 4, 2018, https://www.theguardian.com/environment/2018/jan/04/oceans-suffocating-dead-zones-oxygen-starved
484   "USDA Coexistence Fact Sheets Soybeans," United States Department of Agriculture, February 2015, https://www.usda.gov/sites/default/files/documents/coexistence-soybeans-factsheet.pdf
485   "Certified organic and total U.S. acreage, selected crops and livestock, 1995-2011," United States Department of Agriculture, https://www.ers.usda.gov/webdocs/DataFiles/52407/CertifiedandtotalUSacreageselectedcropslivestock.xls
486   David Pimentel, "Environmental, Energetic, and Economic Comparisons of Organic and Conventional Farming Systems," BioScience, July 2005, vol. 55, no. 7, pp. 573-582, https://doi.org/10.1641/0006-3568(2005)055[0573:EEAECO]2.0.CO;2
487   Walden Bello, Capitalism's Last Stand?: Deglobalization in the Age of Austerity, 2013
488   Mike Barrett, "How Far Does Your Food Travel? 4 Reasons to Choose Local," Natural Society, December 17, 2013, https://naturalsociety.com/how-far-food-travel-food-miles-1500-average
489   Parliamentary Papers, Great Britain. Parliament. House of Commons. Reports from Commissioners. Children's Employment (Mines). Volume 15, 1842, p. 11.
490   "The Modern Union," The Blacksmiths Journal, October 1904, vol. V., no. 10, p. 4.
491   Carlton J. H. Hayes, A Political and Social History of Modern Europe, Volume 2, 1920, The Macmillan Company, p. 85.
492   Negro Slavery in America, The Review of Reviews, January 15, 1900, p. 90.
493   Michael Coulson, The History of Mining - The events, technology, and people involved in the industry that forged the modern world, 2012, Harriman House Ltd., p. 161.
494   Coal Fatalities for 1900 Through 2016, United States Department of Labor - Mine Safety and Health Administration, https://arlweb.msha.gov/stats/centurystats/coalstats.asp
495   Reports on the gases and explosions in collieries, 1847, London.
496   Mines and Quarries. Reports of Arthur H. Stokes Esq., H. M. Inspector of Mines for the Midland District, 1896, London.
497   Parliamentary Papers, Great Britain. Parliament. House of Commons. Reports from Commissioners. Children's Employment (Mines). Volume 15, 1842, p. 136.
498   Frederick Hoffman L., "Fatal Accidents in Coal Mining in 1903," The Engineering and Mining Journal, December 22, 1904, p. 990.
499   Arthur McIvor, and Ronald Johnston, Miner's Lung - A History of Dust Disease in British Coal Mining, 2007, p. 42.
500   William Cavert, The Smoke of London - Energy and Environment in the Early Modern City, 2016, Cambridge University Press, p. 5.
501   Trish Ferguson, Victorian Time - Technologies, Standardizations, Catastrophes, 2013.
502   "The Treatment of Smoke: A Sanitary Parallel," Nature, vol. 66, no. 1722, October 30, 1902, p. 669.
503   American Medicine, "Smoke a Sanitary, Not Only Esthetic Nuisance," American Medicine, May 17, 1902, vol. 3, no. 20, p. 800.
504   Mark Jacobson, Atmospheric Pollution - History, Science, and Regulation, 2002, Cambridge University Press, p. 84.

505 Jayati Gupta, "London Through Alien Eyes," Literary London: Interdisciplinary Studies in the Representation of London, March 2003, vol. 1 no. 1.
506 "Killer smog claims elderly victims," https://www.history.com/this-day-in-history/killer-smog-claims-elderly-victims
507 Angus Gunn, Encylopedia of Disasters - Environmental Catastrophes and Human Tragedies, 2008, p. 425.
508 "1966 New York City smog," https://en.wikipedia.org/wiki/1966_New_York_City_smog
509 Peter Christoff and Robyn Eckersley, Globalization and the Environment, 2013, Rowman & Littlefield Publishers, Inc., p. 86.
510 Lucy Rodgers, "Climate change: The massive CO2 emitter you may not know about," BBC, December 17, 2018, https://www.bbc.com/news/science-environment-46455844
511 Empire State Building Fact Sheet, http://www.esbnyc.com/sites/default/files/esb_fact_sheet_4_9_14_4.pdf
512 Peter Frumhoff, "Global Warming Fact: More than Half of All Industrial CO2 Pollution Has Been Emitted Since 1988," Union of Concerned Scientists Blog, December 15, 2014, http://blog.ucsusa.org/peter-frumhoff/global-warming-fact-co2-emissions-since-1988-764
513 Airbus A380 Specs, http://www.modernairliners.com/airbus-a380/airbus-a380-specs
514 Trends in Global CO2 Emissions 2016 Report, PBL Netherlands Environmental Assessment Agency, The Hague, 2016, p. 43.
515 "Global Gas Flaring Jumps to Levels Last Seen in 2009," Scientific American, July 21, 2020, https://www.worldbank.org/en/news/press-release/2020/07/21/global-gas-flaring-jumps-to-levels-last-seen-in-2009
516 Global Carbon Budget, International Geosphere-Biosphere Programme, December 7, 2015
517 Chloe Farand, "Carbon dioxide levels in Earth's atmosphere reach 'highest level in 800,000 years'," The Independent, May 5, 2018, https://www.independent.co.uk/environment/carbon-dioxide-concentration-atmosphere-highest-level-800000-years-mauna-loa-observatory-hawaii-a8337921.html
518 Marah J. Hardt and Carl Safina, "How Acidification Threatens Oceans from the Inside Out," Scientific American, August 9, 2010
519 O. Hoegh-Guldberg, et al., Coral Reefs Under Rapid Climate Change and Ocean Acidification, Science, December 14, 2007, vol. 318, pp. 1737-1742.
520 John Hawthorne, "How Ordinary People Can Help Ailing Oceans," Truth Dig, December 20, 2017, https://www.truthdig.com/articles/oceans-getting-acidic-heres-can-help
521 "A primer on pH," NOAA, https://www.pmel.noaa.gov/co2/story/A+primer+on+pH
522 Kirsten Jacob, "How Our Bodies Go To Extraordinary Lengths To Maintain Safe pH Levels," Forbes, March 11, 2016, https://www.forbes.com/sites/quora/2016/03/11/how-our-bodies-go-to-extraordinary-lengths-to-maintain-safe-ph-levels
523 UNEP Emerging Issues: Environmental Consequences of Ocean Acidification: A Threat to Food Security, 2010, p. 2.
524 An Updated Synthesis of the Impacts of Ocean Acidification on Marine Biodiversity, United Nations Environmental Programme (UNEP), 2014, p. 11.
525 Marah J. Hardt and Carl Safina, "How Acidification Threatens Oceans from the

Inside Out," Scientific American, August 9, 2010

526   "Ocean's Depth and Volume Revealed," Live Science, May 19, 2010, https://www.livescience.com/6470-ocean-depth-volume-revealed.html

527   Ocean Chemistry, American Chemical Society (ACS), https://www.acs.org/content/acs/en/climatescience/oceansicerocks/oceanchemistry.html

528   Majit S. Kang and Surinder S. Banga, Combating Climate Change - An Agricultural Perspective, 2013, CRC Press, p. 20.

529   "About California Current," NOAA, https://www.integratedecosystemassessment.noaa.gov//regions/california-current-region/about.html

530   F. Chan, J. A. Barth, et al., "Emergence of Anoxia in the California Current Large Marine Ecosystem," Science, February 15, 2008, vol. 319, p. 920.

531   F. Chan, J. A. Barth, et al., "Persistent spatial structuring of coastal ocean acidification in the
California Current System," Scientific Reports, May 2017, doi:10.1038/s41598-017-02777-y

532   An Updated Synthesis of the Impacts of Ocean Acidification on Marine Biodiversity, United Nations Environmental Programme (UNEP), 2014.

533   An Updated Synthesis of the Impacts of Ocean Acidification on Marine Biodiversity, United Nations Environmental Programme (UNEP), 2014, p. 54.

534   K. R. N. Anthony, et al., "Ocean acidification causes bleaching and productivityloss in coral reef builders," PNAS (Proceedings of the National Academy of Sciences), November 11, 2008, vol. 105, no. 45, pp. 17442-17446.

535   An Updated Synthesis of the Impacts of Ocean Acidification on Marine Biodiversity, United Nations Environmental Programme (UNEP), 2014, p. 57.

536   YU KeFu, "Coral reefs in the South China Sea: Their response to and records on past environmental changes," Science China Earth Sciences, August 2012, vol. 55, no. 8, p. 1219, doi: 10.1007/s11430-012-4449-5

537   Glenn De'ath, Janice M. Lough, and Katharina E. Fabricius, "Declining Coral Calcification on the Great Barrier Reef," Science Magazine, January 2, 2009, vol. 323, pp. 116-119.

538   Dorothée Herr, The Ocean and Climate Change: Tools and Guidelines for Action, 2009, p. 14.

539   Dorothée Herr, The Ocean and Climate Change: Tools and Guidelines for Action, 2009, p. 14.

540   "The future is now: Long-term research shows ocean acidification ramping up on the Great Barrier Reef," Phys.org, October 28, 2020, https://phys.org/news/2020-10-future-long-term-ocean-acidification-ramping.html

541   "The future is now: Long-term research shows ocean acidification ramping up on the Great Barrier Reef," Phys.org, October 28, 2020, https://phys.org/news/2020-10-future-long-term-ocean-acidification-ramping.html

542   An Updated Synthesis of the Impacts of Ocean Acidification on Marine Biodiversity, United Nations Environmental Programme (UNEP), 2014, p. 65.

543   Jennifer Chu, "Ocean acidification may cause dramatic changes to phytoplankton - Study finds many species may die out and others may migrate significantly as ocean acidification intensifies," MIT (Massachusetts Institute of Technology) News, July 20, 2015, http://news.mit.edu/2015/ocean-acidification-phytoplankton-0720

544   Jennifer Chu, "Ocean acidification may cause dramatic changes to phytoplankton - Study finds many species may die out and others may migrate significantly as ocean acidification intensifies," MIT (Massachusetts Institute of Technology) News, July 20, 2015, http://news.mit.edu/2015/ocean-acidification-phytoplankton-0720

545 "Key biological mechanism is disrupted by ocean acidification," phys.org, March 14, 2018, https://phys.org/news/2018-03-key-biological-mechanism-disrupted-ocean.html
546 "Key biological mechanism is disrupted by ocean acidification," phys.org, March 14, 2018, https://phys.org/news/2018-03-key-biological-mechanism-disrupted-ocean.html
547 "Canary in the kelp forest: Sea creature dissolves in today's warming, acidic waters," Science Daily, April 19, 2017, https://www.sciencedaily.com/releases/2017/04/170419131935.htm
548 Alex Fox, "Satellite Imagery Shows Northern California Kelp Forests Have Collapsed," Smithsonian Magazine, March 11, 2021, https://www.smithsonianmag.com/smart-news/satellite-imagery-shows-northern-california-kelp-forests-have-collapsed-180977214
549 Roshini Nair, "Ocean acidity increasing along Pacific coast, study finds," CBC News, June 4, 2017, http://www.cbc.ca/news/canada/british-columbia/ocean-acidity-increasing-along-pacific-coast-study-finds-1.4144786
550 Roshini Nair, "Ocean acidity increasing along Pacific coast, study finds," CBC News, June 4, 2017, http://www.cbc.ca/news/canada/british-columbia/ocean-acidity-increasing-along-pacific-coast-study-finds-1.4144786
551 "Study: Ocean acidification killing oysters by inhibiting shell formation," Oregon State University (OSU) News and Research Communications, June 11, 2013, https://today.oregonstate.edu/archives/2013/jun/study-ocean-acidification-killing-oysters-inhibiting-shell-formation-0
552 "Study: Ocean acidification killing oysters by inhibiting shell formation," Oregon State University (OSU) News and Research Communications, June 11, 2013, https://today.oregonstate.edu/archives/2013/jun/study-ocean-acidification-killing-oysters-inhibiting-shell-formation-0
553 An Updated Synthesis of the Impacts of Ocean Acidification on Marine Biodiversity, United Nations Environmental Programme (UNEP), 2014, p. 82.
554 Michael McGowan, "Sydney rock oysters getting smaller as oceans become more acidic," The Guardian, August 15, 2018, https://www.theguardian.com/australia-news/2018/aug/16/sydney-rock-oysters-getting-smaller-as-oceans-become-more-acidic
555 "Ocean Acidification Predicted To Harm Shellfish, Aquaculture," Science News, March 18, 2007, https://www.sciencedaily.com/releases/2007/03/070318133722.htm
556 "Ocean Acidification threatens cod recruitment in the Atlantic," August 24, 2016, https://www.geomar.de/en/news/article/ocean-acidification-threatens-cod-recruitment-in-the-atlantic
557 Roger Harrabin, "More acidic oceans 'will affect all sea life'," BBC News, October 23, 2017, http://www.bbc.com/news/science-environment-41653511
558 "Timeline Of Mass Extinction Events On Earth," http://www.worldatlas.com/articles/the-timeline-of-the-mass-extinction-events-on-earth.html
559 Dana Nuccitelli, "Burning coal may have caused Earth's worst mass extinction," The Guardian, March 12, 2018, https://www.theguardian.com/environment/climate-consensus-97-per-cent/2018/mar/12/burning-coal-may-have-caused-earths-worst-mass-extinction
560 Amina Kahn, "Ocean acidification triggered devastating extinction, study finds," Los Angeles Times, April 9, 2015, http://www.latimes.com/science/sciencenow/la-sci-sn-ocean-acidification-mass-extinction-20150409-story.html

561	Marah J. Hardt and Carl Safina, "How Acidification Threatens Oceans from the Inside Out," Scientific American, August 9, 2010
562	An Updated Synthesis of the Impacts of Ocean Acidification on Marine Biodiversity, United Nations Environmental Programme (UNEP), 2014
563	An Updated Synthesis of the Impacts of Ocean Acidification on Marine Biodiversity, United Nations Environmental Programme (UNEP), 2014, p. 32.
564	UNEP Emerging Issues: Environmental Consequences of Ocean Acidification: A Threat to Food Security, United Nations Environment Programme (UNEP), 2010.
565	Jennifer Fabiano, "How plant-based diets can help reduce greenhouse gas emissions by 70 percent," AccuWeather, https://www.accuweather.com/en/weather-news/how-plant-based-diets-can-help-reduce-greenhouse-gas-emissions-by-70-percent/351781
566	"Exercise: 7 benefits of regular physical activity," The Mayo Clinic, https://www.mayoclinic.org/healthy-lifestyle/fitness/in-depth/exercise/art-20048389
567	Mike Berners-Lee and Duncan Clark, "What's the carbon footprint of … a load of laundry?" The Guardian, November 20, 2010, https://www.theguardian.com/environment/green-living-blog/2010/nov/25/carbon-footprint-load-laundry
568	Tony Long, "Sept. 19, 1991: Hikers stumble upon Otzi, the Alpine iceman," Wired, September 19, 2007, https://www.wired.com/2007/09/dayintech-0919-2
569	Marissa Fessenden, "As Glaciers Retreat, They Give up the Bodies and Artifacts They Swallowed," Smithsonian, May 27, 2015, https://www.smithsonianmag.com/smart-news/glaciers-retreat-they-give-mummies-and-artifacts-they-swallowed-180955399
570	Andrew Curry, "The Big Melt," Archaeology, August 12, 2013, https://www.archaeology.org/issues/105-1309/letter-from/1165-glaciers-ice-patches-norway-global-warming
571	Rebecca Lindsey, "Climate Change: Glacier Mass Balance," , climate.gov, August 1, 2018, https://www.climate.gov/news-features/understanding-climate/climate-change-glacier-mass-balance
572	"Which are the fastest and slowest moving glaciers on Earth?" AntarcticGlaciers.org, http://www.antarcticglaciers.org/question/fastest-slowest-moving-glaciers-earth
573	"Earth's Freshwater," National Geographic, https://www.nationalgeographic.org/media/earths-fresh-water
574	Becky Oskin, "World's Glaciers Have New Size Estimate," Live Science, October 22, 2012, https://www.livescience.com/24168-glacier-volume-sea-level-rise.html
575	"Great Pyramid of Giza," https://en.wikipedia.org/wiki/Great_Pyramid_of_Giza
576	"Water on the Tibetan Plateau Ecological and Strategic Implications," The Hague Centre for Strategic Studies, June 5, 2009, p. 18.
577	"Water on the Tibetan Plateau Ecological and Strategic Implications," The Hague Centre for Strategic Studies, June 5, 2009, p. 59
578	Timothy Gardner, "Tibetan glacial shrink to cut water supply by 2050," Reuters, January 16, 2009, https://www.reuters.com/article/us-glaciers/tibetan-glacial-shrink-to-cut-water-supply-by-2050-idUSTRE50F76420090116
579	Tandong Yao, et al., "Third Pole Environment (TPE)," Environmental Development, April 2012, pp. 52-64, http://dx.doi.org/10.1016/j.envdev.2012.04.002
580	Bruce E. Johansen, The Encyclopedia of Global Warming Science and Technology, 2009, Greenwood Press, pp. 335-336.
581	"Water on the Tibetan Plateau Ecological and Strategic Implications," The

Hague Centre for Strategic Studies, June 5 2009, p. 17.
582   "Water on the Tibetan Plateau Ecological and Strategic Implications," The Hague Centre for Strategic Studies, June 5 2009, p. 19.
583   Donghui, Shangguan, et al., "Monitoring the glacier changes in the Muztag Ata and Konggur mountains, east Pamirs, based on Chinese Glacier Inventory and recent satellite imagery," Annals of Glaciology, 2006, pp. 79-85.
584   "Water on the Tibetan Plateau Ecological and Strategic Implications," The Hague Centre for Strategic Studies, June 5, 2009, p. 64.
585   Bruce E. Johansen, The Encyclopedia of Global Warming Science and Technology, 2009, Greenwood Press, p. 336.
586   Tandong Yao, et al., "Third Pole Environment (TPE)," Environmental Development, April 2012, pp. 52-64, http://dx.doi.org/10.1016/j.envdev.2012.04.002
587   Yang Wei, et al., "Quick ice mass loss and abrupt retreat of the maritime glaciers in the Kangri Karpo Mountains, southeast Tibetan Plateau," Chinese Science Bulletin, August 2008, vol. 54, no. 16. pp. 2547-2551.
588   Shangguan Donghui, et al., "Monitoring the glacier changes in the Muztag Ata and Konggur mountains, east Pamirs, based on Chinese Glacier Inventory and recent satellite imagery," Annals of Glaciology, 2006, pp. 79-85.
589   Bruce E. Johansen, The Encyclopedia of Global Warming Science and Technology, 2009, Greenwood Press, p. 336.
590   "Water on the Tibetan Plateau Ecological and Strategic Implications," The Hague Centre for Strategic Studies, June 5 2009, p. 64.
591   Bruce E. Johansen, The Encyclopedia of Global Warming Science and Technology, 2009, Greenwood Press, p. 336.
592   Yang Wei, et al., "Quick ice mass loss and abrupt retreat of the maritime glaciers in the Kangri Karpo Mountains, southeast Tibetan Plateau," Chinese Science Bulletin, August 2008, vol. 54, no. 16. pp. 2547-2551.
593   Shangguan Donghui, et al., "Monitoring the glacier changes in the Muztag Ata and Konggur mountains, east Pamirs, based on Chinese Glacier Inventory and recent satellite imagery," Annals of Glaciology, 2006, pp. 79-85.
594   Bruce E. Johansen, The Encyclopedia of Global Warming Science and Technology, 2009, Greenwood Press, p. 337.
595   Navin Singh Khadka, "Nepal drains dangerous Everest lake," BBC News, October 21, 2016, https://www.bbc.com/news/world-asia-37797559
596   Yang Wei, et al., "Quick ice mass loss and abrupt retreat of the maritime glaciers in the Kangri Karpo Mountains, southeast Tibetan Plateau," Chinese Science Bulletin, August 2008, vol. 54, no. 16. pp. 2547-2551.
597   "Water on the Tibetan Plateau Ecological and Strategic Implications," The Hague Centre for Strategic Studies, June 5, 2009, p. 32.
598   "Black Carbon Deposits on Himalayan Ice Threaten Earth's "Third Pole"," NASA, https://www.nasa.gov/topics/earth/features/carbon-pole.html
599   Sylvia Downes, "The global cost of China's destruction of the 'roof of the world'," The Ecologist, May 11, 2012, https://theecologist.org/2012/may/11/global-cost-chinas-destruction-roof-world
600   "Water on the Tibetan Plateau Ecological and Strategic Implications," The Hague Centre for Strategic Studies, June 5, 2009, p. 38.
601   "The impact of China's search for wealth on Tibet's environment and its people," Tibet Nature Environmental Conservation Netword, November 25 2014, http://www.tibetnature.net/en/impact-chinas-search-wealth-tibets-environment-people/
602   Timothy Gardner, "Tibetan glacial shrink to cut water supply by 2050," Reu-

ters, January 16, 2009, https://www.reuters.com/article/us-glaciers/tibetan-glacial-shrink-to-cut-water-supply-by-2050-idUSTRE50F76420090116
603   Tandong Yao, et al., "Third Pole Environment (TPE)," Environmental Development, April 2012, pp. 52-64, http://dx.doi.org/10.1016/j.envdev.2012.04.002
604   Daniel Glick, "The Big Thaw," National Geographic, September 2004
605   "Substantial glacier ice loss in Central Asia's largest mountain range," Phys.Org, August 17, 2015, https://phys.org/news/2015-08-substantial-glacier-ice-loss-central.html
606   Elephant, San Diego Zoo, http://animals.sandiegozoo.org/animals/elephant
607   "Substantial glacier ice loss in Central Asia's largest mountain range," Phys.Org, August 17, 2015, https://phys.org/news/2015-08-substantial-glacier-ice-loss-central.html
608   Andy Coghlan, "Much of Asia's Celestial mountain glacier ice could met by 2050," New Scientist, August 17, 2015, https://www.newscientist.com/article/dn28054-much-of-asias-celestial-mountain-glacier-ice-could-melt-by-2050
609   "Water on the Tibetan Plateau Ecological and Strategic Implications," The Hague Centre for Strategic Studies, June 5, 2009, pp. 49-50.
610   "WATER and PEOPLE: whose right is it?" Food and Agriculture Organization of the United Nations (FAO), http://www.fao.org/docrep/005/Y4555E/Y4555E00.HTM
611   Ben Orlove, "Glacier Retreat: Reviewing the Limits of Human Adaptation to Climate Change," Environment, May-June 2009
612   Daniel Glick, , "The Big Thaw," National Geographic, September 2004
613   Daniel Glick, , "The Big Thaw," National Geographic, September 2004
614   Oliver Milman, "US Glacier national park losing its glaciers with just 26 of 150 left," The Guardian, May 11, 2017, https://www.theguardian.com/environment/2017/may/11/us-glacier-national-park-is-losing-its-glaciers-with-just-26-of-150-left
615   Emily Holden, "North American glaciers melting much faster than 10 years ago – study," The Guardian, January 19, 2019, https://www.theguardian.com/environment/2019/jan/18/north-america-glacier-melt-study-climate-change
616   Oliver Milman, "US Glacier national park losing its glaciers with just 26 of 150 left," The Guardian, May 11, 2017, https://www.theguardian.com/environment/2017/may/11/us-glacier-national-park-is-losing-its-glaciers-with-just-26-of-150-left
617   Ed Struzik, "Loss of Snowpack and Glaciers In Rockies Poses Water Threat," Yale Environment 360, July 10, 2014, https://e360.yale.edu/features/loss_of_snowpack_and_glaciers_in_rockies_poses_water_threat
618   Hina Alam, "Majority of glaciers in Western Canada will likely disappear in next 50 years: expert," CTV news, December 28, 2018, https://beta.ctvnews.ca/national/sci-tech/2018/12/27/1_4232003.html
619   Ed Struzik, "Loss of Snowpack and Glaciers In Rockies Poses Water Threat," Yale Environment 360, July 10, 2014, https://e360.yale.edu/features/loss_of_snowpack_and_glaciers_in_rockies_poses_water_threat
620   "Climate change wreaking havoc with Colombia's glaciers: government," Phys.org, July 13, 2018, https://phys.org/news/2018-07-climate-wreaking-havoc-colombia-glaciers.html
621   "Peru's glaciers melting dangerously fast as a result of global warming," Euronews, August 12, 2014, http://www.euronews.com/2014/12/08/peru-s-glaciers-melting-dangerously-fast-as-a-result-of-global-warming
622   "Peru's glaciers melting dangerously fast as a result of global warming," Euronews, August 12, 2014, http://www.euronews.com/2014/12/08/peru-s-glaciers-

melting-dangerously-fast-as-a-result-of-global-warming

623 Marco Aquino, "Glacier breaks in Peru, causing tsunami in Andes," Reuters, April 12, 2010, https://www.reuters.com/article/us-peru-glaciers/glacier-breaks-in-peru-causing-tsunami-in-andes-idUSTRE63B69Y20100412

624 Marco Aquino, "Glacier breaks in Peru, causing tsunami in Andes," Reuters, April 12, 2010, https://www.reuters.com/article/us-peru-glaciers/glacier-breaks-in-peru-causing-tsunami-in-andes-idUSTRE63B69Y20100412

625 "Peru says country's glaciers shrank 40 pct in 4 decades from climate change," Reuters, October 15, 2014, https://www.reuters.com/article/peru-climatechange-glacier-idUSL2N0SA39P20141015

626 Nicholas Case, "In Peru's Deserts, Melting Glaciers Are a Godsend (Until They're Gone)," New York Times, November 26 2017, https://www.nytimes.com/2017/11/26/world/americas/peru-climate-change.html

627 Nicholas Case, "In Peru's Deserts, Melting Glaciers Are a Godsend (Until They're Gone)," New York Times, November 26 2017, https://www.nytimes.com/2017/11/26/world/americas/peru-climate-change.html

628 Dan Collyns, "On thin ice: the farmers adapting to Peru's melting glacier," The Guardian, April 15, 2015, https://www.theguardian.com/global-development-professionals-network/2015/apr/15/peru-glacier-melt-metals-farmers-adaptation-pastoruri

629 Nicholas Case, "In Peru's Deserts, Melting Glaciers Are a Godsend (Until They're Gone)," New York Times, November 26, 2017, https://www.nytimes.com/2017/11/26/world/americas/peru-climate-change.html

630 Andy Coghlan, "Much of Asia's Celestial mountain glacier ice could melt by 2050," New Scientist, August 17, 2015, https://www.newscientist.com/article/dn28054-much-of-asias-celestial-mountain-glacier-ice-could-melt-by-2050

631 Andy Coghlan, "Much of Asia's Celestial mountain glacier ice could melt by 2050," New Scientist, August 17, 2015, https://www.newscientist.com/article/dn28054-much-of-asias-celestial-mountain-glacier-ice-could-melt-by-2050

632 Glaciers, NASA, https://climate.nasa.gov/interactives/global-ice-viewer#/1

633 Empire State Building Fact Sheet, http://www.esbnyc.com/sites/default/files/esb_fact_sheet_4_9_14_4.pdf

634 Rebecca Lindsey, "Climate Change: Glacier Mass Balance," NOAA, August 1, 2018, https://www.climate.gov/news-features/understanding-climate/climate-change-glacier-mass-balance

635 Rebecca Lindsey, "Climate Change: Glacier Mass Balance," NOAA, August 1, 2018, https://www.climate.gov/news-features/understanding-climate/climate-change-glacier-mass-balance

636 "Global ice sheets melting at 'worst-case' rates: UK scientists," Algazeera, January 27, 2021, https://www.aljazeera.com/news/2021/1/27/global-ice-sheets-melting-at-worst-case-rates-uk-scientists

637 Andrew Curry, "The Big Melt," Archaeology, August 12, 2013, https://www.archaeology.org/issues/105-1309/letter-from/1165-glaciers-ice-patches-norway-global-warming

638 Rebecca Lindsey, "Climate Change: Glacier Mass Balance," NOAA, August 1, 2018, https://www.climate.gov/news-features/understanding-climate/climate-change-glacier-mass-balance

639 Damian Carrington, "A third of Himalayan ice cap doomed, finds report," The Guardian, February 4, 2019, https://amp.theguardian.com/environment/2019/feb/04/a-third-of-himalayan-ice-cap-doomed-finds-shocking-report

640	Andy Coghlan, "Much of Asia's Celestial mountain glacier ice could melt by 2050," New Scientist, August 17, 2015, https://www.newscientist.com/article/dn28054-much-of-asias-celestial-mountain-glacier-ice-could-melt-by-2050
641	Brooks Hays, "Glaciers are going to keep melting for decades, research predicts," UPI, March 17, 2018, https://www.upi.com/Glaciers-are-going-to-keep-melting-for-decades-research-predicts/6241521484043
642	Ben Marzeion, et al., "Attribution of global glacier mass loss to anthropogenic and natural causes," Science, 2014, vol. 345, pp. 919-921.
643	"Glacier mass loss: Past the point of no return," Science News, March 19, 2018, https://www.sciencedaily.com/releases/2018/03/180319124258.htm
644	Ben Marzeion, et al., "Attribution of global glacier mass loss to anthropogenic and natural causes," Science, 2014, vol. 345, pp. 919-921.
645	"Melting glaciers raise sea level," Science Daily, November 14, 2012, https://www.sciencedaily.com/releases/2012/11/121114083819.htm
646	Matthias Huss and Daniel Farinotti, "Distributed ice thickness and volume of all glaciers around
the globe," Journal of Geophysical Research, October 11, 2012, vol. 117, doi:10.1029/2012JF002523, 2012
647	Becky Oskin, "World's Glaciers Have New Size Estimate," Live Science, October 22, 2012, https://www.livescience.com/24168-glacier-volume-sea-level-rise.html
648	J. Carl Ganter, "Tibetan Plateau Water Reserves at Risk," Huffington Post, December 6, 2017, https://www.huffingtonpost.com/j-carl-ganter/tibetan-plateau-water-res_b_100818.html
649	Xiaolin Wang, Limin Wang, Yan Wang, The Quality of Growth and Poverty Reduction in China, 2014, The World Bank, Springer, p. 89.
650	Daniel C. Nepstad, "Interactions among Amazon land use, forests and climate: prospects for a near-term forest tipping point," Philosophical Transactions of The Royal Society B, February 2008, pp. 1737–1746, doi:10.1098/rstb.2007.0036
651	Alexis Lassman, "New Theory on How the Amazon Controls the Earth's Climate," Uplift, March 17, 2017, https://upliftconnect.com/amazon-controls-earths-climate
652	Daniel Glick, "Can the Amazon Save the Planet? Scientists climb to perilous heights to gauge how much carbon dioxide the rainforest is absorbing," Scientific American, April 3, 2017, https://www.scientificamerican.com/article/can-the-amazon-save-the-planet
653	Amanda Paulson, "Camp Amazon: Inside the 'lungs of the Earth'," Christian Science Monitor, September 24, 2018, https://www.csmonitor.com/Environment/2018/0924/Camp-Amazon-Inside-the-lungs-of-the-Earth
654	Daniel Glick, "Can the Amazon Save the Planet? Scientists climb to perilous heights to gauge how much carbon dioxide the rainforest is absorbing," Scientific American, April 3, 2017, https://www.scientificamerican.com/article/can-the-amazon-save-the-planet
655	Rodrigo Hierro, et al., "The Amazon basin as a moisture source for an Atlantic Walker-type Circulation," Atmospheric Research, October 2018, DOI: 10.1016/j.atmosres.2018.10.009
656	"Inside the Amazon," WWF - World Wide Fund for Nature, http://wwf.panda.org/knowledge_hub/where_we_work/amazon/about_the_amazon
657	Stephen Leahy, "The REAL Amazon-gate: On the Brink of Collapse Reveals Million $ Study," February 104 2010, https://stephenleahy.net/2010/02/15/the-real-amazon-gate-on-the-brink-of-collapse-million-study

658	Daniel Glick, "Can the Amazon Save the Planet? Scientists climb to perilous heights to gauge how much carbon dioxide the rainforest is absorbing," Scientific American, April 3, 2017, https://www.scientificamerican.com/article/can-the-amazon-save-the-planet
659	Amanda Paulson, "Camp Amazon: Inside the 'lungs of the Earth'," Christian Science Monitor, September 24, 2018, https://www.csmonitor.com/Environment/2018/0924/Camp-Amazon-Inside-the-lungs-of-the-Earth
660	Doug Bennett, "What Are the Reactants & Products in the Equation for Photosynthesis?" Sciencing, April 30, 2018, https://sciencing.com/reactants-products-equation-photosynthesis-8460990.html
661	"Amazon rainforest is home to 16,000 tree species, estimate suggest," The Guardian, October 18, 2013, https://www.theguardian.com/environment/2013/oct/18/amazon-rainforest-tree-species-estimate
662	David Wallace-Wells, "Could One Man Single-Handedly Ruin the Planet?" Intelligencer, October 31, 2018, http://nymag.com/intelligencer/2018/10/bolsanaros-amazon-deforestation-accelerates-climate-change.html
663	Oliver L. Phillips, and Roel J. W. Brienen, "Carbon uptake by mature Amazon forests has mitigated Amazon nations' carbon emissions" Carbon Balance Manage, February 15, 20178, DOI 10.1186/s13021-016-0069-2
664	"Climate Change and Tropical Forests," Global Forest Atlas, https://globalforestatlas.yale.edu/climate-change/climate-change-and-tropical-forests
665	Trends in Global CO2 Emissions 2016 Report, PBL Netherlands Environmental Assessment Agency, The Hague, 2016, p. 43.
666	Rhett Butler, "Rainforest Ecology," Mongabay, January 26, 2017, https://rainforests.mongabay.com/amazon/rainforest_ecology.html
667	Alexis Lassman, "New Theory on How the Amazon Controls the Earth's Climate," Uplift, March 17, 2017, https://upliftconnect.com/amazon-controls-earths-climate
668	Airbus A380 Specs, http://www.modernairliners.com/airbus-a380/airbus-a380-specs
669	"Trees in the Amazon Generate Their Own Clouds and Rain, Study Finds," Yale Environment 360, August 7, 2017, https://e360.yale.edu/digest/trees-in-the-amazon-generate-their-own-clouds-and-rain-study-finds
670	Niklas Boers, et al., "A deforestation-induced tipping point for the South American monsoon system," Scientific Reports, January 25, 2017, DOI: 10.1038/srep41489
671	Alexis Lassman, "New Theory on How the Amazon Controls the Earth's Climate," Uplift, March 17, 2017, https://upliftconnect.com/amazon-controls-earths-climate
672	Peter Bunyard, "Without its rainforest, the Amazon will turn to desert," Ecologist, March 2, 2015, https://theecologist.org/2015/mar/02/without-its-rainforest-amazon-will-turn-desert
673	Fred Pearce, "Rainforests may pump winds worldwide," New Scientist, April 1, 2009
674	Fred Pearce, "Rivers in the Sky: How Deforestation Is Affecting Global Water Cycles," Yale Environment 360, July 24, 2018, https://e360.yale.edu/features/how-deforestation-affecting-global-water-cycles-climate-change
675	Grennan Milliken, "Over Half Of All Amazonian Tree Species Are In Danger," Popular Science, November 20, 2015, https://www.popsci.com/over-half-all-amazonian-tree-species-are-globally-threatened

676   "Rain Forests," National Geographic, https://www.nationalgeographic.com/environment/habitats/rain-forests
677   Aline C. Soterroni, Fernando M. Ramos and Michael Obersteiner, "Fate of the Amazon is on the ballot in Brazil's presidential election (commentary)," Mongabay, October 17, 2018, https://news.mongabay.com/2018/10/fate-of-the-amazon-is-on-the-ballot-in-brazils-presidential-election-commentary
678   Camila Domonoske, "Deforestation Of The Amazon Up 29 Percent From Last Year, Study Finds," NPR, November 30, 2016, https://www.npr.org/sections/thetwo-way/2016/11/30/503867628/deforestation-of-the-amazon-up-29-percent-from-last-year-study-finds
679   "For cattle farmers in the Brazilian Amazon, money can't buy happiness," The Conversation, October 24, 2017, https://theconversation.com/for-cattle-farmers-in-the-brazilian-amazon-money-cant-buy-happiness-85349
680   Nathalie Walker, Barbara Bramble and Sabrina Patel, "From Major Driver of Deforestation and Greenhouse Gas Emissions to Forest Guardians? New Developments in Brazil's Amazon Cattle Industry," National Wildlife Federation, December 2010
681   Julie Kerr Casper, PhD, Forests - More than Trees, 2007, Chelsea House, p. 74.
682   "Amazonian challenges: Cattle ranching and agriculture," Open Learn, May 27, 2014, https://www.open.edu/openlearn/nature-environment/amazonian-challenges-cattle-ranching-and-agriculture
683   "Soybeans," Union of Concerned Scientists, https://www.ucsusa.org/global-warming/stop-deforestation/drivers-of-deforestation-2016-soybeans
684   Rebecca Simmon, "Tropical Deforestation," NASA Earth Observatory, March 30, 2007, https://earthobservatory.nasa.gov/features/Deforestation/deforestation_update.php
685   S. K. Chakravarty, et al. Deforestation: Causes, Effects and Control Strategies, DOI: 10.5772/33342
686   "Amazonian challenges: Cattle ranching and agriculture," Open Learn, May 27, 2014, https://www.open.edu/openlearn/nature-environment/amazonian-challenges-cattle-ranching-and-agriculture
687   "Amazonian challenges: Cattle ranching and agriculture," Open Learn, May 27, 2014, https://www.open.edu/openlearn/nature-environment/amazonian-challenges-cattle-ranching-and-agriculture
688   Elizabeth Palermo, "More Than 30,000 Miles of Roads Built in Amazon in 3 Years," Live Science, November 4, 2013, https://www.livescience.com/40914-amazon-road-building.html
689   Ian Johnston, "Amazon jungle faces death spiral of drought and deforestation, warn scientists," Independent, March 13, 2017, https://www.independent.co.uk/environment/amazon-rainforest-drought-deforestation-jungle-death-spiral-potsdam-institute-a7627931.html
690   David Adam, "Amazon rainforests pay the price as demand for beef soars," The Guardian, May 31, 2009, https://www.theguardian.com/environment/2009/may/31/cattle-trade-brazil-greenpeace-amazon-deforestation
691   Ana Mano, "China-driven Brazil beef bonanza seen lasting: Abiec," Reuters, December 11, 2018, https://www.reuters.com/article/us-brazil-beef-abiec/china-driven-brazil-beef-bonanza-seen-lasting-abiec-idUSKBN1OA1Z8
692   David Adam, "Amazon rainforests pay the price as demand for beef soars," The Guardian, May 31, 2009, https://www.theguardian.com/environment/2009/may/31/cattle-trade-brazil-greenpeace-amazon-deforestation
693   "Amazonian challenges: Cattle ranching and agriculture," Open Learn, May 27,

2014, https://www.open.edu/openlearn/nature-environment/amazonian-challenges-cattle-ranching-and-agriculture

694   David Adam, "Amazon rainforests pay the price as demand for beef soars," The Guardian, May 31, 2009, https://www.theguardian.com/environment/2009/may/31/cattle-trade-brazil-greenpeace-amazon-deforestation

695   Rhett Butler, "Palm oil company destroys 7,000 ha of Amazon rainforest in Peru," Mongabay, March 4, 2013, https://news.mongabay.com/2013/03/palm-oil-company-destroys-7000-ha-of-amazon-rainforest-in-peru

696   James Bargent, "Satellite Images Show Threat of Criminal Activities in Peru's Amazon," InSight Crime, March 20, 2019, https://www.insightcrime.org/news/brief/satellite-images-highlight-threat-to-perus-amazon-forest

697   Rhett Butler, "Population, Poverty, and Deforestation," Mongabay, July 11, 2012, https://rainforests.mongabay.com/0816.htm

698   S. K. Chakravarty, et al. Deforestation: Causes, Effects and Control Strategies, DOI: 10.5772/33342

699   Gregory P. Asnera, et al., "Elevated rates of gold mining in the Amazon revealed through high-resolution monitoring," PNAS, November 12, 2013, vol. 110, no. 46, pp. 18454–18459, https://doi.org/10.1073/pnas.1318271110

700   Camilo Carranza, "Peru Running Out of Ideas to Stop Illegal Mining in Madre de Dios," InSight Crime, March 11, 2019, https://www.insightcrime.org/news/brief/peru-illegal-gold-mining

701   Miriam Wells, "Peru's 'Mining Mafia' Seek To Legalize Their Operations," InSight Crime, August 28, 2013, https://www.insightcrime.org/news/brief/perus-mining-mafias-seek-to-legalize-their-operations

702   Katy Ashe, "Elevated Mercury Concentrations in Humans of Madre de Dios, Peru," PLOS One, March 2012, vol. 7, issue 3, https://doi.org/10.1371/journal.pone.0033305

703   Mac Margolis, "Gold Prices Cause Mining Boom That Threatens Amazon Rainforest," The Daily Beast, August 19, 2011, https://www.thedailybeast.com/gold-prices-cause-mining-boom-that-threatens-amazon-rainforest

704   Rhett Butler, "Environmental impact of mining in the rainforest," Mongabay, July 27, 2012, https://rainforests.mongabay.com/0808.htm

705   Katy Ashe, "Elevated Mercury Concentrations in Humans of Madre de Dios, Peru," PLOS One, March 2012, vol. 7, issue 3, https://doi.org/10.1371/journal.pone.0033305

706   "Mercury in gold mining poses toxic threat," NBC News, January 10, 2009, http://www.nbcnews.com/id/28596948/ns/world_news-world_environment/t/mercury-gold-mining-poses-toxic-threat

707   "Mercury in gold mining poses toxic threat," NBC News, January 10, 2009, http://www.nbcnews.com/id/28596948/ns/world_news-world_environment/t/mercury-gold-mining-poses-toxic-threat

708   Amanda Paulson, "Camp Amazon: Inside the 'lungs of the Earth'," Christian Science Monitor, September 24, 2018, https://www.csmonitor.com/Environment/2018/0924/Camp-Amazon-Inside-the-lungs-of-the-Earth

709   "Amazon has nearly 100,000 km of roads," Mongabay, December 8, 2012, https://news.mongabay.com/2012/12/amazon-has-nearly-100000-km-of-roads

710   Sadia Ahmed, et al., "Temporal patterns of road network development in the Brazilian Amazon," Regional Environmental Change, October 2013, vol. 13, no. 5, pp. 927-937, DOI 10.1007/s10113-012-0397-z

711   Dan Collyns, "Roads are encroaching deeper into the Amazon rainforest,

study says," The Guardian, January 28, 2015, https://www.theguardian.com/environment/2015/jan/28/roads-are-encroaching-deeper-into-the-amazon-rainforest-study-says

712 Liz Kimbroug , "Roads through the rainforest: an overview of South America's 'arc of deforestation'," Mongabay, July 21, 2014, https://news.mongabay.com/2014/07/roads-through-the-rainforest-an-overview-of-south-americas-arc-of-deforestation

713 Liz Kimbroug , "Roads through the rainforest: an overview of South America's 'arc of deforestation'," Mongabay, July 21, 2014, https://news.mongabay.com/2014/07/roads-through-the-rainforest-an-overview-of-south-americas-arc-of-deforestation

714 Philip Fearnside, "Business as Usual: A Resurgence of Deforestation in the Brazilian Amazon," Yale Environment 360, April 18 2017, https://e360.yale.edu/features/business-as-usual-a-resurgence-of-deforestation-in-the-brazilian-amazon

715 Rhett Butler, "Calculating Deforestation Figures for the Amazon," Mongabay, January 26, 2017, https://rainforests.mongabay.com/amazon/deforestation_calculations.html

716 Phillip Fearnside, "Deforestation of the Brazilian Amazon," Oxford Research Encyclopedia of Environmental Science, September 2017, DOI: 10.1093/acrefore/9780199389414.013.102

717 "Amazon deforestation is close to tipping point," Phys.org, March 20, 2018, https://phys.org/news/2018-03-amazon-deforestation.html

718 Chelsea Gohd, "New Study Shows Just How Close The Amazon Rainforest Is to The Brink of Collapse," Science Alert, February 24 2018, https://www.sciencealert.com/deforestation-amazon-collapse-savannah-imminent

719 Lovejoy and Nobre, "Amazon Tipping Point," Science Advances, February 2018, DOI: 10.1126/sciadv.aat2340

720 Stephen Leahy, "The REAL Amazon-gate: On the Brink of Collapse Reveals Million $ Study," February 104 2010, https://stephenleahy.net/2010/02/15/the-real-amazon-gate-on-the-brink-of-collapse-million-study

721 Delphine Zemp, et al., "Self-amplified Amazon forest loss due to vegetation-atmosphere feedbacks," Nature Communications, March 13, 2017, DOI: 10.1038/ncomms14681

722 Delphine Zemp, et al., "Self-amplified Amazon forest loss due to vegetation-atmosphere feedbacks," Nature Communications, March 13, 2017, DOI: 10.1038/ncomms14681

723 Lovejoy and Nobre, "Amazon Tipping Point," Science Advances, February 2018, DOI: 10.1126/sciadv.aat2340

724 "Amazon deforestation is close to tipping point," Phys.org, March 20, 2018, https://phys.org/news/2018-03-amazon-deforestation.html

725 John Hollis, "Mason's Thomas Lovejoy says the time to act for the Amazon is now," George Mason University News, December 10, 2019, https://science.gmu.edu/news/masons-thomas-lovejoy-says-time-act-amazon-now

726 Catherine Brahic, "Parts of Amazon close to tipping point," New Scientist, March 5, 2009, https://www.newscientist.com/article/dn16708-parts-of-amazon-close-to-tipping-point

727 Fred Pearce, "Rivers in the Sky: How Deforestation Is Affecting Global Water Cycles," Yale Environment 360, July 24, 2018, https://e360.yale.edu/features/how-deforestation-affecting-global-water-cycles-climate-change

728 Peter Bunyard, "Without its rainforest, the Amazon will turn to desert," Ecologist, March 2, 2015, https://theecologist.org/2015/mar/02/without-its-rainforest-amazon-will-turn-desert

729   "Stark Beauty: Images of Israel's Negev Desert," Live Science, https://www.livescience.com/31372-israel-negev-desert-photos.html
730   "Forests Precede Us, Deserts Follow," Uncommon Thought, 2015, https://www.uncommonthought.com/mtblog/archives/2015/02/04/forests-precede.php
731   Jim Robbins, "Deforestation and Drought," New York Times, October 9, 2015, https://www.nytimes.com/2015/10/11/opinion/sunday/deforestation-and-drought.html
732   Jonathan Watts, "Amazon rainforest losing ability to regulate climate, scientist warns," The Guardian, October 31, 2014, https://www.theguardian.com/environment/2014/oct/31/amazon-rainforest-deforestation-weather-droughts-report
733   Morgan Kelly, "If a tree falls in Brazil…? Amazon deforestation could mean droughts for western U.S.," Princeton News, November 7, 2013, https://www.princeton.edu/news/2013/11/07/if-tree-falls-brazil-amazon-deforestation-could-mean-droughts-western-us
734   "Tropical Deforestation Affects Rainfall in the U.S. and Around the Globe," NASA, September 13, 2005, https://www.nasa.gov/centers/goddard/news/topstory/2005/deforest_rainfall.html
735   Daniel C. Nepstad, "Interactions among Amazon land use, forests and climate: prospects for a near-term forest tipping point," Philosophical Transactions of The Royal Society B, February 2008, pp. 1737–1746, doi:10.1098/rstb.2007.0036
736   Amanda Paulson, "Camp Amazon: Inside the 'lungs of the Earth'," Christian Science Monitor, September 24, 2018, https://www.csmonitor.com/Environment/2018/0924/Camp-Amazon-Inside-the-lungs-of-the-Earth
737   Jake Spring, "Soy boom devours Brazil's tropical savanna," Reuters, August 28, 2018, https://www.reuters.com/investigates/special-report/brazil-deforestation
738   Philip Fearnside, "Business as Usual: A Resurgence of Deforestation in the Brazilian Amazon," Yale Environment 360, April 18 2017, https://e360.yale.edu/features/business-as-usual-a-resurgence-of-deforestation-in-the-brazilian-amazon
739   Jake Spring, "Soy boom devours Brazil's tropical savanna," Reuters, August 28, 2018, https://www.reuters.com/investigates/special-report/brazil-deforestation
740   Michael Hopkin, "Brazilian savannah 'will disappear by 2030'," Nature, July 20, 2004, doi:10.1038/news040719-6
741   Jonathan Mingle, "The Slow Death of Ecology's Birthplace," Undark.org, December 16, 2016, https://undark.org/article/slow-death-brazil-cerrado-ecology
742   Aline C. Soterroni, Fernando M. Ramos and Michael Obersteiner, "Fate of the Amazon is on the ballot in Brazil's presidential election (commentary)," Mongabay, October 17, 2018, https://news.mongabay.com/2018/10/fate-of-the-amazon-is-on-the-ballot-in-brazils-presidential-election-commentary
743   Chris Fitch, "Deforestation causing São Paulo drought," Geographical, February 5, 2015, http://geographical.co.uk/places/cities/item/761-deforestation-behind-sao-paulo-drought
744   Jim Robbins, "Deforestation and Drought," New York Times, October 9, 2015, https://www.nytimes.com/2015/10/11/opinion/sunday/deforestation-and-drought.html
745   "Brazil reveals plans to privatize key stretches of Amazon highways," The Guardian, January 23, 2019, https://www.theguardian.com/world/2019/jan/22/brazils-government-reveals-plans-to-privatize-key-shipping-routes
746   Aline C. Soterroni, Fernando M. Ramos and Michael Obersteiner, "Fate of the Amazon is on the ballot in Brazil's presidential election (commentary)," Mongabay, October 17, 2018, https://news.mongabay.com/2018/10/fate-of-the-amazon-is-on-

the-ballot-in-brazils-presidential-election-commentary
747 Daniel C. Nepstad, et al., "Interactions among Amazon land use, forests and climate: prospects for a near-term forest tipping point," Philosophical Transactions of the Royal Society B, February 11, 2008, pp. 1737-1746, doi: 10.1098/rstb.2007.0036
748 "Amazon Deforestation and Climate Change," National Geographic, https://www.nationalgeographic.org/media/amazon-deforestation-and-climate-change
749 Amanda Paulson, "Camp Amazon: Inside the 'lungs of the Earth'," Christian Science Monitor, September 24, 2018, https://www.csmonitor.com/Environment/2018/0924/Camp-Amazon-Inside-the-lungs-of-the-Earth
750 Rebecca Simmon, "Tropical Deforestation," NASA Earth Observatory, March 30, 2007, https://earthobservatory.nasa.gov/features/Deforestation/deforestation_update.php
751 David Ellison, et al., "Trees, forests and water: Cool insights for a hot world," Global Environmental Change, March 2017, vol. 43, pp. 51-61, https://doi.org/10.1016/j.gloenvcha.2017.01.002
752 Grennan Milliken, "Over Half Of All Amazonian Tree Species Are In Danger," Popular Science, November 20, 2015, https://www.popsci.com/over-half-all-amazonian-tree-species-are-globally-threatened
753 Michael J. Novacek and Elsa E. Cleland, "The current biodiversity extinction event: Scenarios for mitigation and recovery," PNAS, May 8, 2001, vol. 98, no. 10
754 S. K. Chakravarty, et al. Deforestation: Causes, Effects and Control Strategies, DOI: 10.5772/33342
755 Judith D. Schwartz, "Clearing Forests May Transform Local—and Global—Climate," Scientific American, March 4, 2003, https://www.scientificamerican.com/article/clearing-forests-may-transform-local-and-global-climate
756 Harry CockBurn, "Forest area the size of Italy destroyed last year as trees burned to make way for farms - 'We are trying to put out a house fire with a teaspoon'," Independent, June 27, 2018, https://www.independent.co.uk/environment/deforestation-global-rainforest-trees-forest-amazon-congo-basin-indonesia-a8419621.html
757 Michelle Soto, "The planet loses 40 soccer fields worth of forests every minute," The Tico Times, March 24, 2019, http://www.ticotimes.net/2019/03/24/the-planet-loses-40-soccer-fields-worth-of-forests-every-minute-2
758 Harry CockBurn, "Forest area the size of Italy destroyed last year as trees burned to make way for farms - 'We are trying to put out a house fire with a teaspoon'," Independent, June 27, 2018, https://www.independent.co.uk/environment/deforestation-global-rainforest-trees-forest-amazon-congo-basin-indonesia-a8419621.html
759 Fred Pearce, "Rivers in the Sky: How Deforestation Is Affecting Global Water Cycles," Yale Environment 360, July 24, 2018, https://e360.yale.edu/features/how-deforestation-affecting-global-water-cycles-climate-change
760 "Deforestation in Madagascar," NASA Land-Cover/Land-Use Change Program, https://lcluc.umd.edu/hotspot/deforestation-madagascar
761 Rhett Butler, "Madagascar's Political Chaos Threatens Conservation Gains," Yale Environment 360, January 4, 2010, https://e360.yale.edu/features/madagascars_political_chaos_threatens_conservation_gains
762 Paul Tullis, "How the sapphire trade is driving lemurs toward extinction," National Geographic, March 6, 2019, https://www.nationalgeographic.com/animals/2019/03/sapphire-mining-fuels-lemur-deaths-in-madagascar
763 Judith D. Schwartz, "Clearing Forests May Transform Local—and Global—Climate," Scientific American, March 4, 2003, https://www.scientificamerican.com/article/clearing-forests-may-transform-local-and-global-climate

764 Rita Damary, "Illegal logging, charcoal trade, farming threaten Mau Forest," The Star, Kenya, October 3, 2017, https://www.the-star.co.ke/news/2017/10/03/illegal-logging-charcoal-trade-farming-threaten-mau-forest_c1645309

765 Victor Kiprop, "What next after East Arica's forests vanish?" The East African, March 21, 2018, https://www.theeastafrican.co.ke/scienceandhealth/What-next-after-East-Africa-forests-vanish/3073694-4350820-13y6m5x/index.html

766 Victor Kiprop, "What next after East Arica's forests vanish?" The East African, March 21, 2018, https://www.theeastafrican.co.ke/scienceandhealth/What-next-after-East-Africa-forests-vanish/3073694-4350820-13y6m5x/index.html

767 Justin Murimi, "Mau Forest depletion at the root of worsening droughts and conflicts," The Star, Kenya, March 20, 2017, https://www.the-star.co.ke/news/2017/03/20/mau-forest-depletion-at-the-root-of-worsening-droughts-and-conflicts_c1526197

768 James Morgan, "Kenya's heart stops pumping," BBC News, September 29, 2009, http://news.bbc.co.uk/2/hi/africa/8057316.stm

769 Judith D. Schwartz, "Clearing Forests May Transform Local—and Global—Climate," Scientific American, March 4, 2003, https://www.scientificamerican.com/article/clearing-forests-may-transform-local-and-global-climate

770 "Kenya: Drought - 2014-2019," ReliefWeb, 2017, https://reliefweb.int/disaster/dr-2014-000131-ken

771 Stephen Rutto, "Drought to flood and back again: Cycle that haunts Kenya yearly," The Star, Kenya, April 20, 2018, https://www.the-star.co.ke/news/2018/04/20/drought-to-flood-and-back-again-cycle-that-haunts-kenya-yearly_c1745372

772 Murithi Mutiga, "As drought sweeps Kenya, herders invade farms and old wounds are reopened," The Guardian, March 18, 2017, https://www.theguardian.com/world/2017/mar/19/kenya-range-war-reopens-colonial-wounds

773 Katy Migiro, "Politics of Death: Colonial scars and drought feed Kenya land wars," Reuters, June 22, 2017, https://www.reuters.com/article/us-kenya-landrights-farms/politics-of-death-colonial-scars-and-drought-feed-kenya-land-wars-idUSKBN19E04R

774 Michael Wolosin, and Nancy Harris, "Tropical Forests and Climate Change: The Latest Science," Working Paper. Washington, DC: World Resources Institute. Available online at wri.org/ending-tropical-deforestation.

775 N.M. Mahowald, D.S. Ward, S.C. Doney, P.G. Hess, and J.T. Randerson, "Are the Impacts of Land Use on Warming Underestimated in Climate Policy?" Environmental Research Letters, August 2, 2017, https://doi.org/10.1088/1748-9326/aa836d

776 S. K. Chakravarty, et al. Deforestation: Causes, Effects and Control Strategies, DOI: 10.5772/33342

777 Living Planet Report - 2018: Aiming Higher, World Wildlife Fund, 2018, p. 51.

778 Michelle Soto, "The planet loses 40 soccer fields worth of forests every minute," The Tico Times, March 24, 2019, http://www.ticotimes.net/2019/03/24/the-planet-loses-40-soccer-fields-worth-of-forests-every-minute-2

779 Richard Lydekker, The Royal Natural History: Mammals, 1894, London, p. 192.

780 William Hornaday, Extermination of the American Bison, 1889.

781 Joseph Leidy, MD, "On the extinct species of American ox," 1952, Philadelphia, p. 4.

782 Richard Irving Dodge, The Plains of the Great West and Their Inhabitants, 1877, G. P. Putnam's Sons, New York

783 Weston Phippen, "Kill Every Buffalo You Can! Every Buffalo Dead Is an

Indian Gone," The Atlantic, May 13, 2016, https://www.theatlantic.com/national/archive/2016/05/the-buffalo-killers/482349

784   William Hornaday, Extermination of the American Bison, 1889.

785   Joel Asaph Allen, History of the American Bison, June 1877, pp. 535-536.

786   Joel Asaph Allen, History of the American Bison, June 1877, p. 559.

787   Richard Irving Dodge, The Plains of the Great West and Their Inhabitants, 1877, G. P. Putnam's Sons, New York, pp. xiv, xv.

788   David D. Smits, "The Frontier Army and the Destruction of the Buffalo: 1865-1883," Autumn, 1994, The Western Historical Quarterly, vol. 25, no. 3, p. 327

789   William Hornaday, Extermination of the American Bison, 1889

790   William Denevan, "The Pristine Myth: The Landscape of the Americas in 1492," Annals of the Association of American Geographers, 1992, vol. 82, issue 3.

791   Weston Phippen, "Kill Every Buffalo You Can! Every Buffalo Dead Is an Indian Gone," The Atlantic, May 13, 2016, https://www.theatlantic.com/national/archive/2016/05/the-buffalo-killers/482349

792   Carolyn Merchant, American Environmental History: An Introduction, 2007, Columbia University Press, New York, p. 20.

793   Richard Irving Dodge, The Plains of the Great West and Their Inhabitants, 1877, G. P. Putnam's Sons, New York, pp. xvii-xviii.

794   Carolyn Merchant, American Environmental History: An Introduction, 2007, Columbia University Press, New York, p. 20.

795   The Canadian Naturalist and Quarterly Journal of Science, 1875, vol. 7, p. 199.

796   Joel Asaph Allen, History of the American Bison, June 1877, p. 566.

797   Richard Lydekker, The Royal Natural History: Mammals, 1894, London, p. 191.

798   Hartley Jackson, "Conserving Endangered Wildlife Species," Transactions of the Wisconsin Academy of Science Arts and Letters, 1943, vol. XXXV, p. 80.

799   "5 largest public bison herds in the U.S," Rapid City Journal, November 12, 2014

800   Jed Portman, "5 things you need to know about... The great American Bison," PBS, May 3 ,2011, http://www.pbs.org/wnet/need-to-know/five-things/the-great-american-bison/8950/

801   Kirk Johnson, "Plains Giants Have Foothold on Tables," January 22, 2011, New York Times, http://www.nytimes.com/2011/01/23/us/23buffalo.html

802   John James Audubon, Ornithological biography or an account of the habits of the birds of the United States of America, 1832, Philadelphia. pp. 319-327.

803   Bénédict Henry Révoil, Shooting and Fishing in the Rivers, Prairies, and Backwoods of North America, 1865, London, pp. 93-102.

804   The Passenger Pigeon, Smithsonian, https://www.si.edu/spotlight/passenger-pigeon

805   Barry Yeoman, "Why the Passenger Pigeon Went Extinct," Audubon Magazine, May-June 2014.

806   Barbara Allen, Pigeon, 2009, p. 174.

807   "Flight to Extinction - The Story of the Passenger Pigeon," Maine State Museum, http://www.mainestatemuseum.org/exhibits/flight_to_extinction_-_the_story_of_the_passenger_pigeon

808   Bill Loomis, "Detroit's Delectable Past: Two Centuries of Frog Legs, Pigeon Pie & Drugstore Whiskey," 2012, History Press, Charleston SC.

809   David Biello, "3 Billion to Zero: What Happened to the Passenger Pigeon?" Scientific American, June 27, 2014, https://www.scientificamerican.com/article/3-bil-

lion-to-zero-what-happened-to-the-passenger-pigeon
810   Stephen Taylor, "The 'Bird Bills': A tale of the plume boom," Hoosier State Chronicles, http://blog.newspapers.library.in.gov/the-bird-bills-a-tale-of-the-plume-boom/
811   Sharon Guynup, State of the Wild 2006: A Global Portrait of Wildlife, Wildlands, and Oceans, 2005, Island Press, p. 102.
812   Michael Grunwald, The Swamp: The Everglades, Florida, and the Politics of Paradise, p. 120.
813   B.S. Bowdish, The White Badge of Cruelty, Suburban Life, July 1909, p. 74
814   Robin Doughty, Feather Fashions and Bird Preservation - A Study in Nature Protection, 1975, University of California Press, p. 80.
815   Historical Gold Prices, http://onlygold.com/Info/Historical-Gold-Prices.asp
816   Union Scale of Wages and Hours of Labor, 1907 to 1912: Bulletin of the United States Bureau of Labor Statistics, No. 131, p. 50.
817   Marjory Stoneman Douglas, "Wings," The Saturday Evening Post, March 14, 1931.
818   William Souder, "How Two Women Ended the Deadly Feather Trade," Smithsonian Magazine, March 2013.
819   Sharon Guynup, State of the Wild 2006: A Global Portrait of Wildlife, Wildlands, and Oceans, 2005, Island Press, p. 102.
820   The Audubon Magazine, February 1887 to January 1888
821   821   Robin Doughty, Feather Fashions and Bird Preservation - A Study in Nature Protection, 1975, University of California Press, p. 82.
822   Michael Grunwald, The Swamp: The Everglades, Florida, and the Politics of Paradise, p. 122.
823   Florida Assessment of Coastal Trends, June 1997, D-32,33.
824   Duncan Dayton, The National Parks: America's Best Idea: an Illustrated History, p. 88.
825   Stephen Taylor, "The 'Bird Bills': A tale of the plume boom," Hoosier State Chronicles, http://blog.newspapers.library.in.gov/the-bird-bills-a-tale-of-the-plume-boom/
826   Marjory Stoneman Douglas, "Wings," The Saturday Evening Post, March 14, 1931.
827   William Souder, "How Two Women Ended the Deadly Feather Trade," Smithsonian Magazine, March 2013.
828   "Egret Invasion - A bird which had almost vanished comes north again in big flocks," LIFE, September 27, 1948, pp. 69-75.
829   Ellie Zolfagharifard, "Mankind slaughtered THREE MILLION whales in the 20th Century: Scientists reveal extent of the 'largest hunt in human history'," Daily Mail, March 16, 2015, http://www.dailymail.co.uk/sciencetech/article-2997584/Mankind-slaughtered-THREE-MILLION-whales-20th-Century-Scientists-reveal-extent-largest-hunt-human-history.html
830   R. C. Rocha, P. J. Clapham, & Y. V. Ivashchenko, "Emptying the oceans: A summary of industrial whaling catches in the 20th century," Marine Fisheries Review, 2014, pp. 37–48., doi: dx.doi.org/10.7755/MFR.76.4.3
831   Robert McNamara, "A Brief History of Whaling: The 19th Century Whaling Industry Thrived for Decades," ThoughtCo., March 9, 2017, https://www.thoughtco.com/a-brief-history-of-whaling-1774068
832   Andrew Nikiforuk, "Thar She Blows: Whale Age Teaches Us about Oil: A boom, a bonanza, a decimation, a collapse. Learn from Ahab," May 12, 2011, TheTyee.ca

833   Sir Gerald Elliot, Whaling 1937-1967: The International Control of Whale Stocks, 1997, p. 4.
834   R. C. Rocha, P. J. Clapham, & Y. V. Ivashchenko, "Emptying the oceans: A summary of industrial whaling catches in the 20th century," Marine Fisheries Review, 2014, pp. 37–48., doi: dx.doi.org/10.7755/MFR.76.4.3
835   Johan Nicolay Tønnessen and Arne Odd Johnsen, The History of Modern Whaling, 1982, University of California Press, p. 270.
836   Daniel Cressey, "World's whaling slaughter tallied : Commercial hunting wiped out almost three million animals last century," Nature, March 12 2015, vol. 519, pp. 140-141.
837   R. C Rocha, P. J Clapham, & Y. V, Ivashchenko, "Emptying the oceans: A summary of industrial whaling catches in the 20th century," Marine Fisheries Review, 2014, pp. 37–48., doi: dx.doi.org/10.7755/MFR.76.4.3
838   Ed Yong, "American Whalers Killed Way More Than Just Whales," The Atlantic, September 19, 2016, https://www.theatlantic.com/science/archive/2016/09/the-collateral-damage-of-americas-whaling-fleets/500492
839   R. C Rocha, P. J Clapham, & Y. V, Ivashchenko "Emptying the oceans: A summary of industrial whaling catches in the 20th century," Marine Fisheries Review, 2014, pp. 37–48., doi: dx.doi.org/10.7755/MFR.76.4.3
840   P J. Clapham, Alex Aguilar, Leila Hatch, "Determining spatial and temporal scales for management: lessons from whaling," Marine Mammal Science, January 2008, p. 194, DOI: 10.1111/j.1748-7692.2007.00175.x
841   "Why 12 right whales died in Canadian waters — and why more will if nothing is done," Metro News Canada, December 27, 2017, http://www.metronews.ca/news/canada/2017/12/27/why-12-right-whales-died-in-canadian-waters-and-why-more-will-if.html
842   Robin McKie, "The oceans' last chance: 'It has taken years of negotiations to set this up,'" The Guardian, August 5, 2018, https://www.theguardian.com/environment/2018/aug/05/last-chance-save-oceans-fishing-un-biodiversity-conference
843   Sharon Guynup, State of the Wild 2006: A Global Portrait of Wildlife, Wildlands, and Oceans, 2005, Island Press, p. 101.
844   Ted B. Lyon and Will N. Graves, The Real Wolf: The Science, Politics, and Economics of Co-Existing with Wolves in Modern Times, 2014, p. 278.
845   International Wolf Center, https://wolf.org/wow/united-states
846   Alastair Bland, "Should Trophy Hunting of Lions Be Banned?" Smithsonian, December 7, 2012, hhttps://www.smithsonianmag.com/travel/should-trophy-hunting-of-lions-be-banned-155657735
847   Gerardo Ceballos, Paul R. Ehrlich and Rodolfo Dirzo, "Biological annihilation via the ongoing sixth mass
extinction signaled by vertebrate population losses and declines," Proceedings of the National Academy of Sciences (PNAS), July 10, 2017, www.pnas.org/cgi/doi/10.1073/pnas.1704949114
848   Panthera leo Lion, International Union for Conservation of Nature (IUCN) Red List of Threatened Species, http://www.iucnredlist.org/details/15951/0
849   Panthera leo Lion, International Union for Conservation of Nature (IUCN) Red List of Threatened Species, http://www.iucnredlist.org/details/15951/0
850   Erica Goode, "Lion Population in Africa Likely to Fall by Half, Study Finds," New York Times, October 25, 2015
851   John Platt, "African Lions Face Extinction by 2050, Could Gain Endangered Species Act Protection," Scientific American, October 27, 2014, https://blogs.scientifi-

camerican.com/extinction-countdown/african-lions-face-extinction-by-2050-could-gain-endangered-species-act-protection

852   Giraffe, Giraffa camelopardalis, International Union for Conservation of Nature (IUCN) Red List of Threatened Species, http://www.iucnredlist.org/details/9194/0

853   Giraffe, Giraffa camelopardalis, International Union for Conservation of Nature (IUCN) Red List of Threatened Species, http://www.iucnredlist.org/details/9194/0

854   Damian Carrington, "Giraffes facing extinction after devastating decline, experts warn," The Guardian, December 8, 2016, https://www.theguardian.com/environment/2016/dec/08/giraffe-red-list-vulnerable-species-extinction

855   Black Rhinoceros, Diceros bicornis, International Union for Conservation of Nature (IUCN) Red List of Threatened Species, http://www.iucnredlist.org/details/6557/0

856   Platt, John, "A Record 1,215 Rhinos Were Poached in 2014," Scientific American, January 23, 2015, https://blogs.scientificamerican.com/extinction-countdown/a-record-1-215-rhinos-were-poached-in-2014

857   Faith Karimi, "11 endangered rhinos were moved to start a new population. 10 died," CNN, July 27, 2018, https://www.cnn.com/2018/07/27/africa/black-rhinos-dead-kenya-relocation/index.html

858   "The strange figures behind a secret trade," BBC News, December 4, 2018, http://www.bbc.co.uk/news/resources/idt-sh/rhino_poaching

859   Vikram Dodd, "Crimes against nature: how greed fuels illegal trade in animal parts," The Guardian, January 28, 2018, https://www.theguardian.com/uk-news/2019/jan/28/illegal-trade-animal-parts-scotland-yard-small-wildlife-unit

860   Vikram Dodd, "Crimes against nature: how greed fuels illegal trade in animal parts," The Guardian, January 28, 2018, https://www.theguardian.com/uk-news/2019/jan/28/illegal-trade-animal-parts-scotland-yard-small-wildlife-unit

861   "Endangered listing urged for cheetahs," Science Daily, December 11, 2017, https://www.sciencedaily.com/releases/2017/12/171211092729.htm

862   "Cheetah populations crash as fastest-animal disappears from 91% of its range," Mongabay, December 26, 2016, https://news.mongabay.com/2016/12/cheetah-populations-crash-as-fastest-animal-disappears-from-91-of-its-range

863   Cheetah, Acinonyx jubatus, International Union for Conservation of Nature (IUCN) Red List of Threatened Species, https://www.iucnredlist.org/species/219/50649567

864   Robin McKie, "Asiatic cheetahs on the brink of extinction with only 50 left alive," The Guardian, December 16, 2017, https://www.theguardian.com/environment/2017/dec/16/asiatic-cheetah-brink-extinction-iran-un-funding

865   Chinese Pangolin, Manis pentadactyla, International Union for Conservation of Nature (IUCN) Red List of Threatened Species, http://www.iucnredlist.org/details/12764/0

866   Nick Davies, and Oliver Holmes, "Animal trafficking: the $23bn criminal industry policed by a toothless regulator," The Guardian, September 26, 2016, https://www.theguardian.com/environment/2016/sep/26/animal-trafficking-cites-criminal-industry-policed-toothless-regulator

867   Stephanie Milot, "You've Probably Never Heard of the Most-Trafficked Mammal in the World," Geek.com, April 13, 2018, https://www.geek.com/science/youve-probably-never-heard-of-the-most-trafficked-mammal-in-the-world-1736908

868   Brad Scriber, "100,000 Elephants Killed by Poachers in Just Three Years,

Landmark Analysis Finds," National Geographic, August 18, 2014, https://news.nationalgeographic.com/news/2014/08/140818-elephants-africa-poaching-cites-census

869   "Tigers on verge of extinction in the wild, World Wildlife Fund warns," CNN, February 10, 2010, http://www.cnn.com/2010/TECH/science/02/10/tigers.gone/index.html

870   Tiger, Panthera tigris, International Union for Conservation of Nature (IUCN) Red List of Threatened Species, https://www.iucnredlist.org/species/15955/50659951

871   Courchamp, Franck, et al., "The paradoxical extinction of the most charismatic animals," PLOS Biology, April 12, 2018, https://doi.org/10.1371/journal.pbio.2003997

872   William J. Ripple, et al., "Bushmeat hunting and extinction risk to the world's mammals," Royal Society Open Science, September 20, 2016, http://dx.doi.org/10.1098/rsos.160498

873   Damian Carrington, "World's mammals being eaten into extinction, report warns," The Guardian, October 19, 2017, https://www.theguardian.com/environment/2016/oct/19/worlds-mammals-being-eaten-into-extinction-report-warns

874   Mark Ariel, "Over 9,000 primates killed for single bushmeat market in West Africa every year," Mongabay, March 24, 2014, https://news.mongabay.com/2014/03/over-9000-primates-killed-for-single-bushmeat-market-in-west-africa-every-year

875   Mark Jones, "Is Africa's wildlife being eaten to extinction?" BBC News, August 3, 2010, http://news.bbc.co.uk/2/hi/science/nature/8877062.stm

876   Damian Carrington, "World's mammals being eaten into extinction, report warns," The Guardian, October 19, 2017, https://www.theguardian.com/environment/2016/oct/19/worlds-mammals-being-eaten-into-extinction-report-warns

877   William J. Ripple, et al., "Bushmeat hunting and extinction risk to the world's mammals," Royal Society Open Science, September 20, 2016, http://dx.doi.org/10.1098/rsos.160498

878   R. Nasi, A. Taber, N. Van Vliet, "Empty forests, empty stomachs? Bushmeat and livelihoods in the Congo and Amazon Basins," International Forestry Review, 2011, vol. 13, pp. 355-368.

879   Oliver Milman, "The killing of large species is pushing them towards extinction, study finds," The Guardian, February 6, 2019, https://www.theguardian.com/world/2019/feb/06/the-killing-of-large-species-is-pushing-them-towards-extinction-study-finds

880   Andrew Jacbos, "China's Appetite Pushes Fisheries to the Brink," New York Times, April 30, 2017, https://www.nytimes.com/2017/04/30/world/asia/chinas-appetite-pushes-fisheries-to-the-brink.html

881   Laura Mallonee, "Inside China's Almost-Totally-Legal $400M Fishery in Africa," Wired, March 23, 2017, https://www.wired.com/2017/03/yuyang-liu-drifting-west-africa/

882   Andrew Jacbos, "China's Appetite Pushes Fisheries to the Brink," New York Times, April 30, 2017, https://www.nytimes.com/2017/04/30/world/asia/chinas-appetite-pushes-fisheries-to-the-brink.html

883   R. Nasi, A. Taber, N. Van Vliet, "Empty forests, empty stomachs? Bushmeat and livelihoods in the Congo and Amazon Basins," International Forestry Review, 2011, vol. 13, pp. 355-368.

884   Mark Jones, "Is Africa's wildlife being eaten to extinction?" BBC News, August 3,2010, http://news.bbc.co.uk/2/hi/science/nature/8877062.stm

885   William J. Ripple, et al., "Bushmeat hunting and extinction risk to the

world's mammals," Royal Society Open Science, September 20, 2016, http://dx.doi.org/10.1098/rsos.160498

886 David Biello, "Will Central Africa's Forest Wildlife Be Eaten into Extinction?" Scientific American, September 15, 2008

887 Nick Davies, and Oliver Holmes, "Animal trafficking: the $23bn criminal industry policed by a toothless regulator," The Guardian, September 26, 2016, https://www.theguardian.com/environment/2016/sep/26/animal-trafficking-cites-criminal-industry-policed-toothless-regulator

888 Vikram Dodd, "Crimes against nature: how greed fuels illegal trade in animal parts," The Guardian, January 28, 2018, https://www.theguardian.com/uk-news/2019/jan/28/illegal-trade-animal-parts-scotland-yard-small-wildlife-unit

889 Colin Fernandez, "Cracking down on illegal ivory trade: Tusks from more than 300 elephants hidden inside wooden logs on the Sudan-Uganda border are seized thanks to British scanner technology," Daily Mail, February 11, 2019, https://www.dailymail.co.uk/sciencetech/article-6692083/Tusks-300-elephants-hidden-inside-wooden-logs-seized.html

890 Nick Davies, and Oliver Holmes, "Animal trafficking: the $23bn criminal industry policed by a toothless regulator," The Guardian, September 26, 2016, https://www.theguardian.com/environment/2016/sep/26/animal-trafficking-cites-criminal-industry-policed-toothless-regulator

891 Nick Whigham, "Environmental crimes on the rise as illegal funds pour into conflict zones," news.com.au, September 27, 2018, https://www.news.com.au/technology/environment/environmental-crimes-on-the-rise-as-illegal-funds-pour-into-conflict-zones/news-story/c37644dec9f484f20ac26a62e2092f5d

892 "Shark Fin Soup – what's the scoop?" Stop Shark Finning, http://www.stopsharkfinning.net/shark-fin-soup-whats-the-scoop

893 Matthew Kassel, "Here's What Happens When You Order A $65 Bowl Of Shark Fin Soup," Business Insider, March 19, 2012, http://www.businessinsider.com/new-york-could-ban-shark-fin-trade-2012-3

894 Alexjandra Borunda, "Rare Manta Ray Nursery Discovered," National Geographic, June 18, 2018, https://news.nationalgeographic.com/2018/06/animals-manta-rays-nurseries-babies

895 Dan Levin, "China Weighs Ban on Manta Ray Gills, Sold in Traditional Market as Modern Panacea," New York Times, January 6, 2016, https://www.nytimes.com/2016/01/07/world/asia/china-manta-ray-conservation.html

896 Sylvia Downes, "The global cost of China's destruction of the 'roof of the world'," The Ecologist, May 11, 2012, https://theecologist.org/2012/may/11/global-cost-chinas-destruction-roof-world

897 Benjamin Carlson, "What China's Rich Want: Gold-Plated Cars And Tiger Bits," August 21, 2013, NPR (National Public Radio), https://www.npr.org/sections/parallels/2013/08/21/213841586/what-chinas-rich-want-gold-plated-cars-and-tiger-bits

898 Nick Whigham, "Environmental crimes on the rise as illegal funds pour into conflict zones," news.com.au, September 27, 2018, https://www.news.com.au/technology/environment/environmental-crimes-on-the-rise-as-illegal-funds-pour-into-conflict-zones/news-story/c37644dec9f484f20ac26a62e2092f5d

899 Guy Kelly, "Pangolins: 13 facts about the world's most hunted animal," The Telegraph, January 1, 2015, http://www.telegraph.co.uk/science/2016/03/15/pangolins-13-facts-about-the-worlds-most-hunted-animal

900 Nick Davies, and Oliver Holmes, "Animal trafficking: the $23bn criminal

industry policed by a toothless regulator," The Guardian, September 26, 2016, https://www.theguardian.com/environment/2016/sep/26/animal-trafficking-cites-criminal-industry-policed-toothless-regulator
901   Niyi Aderibigbe, "Wildlife command crazy prices on the black market; snake venom sells at $215,000 per litre," The Nerve Africa, December 12, 2015, http://thenerveafrica.com/1817/wildlife-command-crazy-prices-on-the-black-market-snake-venom-sells-at-215000-per-litre-bear-bile-at-200000-per-pound-and-a-dead-tiger-costs-5000
902   Yemeli Ortega with Joanna Chiu In Guangzhou, China, "Race for Mexico's 'cocaine of the sea' pushes two species toward extinction," phys.org, April 10, 2018, https://phys.org/news/2018-04-mexico-cocaine-sea-species-extinction.html
903   Christopher Joyce, "Chinese Taste For Fish Bladder Threatens Rare Porpoise In Mexico," NPR, February 9, 2016, https://www.npr.org/sections/goatsandsoda/2016/02/09/466185043/chinese-taste-for-fish-bladder-threatens-tiny-porpoise-in-mexico
904   Anthony D. Barnosky, et al., "Has the Earth's sixth mass extinction already arrived?" Nature, March 2011, vol. 471, pp. 51-57.
905   Rodolfo Dirzo, et al., "Defaunation in the Anthropocene," Science, 2014, pp. 401-406, DOI: 10.1126/science.1251817
906   The Balance of Nature and Human Impact, Edited by Klaus Rohde, 2013.
907   Anthony D. Barnosky, et al., "Has the Earth's sixth mass extinction already arrived?" Nature, March 2011, vol. 471, pp. 51-57.
908   The Balance of Nature and Human Impact, Edited by Klaus Rohde, 2013.
909   Jasmine Aguilera, "The Numbers Are Just Horrendous.' Almost 30,000 Species Face Extinction Because of Human Activity," Time, July 18, 2019, https://time.com/5629548/almost-30000-species-face-extinction-new-report/
910   John Muchangi, "Tilapia at risk of extinction due to overfishing - report," The Star, May 2, 2018, https://www.the-star.co.ke/news/2018/05/02/tilapia-at-risk-of-extinction-due-to-overfishing-report_c1752340
911   "Water on the Tibetan Plateau Ecological and Strategic Implications," The Hague Centre for Strategic Studies, June 5, 2009, p. 20.
912   Yuliya Talmazant, "Extinction threatens third of freshwater fish species, report finds," NBC News, February 23, 2021, https://www.nbcnews.com/news/world/extinction-threatens-third-freshwater-fish-species-report-finds-n1258591
913   Michael J. Novacek and Elsa E. Cleland, "The current biodiversity extinction event: Scenarios for mitigation and recovery," PNAS, May 8, 2001, vol. 98, no. 10.
914   William J. Ripple, et al., "Bushmeat hunting and extinction risk to the world's mammals," Royal Society Open Science, September 20, 2016, http://dx.doi.org/10.1098/rsos.160498
915   Michael J. Novacek and Elsa E. Cleland, "The current biodiversity extinction event: Scenarios for mitigation and recovery," PNAS, May 8, 2001, vol. 98, no. 10.
916   Azeen Ghorayshi, "Common bird species such as sparrow and skylark facing decline in Europe," The Guardian, November 2, 2014, https://www.theguardian.com/environment/2014/nov/02/common-bird-species-sparrow-skylark-decline-europe
917   Simon Williams, "Study reveals severe decline of mountain hares," Centre for Ecology & Hydrology, August 13 2018, https://www.ceh.ac.uk/news-and-media/news/study-reveals-severe-decline-mountain-hares
918   Rodolfo Dirzo, et al., "Defaunation in the Anthropocene," Science, 2014, pp. 401-406, DOI: 10.1126/science.1251817
919   Robin McKie, "RSPB boss: Britain has one last chance to save endangered

species," The Guardian, August 18 2018, https://amp.theguardian.com/environment/2018/aug/18/endangered-birds-farming-uk-brexit

920   Rodolfo Dirzo, et al., "Defaunation in the Anthropocene," Science, 2014, pp. 401-406, DOI: 10.1126/science.1251817

921   Jeffrey Kluger, "The Sixth Great Extinction Is Underway—and We're to Blame," Time, April 24, 2017, http://time.com/3035872/sixth-great-extinction

922   Francisco Sánchez-Bayo and Kris A.G. Wyckhuys, "Worldwide decline of the entomofauna: A review of its drivers, " Biological Conservation, April 2019, vol. 232, pp. 8-27, https://doi.org/10.1016/j.biocon.2019.01.020

923   Francisco Sánchez-Bayo and Kris A.G. Wyckhuys, "Worldwide decline of the entomofauna: A review of its drivers, " Biological Conservation, April 2019, vol. 232, pp. 8-27, https://doi.org/10.1016/j.biocon.2019.01.020

924   Nick Haddad, "The Last Butterflies?" September 19, 2019, Scientific American, https://blogs.scientificamerican.com/observations/the-last-butterflies

925   Ben Westcott, "Sixth mass extinction? Two-thirds of wildlife may be gone by 2020: WWF," October 28, 2016, http://www.cnn.com/2016/10/26/world/wild-animals-disappear-report-wwf/index.html

926   Archana Phull, "68% decline in wildlife population since 1970: WWF report," The Statesman, September 11, 2020, https://www.thestatesman.com/cities/shimla/68-decline-wildlife-population-since-1970-wwf-report-1502923161.html

927   Living Planet Report - 2018: Aiming Higher, World Wildlife Fund, 2018

928   Damian Carrington, "World on track to lose two-thirds of wild animals by 2020, major report warns," The Guardian, October 26, 2016, https://www.theguardian.com/environment/2016/oct/27/world-on-track-to-lose-two-thirds-of-wild-animals-by-2020-major-report-warns

929   Damian Carrington, "Humans just 0.01% of all life but have destroyed 83% of wild mammals – study," The Guardian, May 21, 2018, https://www.theguardian.com/environment/2018/may/21/human-race-just-001-of-all-life-but-has-destroyed-over-80-of-wild-mammals-study

930   Gerardo Ceballos, Paul R. Ehrlich and Rodolfo Dirzo, "Biological annihilation via the ongoing sixth mass extinction signaled by vertebrate population losses and declines," Proceedings of the National Academy of Sciences (PNAS), July 10, 2017, www.pnas.org/cgi/doi/10.1073/pnas.1704949114

931   Damian Carrington, "Earth's sixth mass extinction event under way, scientists warn," The Guardian, July 10, 2017, https://www.theguardian.com/environment/2017/jul/10/earths-sixth-mass-extinction-event-already-underway-scientists-warn

932   John D. Sutter, "How to stop the sixth mass extinction," CNN, December 12, 2016, http://www.cnn.com/2016/12/12/world/sutter-vanishing-help

933   Amanda Mascarelli, "What to know before you spray your lawn with pesticides," The Washington Post, July 7, 2014, https://www.washingtonpost.com/national/health-science/what-to-know-before-you-spray-your-lawn-with-pesticides/2014/07/07/77d719a2-f63c-11e3-a606-946fd632f9f1_story.html

934   Francisco Sánchez-Bayo and Kris A.G. Wyckhuys, "Worldwide decline of the entomofauna: A review of its drivers, " Biological Conservation, April 2019, vol. 232, pp. 8-27, https://doi.org/10.1016/j.biocon.2019.01.020

935   Michael C.R. Alavanja, Dr.P.H., "Pesticides Use and Exposure Extensive Worldwide, " Rev Environ Health, Oct-Dec, 2009, pp. 303-309, doi:10.1515/reveh.2009.24.4.303

936   Nutrition and healthy eating, April 8, 2020, https://www.mayoclinic.org/

healthy-lifestyle/nutrition-and-healthy-eating/in-depth/organic-food/art-20043880

937   Sarah van Gelder, "A Brief History of Happiness: How America Lost Track of the Good Life—and Where to Find It Now," Yes Magazine, May 15, 2015, http://www.yesmagazine.org/happiness/how-america-lost-track-of-the-good-life-and-where-to-find-it-now

938   Tim Kasser, "The High Price of Materialism – how our culture of consumerism undermines our well-being," https://newdream.org/videos/high-price-of-materialism

939   John Buell, Politics, Religion, and Culture in an Anxious Age, 2011, Palgrave Macmillan.

940   Sarah van Gelder, "A Brief History of Happiness: How America Lost Track of the Good Life—and Where to Find It Now," Yes Magazine, May 15, 2015, http://www.yesmagazine.org/happiness/how-america-lost-track-of-the-good-life-and-where-to-find-it-now

941   Allen J Frances M.D., "How Psychoanalysis and Behaviorism Helped Create Advertising," Psychology Today, January 9, 2017

942   "The manipulation of the American mind: Edward Bernays and the birth of public relations," The Conversation, July 9, 2015

943   John Buell, Politics, Religion, and Culture in an Anxious Age, 2011, Palgrave Macmillan.

944   Matt Walsh, "If You Shop on Thanksgiving, You Are Part of the Problem," Huffington Post, November 20, 2013, https://www.huffingtonpost.com/matt-walsh/shopping-on-thanksgiving_b_4310109.html

945   Tori DeAngelis, "Consumerism and its discontents - Materialistic values may stem from early insecurities and are linked to lower life satisfaction, psychologists find. Accruing more wealth may provide only a partial fix," American Psychological Association, June 2004, vol. 35, no. 6.

946   "Behind the Ever-Expanding American Dream House," NPR, July 4, 2006, https://www.npr.org/templates/story/story.php?storyId=5525283

947   Tori DeAngelis, "Consumerism and its discontents - Materialistic values may stem from early insecurities and are linked to lower life satisfaction, psychologists find. Accruing more wealth may provide only a partial fix," American Psychological Association, June 2004, vol. 35, no. 6.

948   Carey Goldberg, "Materialism is bad for you, studies say", New York Times, February 8, 2006, http://www.nytimes.com/2006/02/08/health/materialism-is-bad-for-you-studies-say.html

949   Weller, Chris, "16 of history's greatest philosophers reveal the secret to happiness," Business Insider, October 17, 2016, http://www.businessinsider.com/philosophers-quotes-on-happiness-2016-10

950   Professor of Psychology Tim Kasser, "The High Price of Materialism – how our culture of consumerism undermines our well-being," https://newdream.org/videos/high-price-of-materialism

951   Carey Goldberg, "Materialism is bad for you, studies say", New York Times, February 8, 2006, http://www.nytimes.com/2006/02/08/health/materialism-is-bad-for-you-studies-say.html

952   Sarah van Gelder, "A Brief History of Happiness: How America Lost Track of the Good Life—and Where to Find It Now," Yes Magazine, May 15, 2015, http://www.yesmagazine.org/happiness/how-america-lost-track-of-the-good-life-and-where-to-find-it-now

953   Sherrie Bourg Carter Psy.D, "Why Mess Causes Stress: 8 Reasons, 8 Remedies:

The mental cost of clutter," Psychology Today, March 14, 2012, https://www.psychologytoday.com/blog/high-octane-women/201203/why-mess-causes-stress-8-reasons-8-remedies

954   Professor of Psychology Tim Kasser, "The High Price of Materialism – how our culture of consumerism undermines our well-being," https://newdream.org/videos/high-price-of-materialism

955   Emily Holden, "US produces far more waste and recycles far less of it than other developed countries," The Guardian, July 3, 2019, https://www.theguardian.com/us-news/2019/jul/02/us-plastic-waste-recycling

956   Reynard Loki, "America is a wasteland: The U.S. produces a shocking amount of garbage," Salon, July 15, 2016, https://www.salon.com/2016/07/15/america_is_a_wasteland_the_u_s_produces_a_shocking_amount_of_garbage_partner

957   "Waste Not, Want Not – Solid Waste at the Heart of Sustainable Development," The World Bank, March 3, 2016, http://www.worldbank.org/en/news/feature/2016/03/03/waste-not-want-not---solid-waste-at-the-heart-of-sustainable-development

958   A Roadmap for closing Waste Dumpsites The World's most Polluted Places, (ISWA) International Solid Waste Association, September 2012, p. 34.

959   "Global Slavery Index: 2018 Findings," https://www.globalslaveryindex.org/2018/findings/highlights

960   Annie Kelly, "British public bought £14bn of goods made by slaves in 2017, claims report," The Guardian, July 18, 2019, https://www.theguardian.com/global-development/2018/jul/19/british-public-14bn-goods-slaves-global-slavery-index

961   "Global Slavery Index: 2018 Findings Country Studies" United States," https://www.globalslaveryindex.org/2018/findings/country-studies/united-states

962   Suzanne Goldenberg, "Half of all US food produce is thrown away, new research suggests," The Guardian, July 13, 2016, https://www.theguardian.com/environment/2016/jul/13/us-food-waste-ugly-fruit-vegetables-perfect

963   Elizabeth Royte, "How 'Ugly' Fruits and Vegetables Can Help Solve World Hunger," National Geographic, March 2016, https://www.nationalgeographic.com/magazine/2016/03/global-food-waste-statistics/

964   Reynard Loki, "America is a wasteland: The U.S. produces a shocking amount of garbage," Salon, July 15, 2016, https://www.salon.com/2016/07/15/america_is_a_wasteland_the_u_s_produces_a_shocking_amount_of_garbage_partner

965   Dana Gunders, Wasted: How America Is Losing Up to 40 Percent of Its Food from Farm to Fork to Landfill, Natural Resources Defense Council, August 2012.

966   Adam Chandler, "Why Americans Lead the World in Food Waste," The Atlantic, July 15, 2016, https://www.theatlantic.com/business/archive/2016/07/american-food-waste/491513

967   Dana Gunders, Wasted: How America Is Losing Up to 40 Percent of Its Food from Farm to Fork to Landfill, Natural Resources Defense Council, August 2012.

968   Ariana Eunjung Cha, "Why the healthy school lunch program is in trouble. Before/after photos of what students ate," Washington Post, August 26, 2015.

969   Elizabeth Royte, "How 'Ugly' Fruits and Vegetables Can Help Solve World Hunger," National Geographic, March 2016, https://www.nationalgeographic.com/magazine/2016/03/global-food-waste-statistics

970   Water pollution from agriculture: a global review - Executive summary, FAO & IWMI, 2017.

971   United States 2030 Food Loss and Waste Reduction Goal, EPA (Environmental Protection Agency)

972   Elizabeth Royte, "How 'Ugly' Fruits and Vegetables Can Help Solve World Hunger," National Geographic, March 2016, https://www.nationalgeographic.com/magazine/2016/03/global-food-waste-statistics

973   "How the size of an average restaurant meal has QUADRUPLED since the 1950s - with U.S. burgers now three times as big," Daily Mail, May 23 2012, http://www.dailymail.co.uk/news/article-2148970/How-size-average-restaurant-meal-QUADRUPLED-1950s--U-S-burgers-times-big.html

974   Dana Gunders, Wasted: How America Is Losing Up to 40 Percent of Its Food from Farm to Fork to Landfill, Natural Resources Defense Council, August 2012.

975   Lisa F. Berkman, Ichiro Kawachi, and M. Maria Glymour, Social Epidemiology, Oxford University Press, 2014

976   Paul Insel, et al., Nutrition - Fourth Edition, 2011, p. 48.

977   Brian Wansink, Phd and Koert Van Ittersum, Phd, "Portion Size Me: Downsizing Our Consumption Norms," Journal of the American Dieteic Association, July 2007, vol. 107, no. 7, pp. 1103-1106.

978   Zeena Mackerdien PhD, "Obesity Rates Continue to Rise Among Americans," https://www.medpagetoday.org/primarycare/obesity/90456

979   "The Health Effects of Overweight & Obesity," https://www.cdc.gov/healthyweight/effects/index.html

980   https://www.cdc.gov/nchs/fastats/deaths.htm

981   Mahshid Dehghan, Noori Akhtar-Danesh and Anwar T Merchant, "Childhood obesity, prevalence and prevention," Nutrition Journal, September 2005

982   Susan Kelley, "Obesity accounts for 21 percent of U.S. health care costs," Cornell Chronicle, April 4, 2012, http://news.cornell.edu/stories/2012/04/obesity-accounts-21-percent-medical-care-costs

983   Dana Gunders, Wasted: How America Is Losing Up to 40 Percent of Its Food from Farm to Fork to Landfill, Natural Resources Defense Council, August 2012.

984   Dana Gunders, Wasted: How America Is Losing Up to 40 Percent of Its Food from Farm to Fork to Landfill, Natural Resources Defense Council, August 2012.

985   Elizabeth Royte, "How 'Ugly' Fruits and Vegetables Can Help Solve World Hunger," National Geographic, March 2016, https://www.nationalgeographic.com/magazine/2016/03/global-food-waste-statistics/

986   Suzanne Goldenberg, "Half of all US food produce is thrown away, new research suggests," The Guardian, July 13, 2016, https://www.theguardian.com/environment/2016/jul/13/us-food-waste-ugly-fruit-vegetables-perfect

987   Suzanne Goldenberg, "The US throws away as much as half its food produce," Wired, July 14, 2016, https://www.wired.com/2016/07/us-throws-away-much-half-food-produce

988   Reynard Loki, "America is a wasteland: The U.S. produces a shocking amount of garbage," Salon, July 15, 2016, https://www.salon.com/2016/07/15/america_is_a_wasteland_the_u_s_produces_a_shocking_amount_of_garbage_partner

989   United States 2030 Food Loss and Waste Reduction Goal, EPA (Environmental Protection Agency)

990   Elizabeth Royte, "How 'Ugly' Fruits and Vegetables Can Help Solve World Hunger," National Geographic, March 2016, https://www.nationalgeographic.com/magazine/2016/03/global-food-waste-statistics

991   Marc Bain, "Consumer culture has found its perfect match in our mobile-first, fast-fashion lifestyles," Quartz, March 21, 2015, https://qz.com/359040/the-internet-and-cheap-clothes-have-made-us-sport-shoppers/

992   Jamie Feldman, "The Average Woman Owns Over $500 Worth Of Unworn

Clothing, New Survey Reports," Huffington Post, August 27, 2014, https://www.huffingtonpost.com/2014/03/28/unworn-clothing-survey_n_5048486.html
993	Maybelle Morgan, "Throwaway fashion: Women have adopted a 'wear it once culture', binning clothes after only a few wears (so they aren't pictured in same outfit twice on social media)," Daily Mail, June 9, 2015, http://www.dailymail.co.uk/femail/article-3116962/Throwaway-fashion-Women-adopted-wear-culture-binning-clothes-wears-aren-t-pictured-outfit-twice-social-media.html
994	Marc Bain, "Consumer culture has found its perfect match in our mobile-first, fast-fashion lifestyles," Quartz, March 21, 2015, https://qz.com/359040/the-internet-and-cheap-clothes-have-made-us-sport-shoppers/
995	Marc Gunther, "Pressure mounts on retailers to reform throwaway clothing culture," The Guardian, August 10, 2016, https://www.theguardian.com/environment/2016/aug/10/pressure-mounts-on-retailers-to-reform-throwaway-clothing-culture
996	Shannon Whitehead, "5 Truths the Fast Fashion Industry Doesn't Want You to Know," Huffington Post, August 19, 2014, https://www.huffingtonpost.com/shannon-whitehead/5-truths-the-fast-fashion_b_5690575.html
997	M. Leanne Lachman and Deborah L. Brett, Generation Y: Shopping and Entertainment in the Digital Age, Urban Land Institute, 2013.
998	Shuk-Wah Chung, "Fast fashion is "drowning" the world. We need a Fashion Revolution!" Greenpeace, April 21, 2016, https://www.greenpeace.org/international/story/7539/fast-fashion-is-drowning-the-world-we-need-a-fashion-revolution
999	Sam Williams, "The Environmental Impact Of Clothing," Huffington Post UK, December 2016, https://www.huffingtonpost.co.uk/sam-williams1/the-environmental-impact-_1_b_13546078.html
1000	"Waste Couture: Environmental Impact of the Clothing Industry," Environmental Health Perspectives, September 2007, vo. 115, no.9, pp. A449-A454.
1001	Shannon Whitehead, "5 Truths the Fast Fashion Industry Doesn't Want You to Know," Huffington Post, August 19, 2014, https://www.huffingtonpost.com/shannon-whitehead/5-truths-the-fast-fashion_b_5690575.html
1002	Shuk-Wah Chung, "Fast fashion is "drowning" the world. We need a Fashion Revolution!" Greenpeace, April 21, 2016, https://www.greenpeace.org/international/story/7539/fast-fashion-is-drowning-the-world-we-need-a-fashion-revolution
1003	"One garbage truck of textiles wasted every second: report creates vision for change," Ellen MacArthur Foundation, November 28, 2017, https://www.ellenmacarthurfoundation.org/news/one-garbage-truck-of-textiles-wasted-every-second-report-creates-vision-for-change
1004	Marc Bain, "Consumer culture has found its perfect match in our mobile-first, fast-fashion lifestyles," Quartz, March 21 2015, https://qz.com/359040/the-internet-and-cheap-clothes-have-made-us-sport-shoppers/
1005	Marc Gunther, "Pressure mounts on retailers to reform throwaway clothing culture," The Guardian, August 10, 2016, https://www.theguardian.com/environment/2016/aug/10/pressure-mounts-on-retailers-to-reform-throwaway-clothing-culture
1006	Morgan McFall-Johnsen, "The fashion industry emits more carbon than international flights and maritime shipping combined. Here are the biggest ways it impacts the planet," Business Insider,October 21, 2019, https://www.businessinsider.com/fast-fashion-environmental-impact-pollution-emissions-waste-water-2019-10
1007	Morgan McFall-Johnsen, "These facts show how unsustainable the fashion industry is," World Economic Forum, January 21, 2020, https://www.weforum.org/

agenda/2020/01/fashion-industry-carbon-unsustainable-environment-pollution"
1008   Sophie Benson, "It Takes 2,720 Litres Of Water To Make ONE T-Shirt – As Much As You'd Drink In 3 Years," Refinery 29, March 22, 2018, https://www.refinery29.com/en-gb/water-consumption-fashion-industry
1009   Gunther, Marc, "Pressure mounts on retailers to reform throwaway clothing culture," The Guardian, August 10, 2016, https://www.theguardian.com/environment/2016/aug/10/pressure-mounts-on-retailers-to-reform-throwaway-clothing-culture
1010   "Cotton's Water Footprint: How One T-Shirt Makes A Huge Impact On The Environment," Huffington Post, January 27, 2013, https://www.huffingtonpost.com/2013/01/27/cottons-water-footprint-world-wildlife-fund_n_2506076.html
1011   Shuk-Wah Chung, "Fast fashion is "drowning" the world. We need a Fashion Revolution!" Greenpeace, April 21, 2016, https://www.greenpeace.org/international/story/7539/fast-fashion-is-drowning-the-world-we-need-a-fashion-revolution
1012   Olympic-size swimming pool, Wikipedia.
1013   Sophie Benson, "It Takes 2,720 Litres Of Water To Make ONE T-Shirt – As Much As You'd Drink In 3 Years," Refinery 29, March 22, 2018, https://www.refinery29.com/en-gb/water-consumption-fashion-industry
1014   "Get the facts about Organic Cotton," Organic Trade Association, https://ota.com/advocacy/fiber-and-textiles/get-facts-about-organic-cotton
1015   Shuk-Wah Chung, "Fast fashion is "drowning" the world. We need a Fashion Revolution!" Greenpeace, April 21, 2016, https://www.greenpeace.org/international/story/7539/fast-fashion-is-drowning-the-world-we-need-a-fashion-revolution
1016   Bain, Marc, "Consumer culture has found its perfect match in our mobile-first, fast-fashion lifestyles," Quartz, March 21 2015, https://qz.com/359040/the-internet-and-cheap-clothes-have-made-us-sport-shoppers
1017   Wee, Heesun, "'Made in USA' fuels new manufacturing hubs in apparel," CNB, September 23, 2013, https://www.cnbc.com/2013/09/23/inside-made-in-the-usa-showcasing-skilled-garment-workers.html
1018   Yallop, Olivia, "Citarum, the most polluted river in the world?" The Telegraph, April 11, 2014, https://www.telegraph.co.uk/news/earth/environment/10761077/Citarum-the-most-polluted-river-in-the-world.html
1019   Sweeny, Glynis, "Fast Fashion Is the Second Dirtiest Industry in the World, Next to Big Oil," EcoWatch, August 17, 2015, https://www.ecowatch.com/fast-fashion-is-the-second-dirtiest-industry-in-the-world-next-to-big--1882083445.html
1020   "Fast fashion: Rivers turning blue and 500,000 tonnes in landfill," ABC News Australia, March 28 2017, http://www.abc.net.au/news/2017-03-28/the-price-of-fast-fashion-rivers-turn-blue-tonnes-in-landfill/8389156
1021   "One garbage truck of textiles wasted every second: report creates vision for change," Ellen MacArthur Foundation, November 28, 2017, https://www.ellenmacarthurfoundation.org/news/one-garbage-truck-of-textiles-wasted-every-second-report-creates-vision-for-change
1022   Sweeny, Glynis, "Fast Fashion Is the Second Dirtiest Industry in the World, Next to Big Oil," EcoWatch, August 17, 2015, https://www.ecowatch.com/fast-fashion-is-the-second-dirtiest-industry-in-the-world-next-to-big--1882083445.html
1023   Kirchain, Randolph, et al., "Sustainable Apparel Materials," Massachusetts Institute of Technology, September 22, 2015, p. 17.
1024   Williams, Sam, "The Environmental Impact Of Clothing," Huffington Post UK, December 2016, https://www.huffingtonpost.co.uk/sam-williams1/the-environmental-impact-_1_b_13546078.html

1025	Parry, Simon, "The true cost of your cheap clothes: slave wages for Bangladesh factory workers," Post Magazine, June 11, 2016, http://www.scmp.com/magazines/post-magazine/article/1970431/true-cost-your-cheap-clothes-slave-wages-bangladesh-factory
1026	Whitehead, Shannon, "5 Truths the Fast Fashion Industry Doesn't Want You to Know," Huffington Post, August 19, 2014, https://www.huffingtonpost.com/shannon-whitehead/5-truths-the-fast-fashion_b_5690575.html
1027	"Waste Couture: Environmental Impact of the Clothing Industry," Environmental Health Perspectives, September 2007, vo. 115, no.9, pp. A449-A454.
1028	Parry, Simon, "The true cost of your cheap clothes: slave wages for Bangladesh factory workers," Post Magazine, June 11, 2016, http://www.scmp.com/magazines/post-magazine/article/1970431/true-cost-your-cheap-clothes-slave-wages-bangladesh-factory
1029	Winn, Patrick, "The slave labor behind your favorite clothing brands: Gap, H&M and more exposed," Salon, March 22, 2015, https://www.salon.com/2015/03/22/the_slave_labor_behind_your_favorite_clothing_brands_gap_hm_and_more_exposed_partner
1030	Quinn, Shannon, "10 Truly Troubling Facts About The Clothing Industry," List Verse, March 17, 2017, https://listverse.com/2017/03/17/10-truly-troubling-facts-about-the-clothing-industry
1031	Apple iPhone sales worldwide 2007-2017, Statista, https://www.statista.com/statistics/276306/global-apple-iphone-sales-since-fiscal-year-2007
1032	Number of smartphones sold to end users worldwide from 2007 to 2017, Statista, https://www.statista.com/statistics/263437/global-smartphone-sales-to-end-users-since-2007
1033	Juli Clover, "Thousands of Customers Waiting in Line at Apple Retail Stores for iPhone XS, XS Max and Apple Watch Series 4 as Global Launch Continues," MacRumors, September 20, 2018, https://www.macrumors.com/2018/09/20/iphone-xs-apple-watch-launch-day-lines
1034	S. O'Dea, "Number of smartphone users from 2016 to 2021," Statista, August 20, 2020, https://www.statista.com/statistics/330695/number-of-smartphone-users-worldwide
1035	Syed Faraz Ahmed, "The Global Cost of Electronic Waste," The Atlantic, September 29, 2016, https://www.theatlantic.com/technology/archive/2016/09/the-global-cost-of-electronic-waste/502019/
1036	Kyree Leary, "The World's E-Waste Is Piling Up at an Alarming Rate, Says New Report," Science Alert, December 13, 2017, https://www.sciencealert.com/global-electronic-waste-growth-report-2017-significant-increase
1037	"Record 53.6 million tonnes of e-waste dumped globally last year, says UN report," CBC, July 20, 2020, https://www.cbc.ca/news/technology/global-ewaste-monitor-2020-1.5634759
1038	"Electronic waste facts," http://www.theworldcounts.com/counters/waste_pollution_facts/electronic_waste_facts
1039	"World e-waste rises 8 percent by weight in 2 years as incomes rise, prices fall: UN-backed report," EurekAlert!, December 13, 2017, https://www.eurekalert.org/pub_releases/2017-12/tca-wer120817.php
1040	Leestma, David, "Electronic Waste Study Finds $65 Billion in Raw Materials Discarded in Just One Year," EcoWatch, December 14, 2017, https://www.ecowatch.com/electronic-waste-united-nations-2517390624.html
1041	Wilson, Mark, "Smartphones Are Killing The Planet Faster Than Anyone

Expected," Fast Company, March 27, 2018, https://www.fastcompany.com/90165365/smartphones-are-wrecking-the-planet-faster-than-anyone-expected

1042 Leestma, David, "Electronic Waste Study Finds $65 Billion in Raw Materials Discarded in Just One Year," EcoWatch, December 14, 2017, https://www.ecowatch.com/electronic-waste-united-nations-2517390624.html

1043 Doyle, Alister, "Electronic waste at new high, squandering gold, other metals: study," Reuters, December 13, 2017, https://www.reuters.com/article/us-environment-waste/electronic-waste-at-new-high-squandering-gold-other-metals-study-idUSKBN1E721G

1044 Wilson, Mark, "Smartphones Are Killing The Planet Faster Than Anyone Expected," Fast Company, March 27, 2018, https://www.fastcompany.com/90165365/smartphones-are-wrecking-the-planet-faster-than-anyone-expected

1045 "Responsible Mining: The Value of Cobalt Production Outside of the DRC," December 3, 2018, Investing News, https://investingnews.com/innspired/cobalt-production-outside-dr-congo-ethical-mining-battery-metals

1046 "Is my phone powered by child labour?" Amnesty International, https://www.amnesty.org/en/latest/campaigns/2016/06/drc-cobalt-child-labour

1047 Annie Kelly, "Apple and Google named in US lawsuit over Congolese child cobalt mining deaths," The Guardian, December 16, 2019, https://www.theguardian.com/global-development/2019/dec/16/apple-and-google-named-in-us-lawsuit-over-congolese-child-cobalt-mining-deaths

1048 Howard, Emma, "Photocopiers, fridges and flat screen TVs: how second-hand cars are being used to smuggle toxic e-waste," Unearthed, April 19, 2018, https://unearthed.greenpeace.org/2018/04/19/photocopiers-fridges-and-flatscreen-tvs-how-second-hand-cars-are-being-used-to-smuggle-toxic-e-waste

1049 Vidal, John, "Toxic E-Waste Dumped in Poor Nations, Says United Nations," United Nations University, December 16, 2013, https://ourworld.unu.edu/en/toxic-e-waste-dumped-in-poor-nations-says-united-nations

1050 Watson, Ivan, "China: The electronic wastebasket of the world," CNN, May 30, 2013, https://www.cnn.com/2013/05/30/world/asia/china-electronic-waste-e-waste/index.html

1051 "Guiyu: An E-Waste Nightmare," Greenpeace, http://www.greenpeace.org/eastasia/campaigns/toxics/problems/e-waste/guiyu

1052 Watson, Ivan, "China: The electronic wastebasket of the world," CNN, May 30, 2013, https://www.cnn.com/2013/05/30/world/asia/china-electronic-waste-e-waste/index.html

1053 "Guiyu: An E-Waste Nightmare," Greenpeace, http://www.greenpeace.org/eastasia/campaigns/toxics/problems/e-waste/guiyu

1054 Jayapradha Annamalai, "Occupational health hazards related to informal recycling of E-waste in India: An overview," Indian Journal of Environmental Medicine, 2015, pp. 61-65, doi: 10.4103/0019-5278.157013

1055 Pearce, Fred, "How 16 ships create as much pollution as all the cars in the world," Daily Mail, November 21, 2009, http://www.dailymail.co.uk/sciencetech/article-1229857/How-16-ships-create-pollution-cars-world.html

1056 "New NRDC Report: China's Ports Play Major Role in Country's Air Pollution Problems," Natural Resources Defense Council (NRDC), October 28, 2014, https://www.nrdc.org/media/2014/141028

1057 Sofiev, Mikhail, "Cleaner fuels for ships provide public health benefits with climate tradeoffs," Nature Communications, February 6, 2019, DOI: 10.1038/s41467-017-02774-9

1058   Gallucci, Maria, "At Last, the Shipping Industry Begins Cleaning Up Its Dirty Fuels," Yale Environment 360, June 28, 2018, https://e360.yale.edu/features/at-last-the-shipping-industry-begins-cleaning-up-its-dirty-fuels

1059   "Protective ship coatings as an underestimated source of microplastic pollution," Phys.org, February 23, 2021, https://phys.org/news/2021-02-ship-coatings-underestimated-source-microplastic.html

1060   "Protective ship coatings as an underestimated source of microplastic pollution," Phys.org, February 23, 2021, https://phys.org/news/2021-02-ship-coatings-underestimated-source-microplastic.html

1061   Celia Cole, "Overconsumption is costing us the earth and human happiness," The Guardian, June 21, 2010, https://www.theguardian.com/environment/2010/jun/21/overconsumption-environment-relationships-annie-leonard

1062   John Bakeless, America As Seen By Its First Explorers - The Eyes of Discovery, 1950, Dover Publications, New York, p. 102.

1063   John Bakeless, America As Seen By Its First Explorers - The Eyes of Discovery, 1950, Dover Publications, New York, p. 211.

1064   Nicholas Denys, The Description and Natural History of the Coasts of North America, The Champlain Society, Toronto, 1908, p. 189.

1065   Steve Nicholls, Paradise Found: Nature in America at the Time of Discovery, 2009, The University of Chicago Press, Chicago & London.

1066   Simon L. Lewis and Mark A. Maslin, "Defining the Anthropocene," Nature, March 12, 2015, vol. 519, pp. 171-180.

1067   Lois N. Magner, A History of Infectious Diseases and the Microbial World, 2009, Praeger Publishers

1068   William Denevan, "The Pristine Myth: The Landscape of the Americas in 1492," Annals of the Association of American Geographers, 1992, vol. 82, issue 3.

1069   Simon L. Lewis and Mark A. Maslin, "Defining the Anthropocene," Nature, March 12, 2015, vol. 519, pp. 171-180.

1070   Zoë Schlanger, "Did the Anthropocene begin with the deaths of 50 million Native Americans," Newsweek, March 12 2015, http://www.newsweek.com/did-anthropocene-begin-deaths-50-million-native-americans-313319

1071   Jonathan Amos, "America colonisation 'cooled Earth's climate'," BBC News, January 31, 2019, https://www.bbc.com/news/science-environment-47063973

1072   Randall K. Packer, "How Long Can the Average Person Survive Without Water?" Scientific American

1073   Roy Porter, The Greatest Benefit to Mankind, Harper Collins, New York, 1997, p. 399.

1074   Report of the Unveiling and Dedication of Indiana Monument at Andersonville, Georgia (National Cemetery), Thursday, November 26, 1908, pp. 73–102.

1075   "Cholera's seven pandemics, " CBC News, May 9 2008, http://www.cbc.ca/news/technology/cholera-s-seven-pandemics-1.758504

1076   David L. Streiner, Douglas W. MacPherson, and Brian D. Gushulak, PDQ Public Health, 2010, p. 198.

1077   Dr. Dalong Hu, et al., "Origins of the current seventh cholera pandemic," Proceedings of the National Academy of Sciences (PNAS), November 29, 2016, https://doi.org/10.1073/pnas.1608732113

1078   "Cholera's seven pandemics, " CBC News, May 9, 2008, http://www.cbc.ca/news/technology/cholera-s-seven-pandemics-1.758504

1079   Dr. Dalong Hu, et al., "Origins of the current seventh cholera pandemic," Proceedings of the National Academy of Sciences (PNAS), November 29, 2016, https://doi.

org/10.1073/pnas.1608732113

1080   Ending Cholera: A Global Roadmap to 2030, Global Task Force on Cholera Control, 2017.

1081   "Water, Sanitation and Hygiene," UN Water, http://www.unwater.org/water-facts/water-sanitation-and-hygiene

1082   "Water, Sanitation and Hygiene (WASH)," Unicef, https://www.unicef.org/wash/index_wes_related.html

1083   "Polluted water kills more people than war: UN," The Hindu, March 23, 2010, https://www.thehindu.com/sci-tech/health/Polluted-water-kills-more-people-than-war-UN/article16588753.ece

1084   John Vidal, "Cleaning the world's water: 'We are now more polluted than we have ever been,'" The Guardian, August 31, 2016, https://www.theguardian.com/environment/2016/aug/31/cleaning-the-worlds-water-we-are-now-more-polluted-than-we-have-ever-been

1085   Justin Rowlatt, "India's dying mother," BBC News, May 12, 2016, https://www.bbc.co.uk/news/resources/idt-aad46fca-734a-45f9-8721-61404cc12a39

1086   "Water on the Tibetan Plateau Ecological and Strategic Implications," The Hague Centre for Strategic Studies, June 5, 2009, p. 19.

1087   "The National Ganga River Basin Project," The World Bank, March 23, 2015, http://www.worldbank.org/en/news/feature/2015/03/23/india-the-national-ganga-river-basin-project

1088   Justin Rowlatt, "India's dying mother," BBC News, May 12 2016, https://www.bbc.co.uk/news/resources/idt-aad46fca-734a-45f9-8721-61404cc12a39

1089   Eric Zerkel, "World's Most Polluted Rivers," The Weather Channel, https://weather.com/news/news/worlds-most-polluted-rivers-20130627

1090   Michael Safi, "Murder most foul: polluted Indian river reported dead despite 'living entity' status," The Guardian, July 7, 2017, https://www.theguardian.com/world/2017/jul/07/indian-yamuna-river-living-entity-ganges

1091   Eric Zerkel, "World's Most Polluted Rivers," The Weather Channel, https://weather.com/news/news/worlds-most-polluted-rivers-20130627

1092   Julie McCarthy, "Can India's Sacred But 'Dead' Yamuna River Be Saved?" NPR, May 11, 2016, https://www.npr.org/2016/05/11/477415686/can-indias-sacred-but-dead-yamuna-river-be-saved

1093   Julie McCarthy, "Can India's Sacred But 'Dead' Yamuna River Be Saved?" NPR, May 11, 2016, https://www.npr.org/2016/05/11/477415686/can-indias-sacred-but-dead-yamuna-river-be-saved

1094   Arshard R. Zargar, "Hindu festival in India marred by a river of toxic foam and a blanket of killer smog," CBS News, November 5, 2019, https://www.cbsnews.com/news/yamuna-rivers-toxic-foam-and-delhi-air-pollution-greet-india-hindu-devotees-for-chhath-puja-festival/

1095   Rona Rui, "Heavily Polluted Yangtze River Endangers 400 Million Chinese," The Epoch Times, September 2, 2010,
https://www.theepochtimes.com/heavily-polluted-yangtze-river-endangers-400-million-chinese_1508892.html

1096   "Swimming in Poison," Greenpeace, August 2010, p. 13.

1097   "Threat of Pollution in the Yangtze," World Wildlife Federation, http://wwf.panda.org/our_work/water/freshwater_problems/river_decline/10_rivers_risk/yangtze/yangtze_threats

1098   Nicola Davison, "Rivers of blood: the dead pigs rotting in China's water supply," The Guardian, March 29, 2013, https://www.theguardian.com/world/2013/

mar/29/dead-pigs-china-water-supply
1099   "Water pollution from agriculture: a global review," Food and Agriculture Organization of the United Nations and the International Water Management Institute, 2017, p. 2.
1100   "Polluted water kills more people than war: UN," The Hindu, March 23, 2010, http://www.thehindu.com/sci-tech/health/Polluted-water-kills-more-people-than-war-UN/article16588753.ece
1101   John Vidal, "Cleaning the world's water: 'We are now more polluted than we have ever been,'" The Guardian, August 31, 2016, https://www.theguardian.com/environment/2016/aug/31/cleaning-the-worlds-water-we-are-now-more-polluted-than-we-have-ever-been
1102   The murky future of global water quality, International Food Policy Research Institute (IFPRI), 2015.
1103   "Water pollution from agriculture: a global review," Food and Agriculture Organization of the United Nations and the International Water Management Institute, 2017, p. 9.
1104   Matt Rosenberg, "Current and Historical World Population, " ThoughtCo, March 25, 2018, https://www.thoughtco.com/current-world-population-1435270
1105   Water pollution from agriculture: a global review - Executive summary, FAO & IWMI, 2017.
1106   "Water pollution from agriculture: a global review," Food and Agriculture Organization of the United Nations and the International Water Management Institute, 2017, p. 3.
1107   "Water pollution from agriculture: a global review," Food and Agriculture Organization of the United Nations and the International Water Management Institute, 2017, pp. 5-7.
1108   Damian Carrington, "UN experts denounce 'myth' pesticides are necessary to feed the world," The Guardian, March 7, 2017, https://www.theguardian.com/environment/2017/mar/07/un-experts-denounce-myth-pesticides-are-necessary-to-feed-the-world
1109   Philip J Landrigan, et al., "The Lancet Commission on pollution and health," The Lancet, October 19, 2017, http://dx.doi.org/10.1016/S0140-6736(17)32345-0
1110   Eric DuVall, "U.N. report estimates pesticides kill 200,000 people per year," UPI, March 9, 2017, https://www.upi.com/UN-report-estimates-pesticides-kill-200000-people-per-year/1161489037649
1111   Robert Hunziker, "Is the EPA Hazardous to Your Health?" Counterpunch, March 23, 2018, https://www.counterpunch.org/2018/03/23/is-the-epa-hazardous
1112   Damian Carrington, "Loss of wild pollinators serious threat to crop yields, study finds," The Guardian, February 28, 2013, https://www.theguardian.com/environment/2013/feb/28/wild-bees-pollinators-crop-yields
1113   Philip J Landrigan, et al., "The Lancet Commission on pollution and health," TheLancet, October 19, 2017, http://dx.doi.org/10.1016/S0140-6736(17)32345-0
1114   "Drug waste clogs rivers around the world, scientists say," The Guardian, April 10, 2018, https://www.theguardian.com/environment/2018/apr/11/drug-waste-clogs-rivers-around-the-world-scientists-say
1115   Matt Harvey, "Your tap water is probably laced with antidepressants," Salon, March 14, 2013, https://www.salon.com/2013/03/14/your_tap_water_is_probably_laced_with_anti_depressants_partner
1116   "New scientific study finds coral reefs under attack from chemical in sunscreen as global bleaching event hits," MarineSafe.org, http://www.marinesafe.org/

blog/2015/10/20/here-is-another-post

1117   Matt Harvey, "Your tap water is probably laced with antidepressants," Salon, March 14, 2013, https://www.salon.com/2013/03/14/your_tap_water_is_probably_laced_with_anti_depressants_partner

1118   "Prescription Drugs Found In Large Concentrations In Water Near Manufacturing Plants," NPR All Things Considered, December 11, 2019, https://www.npr.org/2019/12/11/787192604/prescription-drugs-found-in-large-concentrations-in-water-near-manufacturing-pla

1119   "Drugs in the water," Harvard Health Letter, June 2011.

1120   Ron Meador, "New findings suggest serious threat to Great Lakes fish from, yes, Prozac," Minnesota Post, September 8, 2017, https://www.minnpost.com/earth-journal/2017/09/new-findings-suggest-serious-threat-great-lakes-fish-yes-prozac

1121   Katie Hastings, "Antidepressants found in Niagara River fish," Great Lakes Center Newsletter, Issue 11, Fall 2017, p. 2.

1122   Matt Harvey, "Your tap water is probably laced with antidepressants," Salon, March 14, 2013, https://www.salon.com/2013/03/14/your_tap_water_is_probably_laced_with_anti_depressants_partner

1123   "Antidepressants are finding their way into fish brains," The Economist, February 8, 2018, https://www.economist.com/news/united-states/21736558-thats-bad-news-fish-antidepressants-are-finding-their-way-fish-brains

1124   Matt Harvey, "Your tap water is probably laced with antidepressants," Salon, March 14, 2013, https://www.salon.com/2013/03/14/your_tap_water_is_probably_laced_with_anti_depressants_partner

1125   E. J. Mundell, "Antidepressant use in U.S. soars by 65 percent in 15 years," CBS News, August 16, 2017, https://www.cbsnews.com/news/antidepressant-use-soars-65-percent-in-15-years/

1126   James Gorman, "A Drug Used for Cattle Is Said to Be Killing Vultures," New York Times, January 29, 2004, https://www.nytimes.com/2004/01/29/world/a-drug-used-for-cattle-is-said-to-be-killing-vultures.html

1127   Rachel Becker, "Cattle drug threatens thousands of vultures," Nature, April 29, 2016, https://www.nature.com/news/cattle-drug-threatens-thousands-of-vultures-1.19839

1128   Shreya Dasgupta, "Ban of vulture-killing drug in India is working," New Scientist, October 22, 2014, https://www.newscientist.com/article/dn26431-ban-of-vulture-killing-drug-in-india-is-working/

1129   Damian Carrington, "Plastic fibres found in tap water around the world, study reveals," The Guardian, September 5, 2017, https://www.theguardian.com/environment/2017/sep/06/plastic-fibres-found-tap-water-around-world-study-reveals

1130   Greg Beach, "Study suggests the average person consumes 70,000 microplastic bits every year," Inhabitat, April 10, 2018, https://inhabitat.com/study-suggests-the-average-person-consumes-70000-microplastic-bits-every-year/

1131   Sherri A. Mason, Victoria Welch, Joseph Neratko, "Synthetic Polymer Contamination in Bottled Water," State University of New York at Fredonia, Department of Geology & Environmental Sciences

1132   Taylor Rock, "Popular bottle water brands test positive for plastic contamination in new study," The Baltimore Sun, March 15, 2018, http://www.baltimoresun.com/entertainment/dining/dailymeal-popular-bottle-water-brands-test-positive-for-plastic-contamination-in-new-study-20180315-story.html

1133   "We Could Be Eating More Than 100 Pieces Of Plastic In Every Meal, New

Study Says," IFL Science! April 9, 2018, http://www.iflscience.com/health-and-medicine/we-could-be-eating-more-than-100-pieces-of-plastic-in-every-meal-new-study-says

1134 "High levels of microplastics released from infant feeding bottles during formula prep," Phys.org, October 19, 2020, https://phys.org/news/2020-10-high-microplastics-infant-bottles-formula.html

1135 Doyle Rice, "Forget acid rain. Plastic rain is now falling across the U.S.," USA Today, June 12, 2020, https://www.usatoday.com/story/news/nation/2020/06/12/plastic-rain-now-falling-across-u-s-new-study-found/3174549001

1136 Maanvi Singh, "It's raining plastic: microscopic fibers fall from the sky in Rocky Mountains," The Guardian, August 13, 2019, https://www.theguardian.com/us-news/2019/aug/12/raining-plastic-colorado-usgs-microplastics

1137 Ian Johnston, "Scientist warns we could be breathing in microplastic particles laden with chemicals," The Independent, July 13, 2016, https://www.independent.co.uk/news/science/microplastic-microbeads-microfibres-pollution-environment-audit-committee-mps-evidence-a7021051.html

1138 John L. Pauly, et al., "Inhaled Cellulosic and Plastic Fibers Found in Human Lung Tissue," Cancer Epidemiology, Biomarkers & Prevention, May 1998, vol. 7, pp. 419-428.

1139 Dennis Thompson, "Autopsies show microplastics in all major human organs, Medical Xpress, August 17, 2020, https://medicalxpress.com/news/2020-08-autopsies-microplastics-major-human.html

1140 https://www.statista.com/statistics/282732/global-production-of-plastics-since-1950

1141 Corey Charlton, "The OTHER Chinese creation you can see from space - SMOG: Giant pollution cloud revealed in satellite images," The Daily Mail, December 2, 2015, http://www.dailymail.co.uk/news/article-3342634/The-Chinese-landmark-space-SMOG-Giant-pollution-cloud-revealed-satellite-images.html

1142 "Particle Pollution," The American Lung Association, http://www.lung.org/our-initiatives/healthy-air/outdoor/air-pollution/particle-pollution.html

1143 Corey Charlton, "The OTHER Chinese creation you can see from space - SMOG: Giant pollution cloud revealed in satellite images," The Daily Mail, December 2, 2015, http://www.dailymail.co.uk/news/article-3342634/The-Chinese-landmark-space-SMOG-Giant-pollution-cloud-revealed-satellite-images.html

1144 Corey Charlton, "Delhi reels under air pollution: PM2.5 levels reach over 20 times the safe limit; Arvind Kejriwal blames stubble burning," The First Post, November 9, 2017, hhttps://www.firstpost.com/india/delhi-reels-under-air-pollution-pm2-5-levels-reach-over-20-times-the-safe-limit-arvind-kejriwal-blames-stubble-burning-4201759.html

1145 Kathryn Doyle, "Pollution particles damage blood vessels, may lead to heart disease," Reuters, October 26, 2016, https://www.reuters.com/article/us-health-cardiovascular-pm2-5-pollution-idUSKCN12Q2LM

1146 RD Brooks et al., "Particulate matter air pollution and cardiovascular disease: An update to the scientific statement from the American Heart Association," Circulation, June 1, 2010, DOI: 10.1161/CIR.0b013e3181dbece1

1147 Jay Apt, "The Other Reason to Shift away from Coal: Air Pollution That Kills Thousands Every Year," Scientific American, June 7, 2017, https://www.scientificamerican.com/article/the-other-reason-to-shift-away-from-coal-air-pollution-that-kills-thousands-every-year

1148 Dr. Aaron J. Cohen DSc, et al., "Estimates and 25-year trends of the global burden of disease attributable to ambient air pollution: an analysis of data from the Global Burden of Diseases Study 2015," The Lancet, vol. 389, issue 10082, May 2017, pp. 1907-1918.

1149   Sam Meredith, "Pollution linked to one in six deaths worldwide — and threatens 'survival of human societies'," CNBC, October 20, 2017, https://www.cnbc.com/2017/10/20/pollution-linked-to-one-in-six-deaths-worldwide-and-threatens-survival-of-human-societies.html

1150   John Vidal, "Air pollution more deadly in Africa than malnutrition or dirty water, study warns," The Guardian, October 20, 2016, https://www.theguardian.com/global-development/2016/oct/20/air-pollution-deadlier-africa-than-dirty-water-or-malnutrition-oecd

1151   Philip J Landrigan, et al., "The Lancet Commission on pollution and health," The Lancet, October 19, 2017, http://dx.doi.org/10.1016/S0140-6736(17)32345-0

1152   "Fukushima Daiichi Accident," World Nuclear Association, April, 2021, London, WC2E 7HA, United Kingdom, https://www.world-nuclear.org/information-library/safety-and-security/safety-of-plants/fukushima-daiichi-accident.aspx

1153   "Fukushima Daiichi Accident," World Nuclear Association, April, 2021, London, WC2E 7HA, United Kingdom, https://www.world-nuclear.org/information-library/safety-and-security/safety-of-plants/fukushima-daiichi-accident.aspx

1154   "Fukushima Daiichi Accident," World Nuclear Association, April, 202, London, WC2E 7HA, United Kingdom, https://www.world-nuclear.org/information-library/safety-and-security/safety-of-plants/fukushima-daiichi-accident.aspx

1155   "Fukushima Daiichi Accident," World Nuclear Association, April, 2021, London, WC2E 7HA, United Kingdom, https://www.world-nuclear.org/information-library/safety-and-security/safety-of-plants/fukushima-daiichi-accident.aspx

1156   Kenneth Pletcher and John P.Rafferty, "Japan earthquake and tsunami of 2011," Encyclopedia Britannica, March 4, 2021, https://www.britannica.com/event/Japan-earthquake-and-tsunami-of-2011

1157   "Fukushima Nuclear Power Station, Japan," Power Technology, 2021, https://www.power-technology.com/projects/fukushima-daiichi

1158   "Australia is smashing solar records!" Climate Council, February 16, 2018, https://www.climatecouncil.org.au/australia-is-smashing-solar-records

1159   "Nuclear Energy Factsheet," University of Michigan, Center for Sustainable Systems, 2020, Pub. No. CSS11-15, http://css.umich.edu/sites/default/files/Nuclear%20Energy_CSS11-15_e2020.pdf

1160   "Backgrounder on Radioactive Waste," United States Nuclear Regulatory Commission (U.S.NRC), Washington, DC., https://www.nrc.gov/reading-rm/doc-collections/fact-sheets/radwaste.html

1161   "Backgrounder on Radioactive Waste," United States Nuclear Regulatory Commission (U.S.NRC), Washington, DC., https://www.nrc.gov/reading-rm/doc-collections/fact-sheets/radwaste.html

1162   "Radioactive Decay," United States Environmental Protection Agency (EPA), 2019, Washington, DC., https://www.epa.gov/radiation/radioactive-decay

1163   Kenneth Pletcher and John P.Rafferty, "Japan earthquake and tsunami of 2011," Encyclopedia Britannica, March 4, 2021, https://www.britannica.com/event/Japan-earthquake-and-tsunami-of-2011

1164   "The Fukushima Disaster," World Information Service on Energy (WISE), 2019, Washington, Netherlands, https://wiseinternational.org/campaign/fukushima-disaster

1165   "The Fukushima Disaster," World Information Service on Energy (WISE),

2019, Washington, Netherlands, https://wiseinternational.org/campaign/fukushima-disaster

1166 "The Fukushima Disaster," World Information Service on Energy (WISE), 2019, Washington, Netherlands, https://wiseinternational.org/campaign/fukushima-disaster

1167 "Fukushima Radiation in U.S West Coast Tuna," National Oceanic and Atmospheric Association (NOAA), Silver Spring, MD., https://www.fisheries.noaa.gov/west-coast/science-data/fukushima-radiation-us-west-coast-tuna

1168 "New type of fallout from Fukushima Daiichi found a decade after nuclear disaster," PhysicsWorld, March 15, 2021, https://physicsworld.com/a/new-type-of-fallout-from-fukushima-daiichi-found-a-decade-after-nuclear-disaster

1169 "New type of fallout from Fukushima Daiichi found a decade after nuclear disaster," PhysicsWorld, March 15, 2021, https://physicsworld.com/a/new-type-of-fallout-from-fukushima-daiichi-found-a-decade-after-nuclear-disaster

1170 "Fukushima Daiichi 2011-2121. The decontamination myth and a decade of human rights violations," Greenpeace, March 2021, https://www.greenpeace.org/static/planet4-japan-stateless/2021/03/ff71ab0b-finalfukushima2011-2020_web.pdf

1171 "Fukushima Daiichi 2011-2121. The decontamination myth and a decade of human rights violations," Greenpeace, March 2021, https://www.greenpeace.org/static/planet4-japan-stateless/2021/03/ff71ab0b-finalfukushima2011-2020_web.pdf

1172 "Water leaks indicate new damage at fukushima nuclear plant," U.S. News & World Report, February 19, 2021, https://www.usnews.com/news/world/articles/2021-02-19/water-leaks-indicate-new-damage-at-fukushima-nuclear-plant

1173 "Water leaks indicate new damage at fukushima nuclear plant," U.S. News & World Report, February 19, 2021, https://www.usnews.com/news/world/articles/2021-02-19/water-leaks-indicate-new-damage-at-fukushima-nuclear-plant

1174 "Water leaks indicate new damage at fukushima nuclear plant," U.S. News & World Report, February 19, 2021, https://www.usnews.com/news/world/articles/2021-02-19/water-leaks-indicate-new-damage-at-fukushima-nuclear-plant

1175 Dennis Normile, "Japan plans to release Fukushima's wastewater into the ocean," American Association for the Advancement of Science, February 19, 2021, https://www.sciencemag.org/news/2021/04/japan-plans-release-fukushima-s-contaminated-water-ocean

1176 Dennis Normile, "Japan plans to release Fukushima's wastewater into the ocean," American Association for the Advancement of Science, February 19, 2021, https://www.sciencemag.org/news/2021/04/japan-plans-release-fukushima-s-contaminated-water-ocean

1176 Dennis Normile, "Japan plans to release Fukushima's wastewater into the ocean," American Association for the Advancement of Science, February 19, 2021, https://www.sciencemag.org/news/2021/04/japan-plans-release-fukushima-s-contaminated-water-ocean

1177 Justin McCurry, "Fukushima: Japan announces it will dump contaminated water into sea," The Guardian, April 13, 2021, https://www.theguardian.com/environment/2021/apr/13/fukushima-japan-to-start-dumping-contaminated-water-pacific-ocean

1178 Aristyo Rizka Darmawan, "Toxic reaction to Japan's Fukushima water dump," Lowy Institute, April 22, 2021, https://www.lowyinstitute.org/the-interpreter/toxic-reaction-japan-s-fukushima-water-dump

1179 "Nuclear explained, Nuclear power and the environment," U.S. Energy Information Administration, 2020, https://www.eia.gov/energyexplained/nuclear/nuclear-power-and-the-environment.php

1180   "What is nuclear waste, and what do we do with it?" World Nuclear Association, https://world-nuclear.org/nuclear-essentials/what-is-nuclear-waste-and-what-do-we-do-with-it.aspx

1181   "Nuclear explained, Nuclear power and the environment," U.S. Energy Information Administration, 2020, https://www.eia.gov/energyexplained/nuclear/nuclear-power-and-the-environment.php

1182   Padmaparna Ghosh, "Nuclear Power 101," National Resources Defence Council, May 14, 2020, https://www.nrdc.org/stories/nuclear-power-101

1183   Padmaparna Ghosh, "Nuclear Power 101," National Resources Defence Council, May 14, 2020, https://www.nrdc.org/stories/nuclear-power-101

1184   Vicky Brown Varela, "What's Next for Fracking Under Biden?" Council on Foreign Relations, December 18, 2020, https://www.cfr.org/in-brief/whats-next-fracking-under-biden

1185   "What is fracking and why is it controversial?" BBC News, October 15, 2018, https://www.bbc.com/news/uk-14432401

1186   "What is fracking and why is it controversial?" BBC News, October 15, 2018, https://www.bbc.com/news/uk-14432401

1187   John Manfreda, "The origin of fracking actually dates back to the Civil War," OilPrice.com, April 14, 2015, https://www.businessinsider.com/the-history-of-fracking-2015-4

1188   U.S. EPA. Hydraulic Fracturing for Oil and Gas: Impacts from the Hydraulic Fracturing Water Cycle on Drinking Water Resources in the United States (Final Report), U.S. Environmental Protection Agency, EPA/600/R-16/236F, https://cfpub.epa.gov/ncea/hfstudy/recordisplay.cfm?deid=332990

1189   Melissa Denchak, Fracking 101. Natural Resources Defence Council, April 19, 2019, https://www.nrdc.org/stories/fracking-101

1190   Melissa Denchak, Fracking 101. Natural Resources Defence Council, April 19, 2019, https://www.nrdc.org/stories/fracking-101

1191   Hydraulic Fracturing for Oil and Gas: Impacts from the Hydraulic Fracturing Water Cycle on Drinking Water Resources in the United States: Washington DC, United States Environmental Protection Agency (EPA), December 2016, EPA-600-R-16-236ES, https://www.epa.gov/sites/production/files/2016-12/documents/hfdwa_executive_summary.pdf

1192   "What is Bauxite?" Geoscience News and Information, 2021, https://geology.com/minerals/bauxite.shtml

1193   "Hydraulic Fracturing Proppants," Wikipedia, 2020, https://en.wikipedia.org/wiki/Hydraulic_fracturing_proppants

1194   Hydraulic Fracturing for Oil and Gas: Impacts from the Hydraulic Fracturing Water Cycle on Drinking Water Resources in the United States: Washington DC, United States Environmental Protection Agency (EPA), December 2016, EPA-600-R-16-236ES, https://www.epa.gov/sites/production/files/2016-12/documents/hfdwa_executive_summary.pdf

1195   Hydraulic Fracturing for Oil and Gas: Impacts from the Hydraulic Fracturing Water Cycle on Drinking Water Resources in the United States: Washington DC, United States Environmental Protection Agency (EPA), December 2016, EPA-600-R-16-236ES, https://www.epa.gov/sites/production/files/2016-12/documents/hfdwa_executive_summary.pdf

1196   Deirdre Lockwood, "Toxic chemicals from fracking wastewater spills can persist for years," Chemical & Engineering News, https://cen.acs.org/articles/94/

web/2016/05/Toxic-chemicals-fracking-wastewater-spills.html
1197   Ahsen Soomrr, "9 Most Dangerous Chemicals Used in Fracking," October 22, 2019, https://www.environmentbuddy.com/environment/most-dangerous-chemicals-used-in-fracking
1198   Hydraulic Fracturing for Oil and Gas: Impacts from the Hydraulic Fracturing Water Cycle on Drinking Water Resources in the United States: Washington DC, United States Environmental Protection Agency (EPA), December 2016, EPA-600-R-16-236ES, https://www.epa.gov/sites/production/files/2016-12/documents/hfdwa_executive_summary.pdf
1199   Hydraulic Fracturing for Oil and Gas: Impacts from the Hydraulic Fracturing Water Cycle on Drinking Water Resources in the United States: Washington DC, United States Environmental Protection Agency (EPA), December 2016, EPA-600-R-16-236ES, https://www.epa.gov/sites/production/files/2016-12/documents/hfdwa_executive_summary.pdf
1200   "Chemicals used in hydraulic fracturing," Government of Western Australia Department of Mines, Industry Regulation and Safety, https://www.dmp.wa.gov.au/Petroleum/Chemicals-used-in-hydraulic-25615.aspx
1201   Hydraulic Fracturing for Oil and Gas: Impacts from the Hydraulic Fracturing Water Cycle on Drinking Water Resources in the United States: Washington DC, United States Environmental Protection Agency (EPA), December 2016, EPA-600-R-16-236ES, https://www.epa.gov/sites/production/files/2016-12/documents/hfdwa_executive_summary.pdf
1202   Hydraulic Fracturing for Oil and Gas: Impacts from the Hydraulic Fracturing Water Cycle on Drinking Water Resources in the United States: Washington DC, United States Environmental Protection Agency (EPA), December 2016, EPA-600-R-16-236ES, https://www.epa.gov/sites/production/files/2016-12/documents/hfdwa_executive_summary.pdf
1203   Hydraulic Fracturing for Oil and Gas: Impacts from the Hydraulic Fracturing Water Cycle on Drinking Water Resources in the United States: Washington DC, United States Environmental Protection Agency (EPA), December 2016, EPA-600-R-16-236ES, https://www.epa.gov/sites/production/files/2016-12/documents/hfdwa_executive_summary.pdf
1204   Melissa Denchak, Fracking 101. Natural Resources Defence Council, April 19, 2019, https://www.nrdc.org/stories/fracking-101
1205   M. Weingarten, et al., "High injection rates of wastewater into deep wells increase the risk of earthquakes in regions prone to induced seismicity," Science, June 2015, pp. 1336-1340, DOI: 10.1126/science.aab1345, https://science.sciencemag.org/content/348/6241/1336.full
1206   M. Weingarten, et al., "High injection rates of wastewater into deep wells increase the risk of earthquakes in regions prone to induced seismicity," Science, June 2015, pp. 1336-1340, DOI: 10.1126/science.aab1345, https://science.sciencemag.org/content/348/6241/1336.full
1207   U.S. Department of Energy. Economic and National Security Impacts under a Hydraulic Fracturing Ban. January 2021. https://www.energy.gov/sites/prod/files/2021/01/f82/economic-and-national-security-impacts-under-a-hydraulic-fracturing-ban.pdf
1208   Katie Rogers, "The White House warns that the U.S. lags behind china on developing clean technologies," April 23, 2021, The New York Times, https://www.nytimes.com/2021/04/23/us/white-house-clean-technology-china.html
1209   Katie Rogers, "The White House warns that the U.S. lags behind china on

developing clean technologies," April 23, 2021, The New York Times, https://www.nytimes.com/2021/04/23/us/white-house-clean-technology-china.html

1210   "The neglected menace of pollution across world," Gulf Times, March 24, 2018, https://www.gulf-times.com/story/586275/The-neglected-menace-of-pollution-across-world

1211   Sam Meredith, "Pollution linked to one in six deaths worldwide — and threatens 'survival of human societies'," CNBC, October 20, 2017, https://www.cnbc.com/2017/10/20/pollution-linked-to-one-in-six-deaths-worldwide-and-threatens-survival-of-human-societies.html

1212   Damian Carrington, "Global pollution kills 9m a year and threatens 'surival of human societies'," The Guardian, October 20, 2016,

1213   Damian Carrington, "Global pollution kills 9m a year and threatens 'survival of human societies'," The Guardian, October 20, 2016,

1214   Pope Francis, "Pope Francis: The Earth, our home, is beginning to look like an immense pile of filth," The Guardian, June 18, 2015, https://www.theguardian.com/commentisfree/2015/jun/18/pope-francis-encyclical-extract

1215   Ruben Castaneda, "7 Alternatives to Toxic Cleaning U.S. News and World Report, April 26 2018, https://health.usnews.com/wellness/slideshows/7-alternatives-to-toxic-cleaning-products

1216   "Recognizing a Dehydration Headache," Healthline, https://www.healthline.com/health/dehydration-headache

1217   Collene Laswhorn, "Hypoglycemia and Headaches," Very well health, January 20, 2020, https://www.verywellhealth.com/hypoglycemia-and-headaches-1719537

1218   "4 ways to tame tension headaches," Harvard Medical School, April 18, 2020, https://www.healthline.com/health/dehydration-headache

1219   "What to know about caffeine withdrawal headaches," Medical News Today, November 9, 2019, https://www.medicalnewstoday.com/articles/326950

1220   Doyle Rice, "200 years ago, we endured a 'year without a summer'," USA Today, May 26, 2016, https://www.usatoday.com/story/weather/2016/05/26/year-without-a-summer-1816-mount-tambora/84855694

1221   Keith Veronese, "The Year Without a Summer, and How It Spawned Frankenstein," Gizmod, February 17, 2012, https://io9.gizmodo.com/the-year-without-a-summer-and-how-it-spawned-frankenst-5885668

1222   Felix J. Koch, "A Year Without A Summer," The Farm Journal, October 1909, vol. XXXIII, no. 10, p. 432.

1223   Gillen D'Arcy Wood, "The Volcano That Shrouded the Earth and Gave Birth to a Monster," Nautilus, October 5, 2017, http://nautil.us/issue/53/monsters/the-volcano-that-shrouded-the-earth-and-gave-birth-to-a-monster-rp

1224   The Historical Record A Monthly Publication Devoted Principally to the Early History of Wyoming Valley, March 1887, vol. 1, no. 7, p. 107.

1225   Clive Oppenheimer, "Climatic, environmental and human consequences of the largest known historic eruption: Tambora volcano (Indonesia) 1815," Progress in Physical Geography, 2003, pp. 230-259, DOI: 10.1191/0309133303pp379ra

1226   Gillen D'Arcy Wood, "The Volcano That Shrouded the Earth and Gave Birth to a Monster," Nautilus, October 5, 2017, http://nautil.us/issue/53/monsters/the-volcano-that-shrouded-the-earth-and-gave-birth-to-a-monster-rp

1227   Clive Oppenheimer, "Climatic, environmental and human consequences of the largest known historic eruption: Tambora volcano (Indonesia) 1815," Progress in Physical Geography, 2003, pp. 230-259, DOI: 10.1191/0309133303pp379ra

1228   Emma Johnston, "Up From The Ashes," Popular Archaeology, May 31, 2012.

1229   Gillen D'Arcy Wood, "The Volcano That Shrouded the Earth and Gave Birth to a Monster," Nautilus, October 5, 2017, http://nautil.us/issue/53/monsters/the-volcano-that-shrouded-the-earth-and-gave-birth-to-a-monster-rp
1230   Jane J. Lee, "Volcanic Eruption That Changed World Marks 200th Anniversary," National Geographic, April 10, 2015, https://news.nationalgeographic.com/2015/04/150410-tambora-volcano-eruption-climate-change-famine-earth-science
1231   Keith Veronese, "The Year Without a Summer, and How It Spawned Frankenstein," Gizmod, February 17, 2012, https://io9.gizmodo.com/the-year-without-a-summer-and-how-it-spawned-frankenst-5885668
1232   Emma Johnston, "Up From The Ashes," Popular Archaeology, May 31, 2012.
1233   Doyle Rice, "200 years ago, we endured a 'year without a summer'," USA Today, May 26, 2016, https://www.usatoday.com/story/weather/2016/05/26/year-without-a-summer-1816-mount-tambora/84855694/
1234   Jane J. Lee, "Volcanic Eruption That Changed World Marks 200th Anniversary," National Geographic, April 10, 2015, https://news.nationalgeographic.com/2015/04/150410-tambora-volcano-eruption-climate-change-famine-earth-science
1235   Jack Williams, "The epic volcano eruption that led to the 'Year Without a Summer'," The Washington Post, June 10, 2016, https://www.washingtonpost.com/news/capital-weather-gang/wp/2015/04/24/the-epic-volcano-eruption-that-led-to-the-year-without-a-summer
1236   William K. Klingaman and Nicholas P. Klingaman, "Tambora Erupts in 1815 and Changes World History [Excerpt]," Scientific American, March 1, 2013.
1237   Achmad Djumarma Wirakusumah and Heryadi Rachmat, "Impact of the 1815 Tambora Eruption to global climate change," IOP Conference Series: Earth and Environmental Science, 2017, doi:10.1088/1755-1315/71/1/012007
1238   Clive Oppenheimer, "Climatic, environmental and human consequences of the largest known historic eruption: Tambora volcano (Indonesia) 1815," Progress in Physical Geography, 2003, pp. 230-259, DOI: 10.1191/0309133303pp379ra
1239   Gillen D'Arcy Wood, "The Volcano That Shrouded the Earth and Gave Birth to a Monster," Nautilus, October 5, 2017, http://nautil.us/issue/53/monsters/the-volcano-that-shrouded-the-earth-and-gave-birth-to-a-monster-rp
1240   Christoph C. Raible, et al., "Tambora 1815 as a test case for high impact volcanic eruptions: Earth system effects," WIREs Climate Change, July/August 2016, vol. 7, pp. 569-589, doi: 10.1002/wcc.407
1241   Jane J. Lee, "Volcanic Eruption That Changed World Marks 200th Anniversary," National Geographic, April 10, 2015, https://news.nationalgeographic.com/2015/04/150410-tambora-volcano-eruption-climate-change-famine-earth-science
1242   Jack Williams, "The epic volcano eruption that led to the 'Year Without a Summer'," The Washington Post, June 10, 2016, https://www.washingtonpost.com/news/capital-weather-gang/wp/2015/04/24/the-epic-volcano-eruption-that-led-to-the-year-without-a-summer
1243   "Atmospheric Aerosols: What Are They, and Why Are They So Important?" NASA, August 7, 2017, https://www.nasa.gov/centers/langley/news/factsheets/Aerosols.html (accessed September 20, 2008).
1244   "Atmospheric Aerosols: What Are They, and Why Are They So Important?" NASA, August 7, 2017, https://www.nasa.gov/centers/langley/news/factsheets/Aerosols.html (accessed September 20, 2008).

1245   "Volcano World: Measuring Volcanic Gases," Oregon State University, http://volcano.oregonstate.edu/measuring-volcanic-gases

1246   William K. Klingaman and Nicholas P. Klingaman, "Tambora Erupts in 1815 and Changes World History [Excerpt]," Scientific American, March 1, 2013.

1247   Jack Williams, "The epic volcano eruption that led to the 'Year Without a Summer'," The Washington Post, June 10 2016, https://www.washingtonpost.com/news/capital-weather-gang/wp/2015/04/24/the-epic-volcano-eruption-that-led-to-the-year-without-a-summer

1248   Christoph C. Raible, et al., "Tambora 1815 as a test case for high impact volcanic eruptions: Earth system effects," WIREs Climate Change, July/August 2016, vol. 7, pp. 569-589, doi: 10.1002/wcc.407

1249   Empire State Building Fact Sheet, http://www.esbnyc.com/sites/default/files/esb_fact_sheet_4_9_14_4.pdf

1250   Jack Williams, "The epic volcano eruption that led to the 'Year Without a Summer'," The Washington Post, June 10, 2016, https://www.washingtonpost.com/news/capital-weather-gang/wp/2015/04/24/the-epic-volcano-eruption-that-led-to-the-year-without-a-summer

1251   G. P. Brasseur and E. Roeckner, "Impact of improved air quality on the future evolution of climate" Geophysical Research Letters, August, 2005, vol. 32, doi:10.1029/2005GL023902

1252   Hannah Ritchie and Max Roser, "Air Pollution," 2018, Published online at OurWorldInData.org. https://ourworldindata.org/air-pollution

1253   "Volcano World: Climate Cooling," Oregon State University, http://volcano.ore3onstate.edu/climate-cooling

1254   "Atmospheric Aerosols: What Are They, and Why Are They So Important?" NASA, August 7, 2017, https://www.nasa.gov/centers/langley/news/factsheets/Aerosols.html (accessed September 20, 2008).

1255   "Volcano World: Man Versus the Volcano," Oregon State University, http://volcano.oregonstate.edu/man-versus-volcanos

1256   Alister Doyle, "Sun-dimming volcanoes partly explain global warming hiatus-study," Reuters, February 23, 2014, https://www.reuters.com/article/us-climate-volcanoes/sun-dimming-volcanoes-partly-explain-global-warming-hiatus-study-idUSBREA1M0W920140223

1257   "Scientists part the clouds on how droplets form," Science News, March 24, 2016, https://www.sciencedaily.com/releases/2016/03/160324145422.htm

1258   Daniel Cressey, "Brown clouds boost global warming," Nature, August 1, 2007, https://www.nature.com/news/2007/070730/full/news070730-6.html

1259   Amarnath Amarasingam, "What Is Global Dimming?" Huffington Post, May 25, 2011, https://www.huffingtonpost.com/amarnath-amarasingam/what-is-global-dimming_b_740996.html

1260   Amarnath Amarasingam, "What Is Global Dimming?" Huffington Post, May 25, 2011, https://www.huffingtonpost.com/amarnath-amarasingam/what-is-global-dimming_b_740996.html

1261   Martin Wild, Atsumu Ohmura and Knut Makowski, "Impact of global dimming and brightening on global warming" Geophysical Research Letters, February 2007, vol. 34, doi:10.1029/2006GL028031

1262   T. Storelvmo, et al., "Disentangling greenhouse warming and aerosol cooling to reveal Earth's climate sensitivity," Nature Geoscience, March 14, 2016, pp. 286–289, http://doi.org/10.1038/NGEO2670

1263   "Dimming the Sun," Nova, April 18, 2006, http://www.pbs.org/wgbh/nova/

transcripts/3310_sun.html
1264   "Dimming the Sun," Nova, April 18, 2006, http://www.pbs.org/wgbh/nova/transcripts/3310_sun.html
1265   T. Storelvmo, et al., "Disentangling greenhouse warming and aerosol cooling to reveal Earth's climate sensitivity," Nature Geoscience, March 14, 2016, pp. 286–289, http://doi.org/10.1038/NGEO2670
1266   Martin Wild, Atsumu Ohmura and Knut Makowski, "Impact of global dimming and brightening on global warming" Geophysical Research Letters, February 2007, vol. 34, doi:10.1029/2006GL028031
1267   Martin Wild, "Enlightening global dimming and brightening," American Meteorological Society, January 2010, pp. 27-37.
1268   "How Can We Stop Global Dimming?" Forbes, May 18 2017, https://www.forbes.com/sites/quora/2017/05/18/how-can-we-stop-global-dimming
1269   John Abraham, "Pollution is slowing the melting of Arctic sea ice, for now," The Guardian, August 3, 2018, https://www.theguardian.com/environment/climate-consensus-97-per-cent/2018/aug/03/pollution-is-slowing-the-melting-of-arctic-sea-ice-for-now
1270   "How Can We Stop Global Dimming?" Forbes, May 18, 2017, https://www.forbes.com/sites/quora/2017/05/18/how-can-we-stop-global-dimming
1271   John Abraham, "Pollution is slowing the melting of Arctic sea ice, for now," The Guardian, August 3, 2018, https://www.theguardian.com/environment/climate-consensus-97-per-cent/2018/aug/03/pollution-is-slowing-the-melting-of-arctic-sea-ice-for-now
1272   "Global warming levels masked by aerosols: study," Cosmos, March 15, 2018, https://cosmosmagazine.com/climate/global-warming-levels-masked-aerosols-study
1273   "Dimming the Sun," Nova, April 18, 2006, http://www.pbs.org/wgbh/nova/transcripts/3310_sun.html
1274   G. P. Brasseur and E. Roeckner, "Impact of improved air quality on the future evolution of climate" Geophysical Research Letters, August, 2005, vol. 32, doi:10.1029/2005GL023902
1275   James Rainey, "A last-ditch global warming fix? A man-made 'volcanic' eruption," NBC News, October 11, 2018, https://www.nbcnews.com/news/us-news/last-ditch-global-warming-fix-man-made-volcanic-eruption-n918826
1276   James Rainey, "A last-ditch global warming fix? A man-made 'volcanic' eruption," NBC News, October 11, 2018, https://www.nbcnews.com/news/us-news/last-ditch-global-warming-fix-man-made-volcanic-eruption-n918826
1277   Alister Doyle, "Sun-dimming volcanoes partly explain global warming hiatus-study," Reuters, February 23, 2014, https://www.reuters.com/article/us-climate-volcanoes/sun-dimming-volcanoes-partly-explain-global-warming-hiatus-study-idUSBREA1M0W920140223
1278   "Dimming the Sun to cool Earth could ravage wildlife: Study," The Straits Times, January 23, 2018, https://www.straitstimes.com/world/dimming-the-sun-to-cool-earth-could-ravage-wildlife-study
1279   "2019 was 2nd hottest year on record for Earth say NOAA, NASA, NOAA finds ocean heat content was the highest in recorded history," National Oceanic and Atmospheric Administration (NOAA), January 15, 2020, https://www.noaa.gov/news/2019-was-2nd-hottest-year-on-record-for-earth-say-noaa-nasa
1280   (Professor Peter Cox, personal communication, October 4, 2018)
1281   " Sumbawa Tambora, Indonesia," Oregon State University, http://volcano.oregonstate.edu/vwdocs/volc_images/southeast_asia/indonesia/tambora.html

1282   Kevin E. Trenberth and Lesley Smith, "The Mass of the Atmosphere: A Constraint on Global Analyses," National Center for Atmospheric Research, June 29 2004, pp. 864-875.
1283   G. P. Brasseur and E. Roeckner, "Impact of improved air quality on the future evolution of climate" Geophysical Research Letters, August, 2005, vol. 32, doi:10.1029/2005GL023902
1284   Pam Wright, "First Sun-Dimming Experiment Using Ingredient in Antacid Planned for 2019," The Weather Channel, December 5, 2018, https://weather.com/science/environment/news/2018-12-05-sun-dimming-experiment-planned-2019-harvard-scientists
1285   QER Quadrennial Energy Review) Report: Energy Transmission, Storage, and Distribution Infrastructure, April 2015, p. 3-4.
1286   Seth Blumsack, "How complexity science can help keep the lights on," Christian Science Monitor, March 2, 2017, http://www.csmonitor.com/Science/Complexity/2017/0302/How-complexity-science-can-help-keep-the-lights-on
1287   "Electricity storage can smooth out moment-to-moment variations in electricity demand," U.S. Energy Information Administration, May 22, 2012, https://www.eia.gov/todayinenergy/detail.php?id=6370
1288   "Aging And Unstable, The Nation's Electrical Grid Is The Weakest Link" NPR, August 22, 2016, http://www.npr.org/templates/transcript/transcript.php?storyId=490932307
1289   Seth Blumsack, "How complexity science can help keep the lights on," Christian Science Monitor, March 2, 2017, http://www.csmonitor.com/Science/Complexity/2017/0302/How-complexity-science-can-help-keep-the-lights-on
1290   Steve Reilly, and Ryan Sabalow, "Power grid security fears surge since 2003 blackout," USA Today, March 24, 2015, https://www.usatoday.com/story/news/2015/03/24/power-grid-security-solutions-and-ideas-arose-after-2003-blackout/24892721
1291   "The Sun," Northern Illinois Center for Accelerator and Detector Development (NICADD), http://nicadd.niu.edu/~macc/162/class_5b.pdf
1292   "NASA: Understanding the Magnetic Sun", National Aeronautics and Space Administration (NASA), January 29, 2016, https://www.nasa.gov/feature/goddard/2016/understanding-the-magnetic-sun
1293   Magnetic Field Lines Tangle as Sun Rotates, The University Corporation for Atmospheric Research (UCAR) Center for Science Education, https://scied.ucar.edu/magnetic-field-lines-tangle-sun-rotates
1294   "Solar Flares: What Does It Take to Be X-Class?" NASA, August 9, 2011, https://www.nasa.gov/mission_pages/sunearth/news/X-class-flares.html
1295   National Oceanic and Atmospheric Administration (NOAA) Space Weather Scales, http://www.swpc.noaa.gov/noaa-scales-explanation
1296   Coronal Mass Ejection Prediction Page, Montana State University, http://solar.physics.montana.edu/press/faq.html
1297   K.M. Omatola, and I.C. Okeme, "Impacts of solar storms on energy and communications technologies," Archives of Applied Science Research, 2012, vol. 4, issue 4, pp. 1825-1832.
1298   Anthony L. Peratt, Physics of the Plasma, Second Edition, 2015, Springer, p. 355.
1299   Ellen Clarke, et al., "An estimation of the Carrington flare magnitude from solar flare effects (sfe) in the geomagnetic records," RAS National Astronomy Meeting, Glasgow, Scotland, April 2010

1300   Juan José Curto, Josep Castell, and Ferran Del Moral, "Sfe: waiting for the big one," Journal of Space Weather and Space Climate, 2016, vol. 6, DOI: 10.1051/swsc/2016018
1301   Lewis H. Latimer, MIT (Massachusetts Institute of Technology), http://lemelson.mit.edu/resources/lewis-h-latimer
1302   "Magnetic Storms and 'Earth-Currents,'" Proceedings of the British Meteorological Society, Vol I. 1861, November 20, to 1863, June 17, p. 76.
1303   K.M. Omatola, and I.C. Okeme, "Impacts of solar storms on energy and communications technologies," Archives of Applied Science Research, 2012, vol. 4, issue 4, p. 1829.
1304   "Solar Cycle Primer", National Aeronautics and Space Administration (NASA), https://www.nasa.gov/mission_pages/sunearth/news/solarcycle-primer.html
1305   John Kappenman, "Geomagnetic Storms and Their Impacts on the U.S. Power Grid," Metatech Corporation, January 2010
1306   Doris Elin, "Monster Solar Flare Marks 7th Powerful Sun Storm in 7 Days," space.com, September 11, 2017, https://www.space.com/38115-sun-monster-solar-flares-seven-days.html
1307   K.M. Omatola, and I.C. Okeme, "Impacts of solar storms on energy and communications technologies," Archives of Applied Science Research, 2012, vol. 4, issue 4, p. 1829.
1308   John Kappenman, "Geomagnetic Storms and Their Impacts on the U.S. Power Grid," Metatech Corporation, January 2010
1309   "Sunspot credited with rail tie-up," New York Times, May 16, 1921, p. 2.
1310   Greg White, "A solar event that already happened in 1921 will kill 280 million Americans when it happens again… NASA says it's inevitable," Newstarget, February 29, 2016, http://www.newstarget.com/2016-02-29-a-solar-event-that-already-happened-in-1921-will-kill-280-million-americans-when-it-happens-again-nasa-says-its-inevitable.html
1311   "Sunspot credited with rail tie-up," New York Times, May 16, 1921, p. 2.
1312   Securing the U.S. Electrical Grid, The Center for the Study of the Presidency and Congress, 2014, p. 70.
1313   John Kappenman, "Geomagnetic Storms and Their Impacts on the U.S. Power Grid," Metatech Corporation, January 2010
1314   Securing the U.S. Electrical Grid, The Center for the Study of the Presidency and Congress, 2014, p. 50.
1315   Jay Lindsay, "25% of New England power generators are more than 40 years old," Bangor Daily News, October 6, 2011, http://bangordailynews.com/2011/10/06/business/25-of-new-england-power-generators-are-more-than-40-years-old
1316   Joshua D. Rhodes, "The outdated US electric grid is going to cost $5 trillion to replace," Business Insider, March 16, 2017, http://www.businessinsider.com/replacing-us-electrical-grid-cost-2017-3
1317   2017 Infrastructure Report Card, American Society of Civil Engineers, 2017.
1318   Securing the U.S. Electrical Grid, The Center for the Study of the Presidency and Congress, 2014
1319   John Kappenman, "Geomagnetic Storms and Their Impacts on the U.S. Power Grid," Metatech Corporation, January 2010
1320   Brid-Aine Parnell, "Massive Solar Superstorm Narrowly Missed Blasting The Earth Back Into The Dark Ages," Forbes, March 19, 2014, https://www.forbes.com/sites/bridaineparnell/2014/03/19/massive-solar-superstorm-narrowly-missed-blasting-the-earth-back-into-the-dark-ages

1321   Severe space weather events – understanding societal and economic impacts, 2008, National Research Council, The National Academies Press, p. 77.
1322   David Graham, "When the Lights Go Out," The Atlantic, September 9, 2015, https://www.theatlantic.com/technology/archive/2015/09/how-safe-is-the-us-electrical-grid-really/402640
1323   Elizabeth Harrington, "Hearing: Electric Grid Vulnerable to EMP Witness: Could kill 9 in 10 Americans," Washington Free Beacon, May 8 ,2014, http://freebeacon.com/national-security/hearing-electric-grid-vulnerable-to-emp
1324   Elizabeth Harrington, "Hearing: Electric Grid Vulnerable to EMP Witness: Could kill 9 in 10 Americans," Washington Free Beacon, May 8, 2014, http://freebeacon.com/national-security/hearing-electric-grid-vulnerable-to-emp
1325   2017 Infrastructure Report Card, American Society of Civil Engineers, 2017.
1326   Kathryn Schulz, "The Really Big One - An earthquake will destroy a sizable portion of the coastal Northwest. The question is when," The New Yorker, July 20, 2015.
1327   Jessica Leader, "Will The Next Natural Disaster Doom The Nation's Energy Grid?," Huffington Post, July 14, 2016, http://www.huffingtonpost.com/jessica-leader/will-the-next-natural-disaster_b_10972780.html
1328   Brid-Aine Parnell, "Massive Solar Superstorm Narrowly Missed Blasting The Earth Back Into The Dark Ages," Forbes, March 19, 2014, https://www.forbes.com/sites/bridaineparnell/2014/03/19/massive-solar-superstorm-narrowly-missed-blasting-the-earth-back-into-the-dark-ages
1329   "Near Miss: The Solar Superstorm of July 2012," National Aeronautics and Space Administration (NASA), July 23, 2014, https://science.nasa.gov/science-news/science-at-nasa/2014/23jul_superstorm
1330   Juan José Curto, Josep Castell, and Ferran Del Moral, "Sfe: waiting for the big one," Journal of Space Weather and Space Climate, 2016, vol. 6, DOI: 10.1051/swsc/2016018
1331   "Near Miss: The Solar Superstorm of July 2012," National Aeronautics and Space Administration (NASA), July 23, 2014, https://science.nasa.gov/science-news/science-at-nasa/2014/23jul_superstorm
1332   "Solar Power & Other Renewable Energy For Your Home," https://www.thisoldhouse.com/solar-alternative-energy/21072584/all-about-home-solar-alternative-energy
1333   "Seven ways to power your home with renewable energy," https://www.eonenergy.com/spark/ways-to-power-your-home-with-renewable-energy.html
1334   "Planning for Home Renewable Energy Systems," https://www.energy.gov/energysaver/buying-and-making-electricity/planning-home-renewable-energy-systems
1335   Zeena Mackerdien PhD, "Obesity Rates Continue to Rise Among Americans," https://www.medpagetoday.org/primarycare/obesity/90456
1336   "Prevalence of Childhood Obesity in the United States," https://www.cdc.gov/obesity/data/childhood.html
1337   "The Health Effects of Overweight & Obesity," https://www.cdc.gov/healthyweight/effects/index.html
1338   https://www.cdc.gov/nchs/fastats/deaths.htm
1339   Robert Preidt, "Americans Taking More Prescription Drugs Than Ever," https://www.webmd.com/drug-medication/news/20170803/americans-taking-more-prescription-drugs-than-ever-survey
1340   "Death By Prescription," U.S. News and World Report, 2016, https://health.usnews.com/health-news/patient-advice/articles/2016-09-27/the-danger-in-taking-prescribed-medications

# Images

Plastic Oceans

Floating garbage off the shore of Manila Bay – Dunham Will, "Clogged by plastic - World seas polluted by tons of waste," Times of Malta, February 14, 2015, http://www.timesofmalta.com/articles/view/20150214/environment/clogged-by-plastic.556026

Diagram showing the main sources and movement pathways for plastics in the marine environment – Peter G. Ryan et al., "Monitoring the abundance of plastic debris in the marine environment," Philosophical Transaction of the Royal Society B, 2009, vol. 364, p. 1999, doi:10.1098/rstb.2008.0207.

Global plastic production by industry in millions of tons. – Parker Laura, "Planet or Plastic?" National Geographic, May 15, 2018, https://www.nationalgeographic.com/magazine/2018/06/plastic-planet-waste-pollution-trash-crisis
Lake Erie microplastics – Ehrenberg Rachel, "Puny plastic particles mar Lake Erie's waters," Science News, April 12, 2013, https://www.sciencenews.org/article/puny-plastic-particles-mar-lake-eries-waters

The scourge of microplastics – "UN's mission to keep plastics out of oceans and marine life," UN News Centre, April 27, 2017, http://www.un.org/apps/news/story.asp?NewsID=56638

A baby albatross chick can have an ounce of plastic in its belly and remain healthy; the dead chicks have twice as much – "Laysan Albatrosses' Plastic Problem," Smithsonian, https://ocean.si.edu/ocean-life/seabirds/laysan-albatrosses-plastic-problem

Fished Out

Collapse of Northwest Atlantic Cod – Fishing For The Future: Aquaculture and Aquaponics, http://kanat.jsc.vsc.edu/student/grzyb/main.htm
Atlantic halibut total catch in Metric Tons from the Gulf of Maine-Georges Bank region – Laurel A. Col and Christopher M. Legault, The 2008 Assessment of

Atlantic Halibut in the Gulf of Maine-Georges Bank Region, April 2009, Northeast Fisheries Science Center, 166 Water Street, Woods Hole, MA 02543

Total global catches, separated into reported landings, unreported landings and estimated discards from 1950 to 2014. Note these data exclude marine mammals, reptiles and plant material, as well as all freshwater catches. – Dirk Zeller, et al., "Global marine fisheries discards: A synthesis of reconstructed data," Fish and Fisheries, 2017, DOI: 10.1111/faf.12233

Global trends in the state of world marine fish stocks since, 1974 – 2017 – "The State of World Fisheries and Aquaculture 2020," Food and Agriculture Organization of the United Nations (FAO), www.fao.org

Getting Warmer: Sea surface temperatures in the Gulf of Maine have been rising over the last 35 years, and at nearly the fastest rate on the planet over the last 10. 2012 had the warmest readings in the 150 years humans have been collecting them. – Woodard, Colin, "Big changes are occurring in one of the fastest-warming spots on Earth," Portland Press Herald, October 25, 2015, https://www.pressherald.com/2015/10/25/climate-change-imperils-gulf-maine-people-plants-species-rely

World capture fisheries and aquaculture production – "The State of World Fisheries and Aquaculture 2012," Food and Agriculture Organization of the United Nations (FAO), www.fao.org

Projection to 2050 of ocean species collapse – https://sciencenotes.files.wordpress.com/2008/04/2006-fish-trend-small.jpg

Coral Reef Carnage

Decline in percent coral cover on Caribbean coral reefs from 1963 to 2012 – Jeremy Jackson, et al., "Tropical Americas Coral Reef Resilience Workshop," Global Coral Reef Monitoring Network (GCRMN), April-May 2012, p. 8.

Carysfort Reef, Florida Keys 1975-2014 – Caribbean Coral Reefs Through Time: 1972-2013, http://biospherefoundation.org/project/coral-reef-change

Decline in percentage of reef surface covered by live coral across the Great Barrier Reef, Australia, 1985-2012 – The International Society for Reef Studies (ISRS) Consensus Statement on Climate Change and Coral Bleaching, October 2015, www.coralreefs.org

Live coral cover declines in the South China Sea in comparison to long-term trends in the Great Barrier Reef and the Caribbean. – YU KeFu, "Coral reefs in the South China Sea: Their response to and records on past environmental changes," Science China Earth Sciences, August 2012, vol. 55, no. 8, p. 1219, doi: 10.1007/s11430-012-4449-5

Results of aerial surveys of bleaching of the Great Barrier Reef. – Mooney, Chris, "Global warming has changed the Great Barrier Reef 'forever,' scientists say," Washington Post, April 18, 2018, https://www.washingtonpost.com/news/energy-environment/wp/2018/04/18/global-warming-has-changed-the-great-barrier-reef-forever-scientists-say

Dead Zones

Gulf of Mexico drainage basin – S. S. Rabotyagov, et al., "The Economics of Dead Zones: Causes, Impacts, Policy Challenges, and a Model of the Gulf of Mexico Hypoxic Zone," Review of Environmental Economics and Policy, January 4, 2014, pp. 58-79, https://doi.org/10.1093/reep/ret024

Dead zones are on the rise globally, with 95,000 square miles affected across the world's oceans. Pictured is a map showing coastal (red) and open ocean (blue) dead zones worldwide – Pettit, Harry, "'The ocean is suffocating': Fish-killing dead zone is found growing in the Arabian Sea - and it is already bigger than SCOTLAND," Daily Mail, April 27, 2018, https://www.dailymail.co.uk/science-tech/article-5664999/Dead-zone-bigger-Scotland-suffocating-Arabian-Sea.html

Lucy Ngatia, et al., "Nitrogen and Phosphorus Eutrophication in Marine Ecosystems," IntechOpen, April 28, 2018, DOI: 10.5772/intechopen.81869 – Period of the explosive increase in coastal eutrophication in relation to global additions of nitrogen.

Illustration of the coastal upwelling process, in which winds blowing along the shore cause nutrient-poor surface waters to be replaced with nutrient-rich, cold water from deep in the ocean. – Tony Barboza, "Coastal winds intensifying with climate change, study says," Los Angeles Times, July 3, 2014, http://www.latimes.com/science/sciencenow/la-sci-sn-coastal-upwelling-winds-climate-change-20140701-story.html

A number of greenhouse gas concentration trajectories scenarios (Representative Concentration Pathways or RCPs) show global ocean oxygen concentration continuing to decrease. – Scott C. Doney, Laurent Bopp, and Matthwe C. Long, "Historical and Future Trends in Ocean Climate and Biogeochemistry," Oceanography, 2014, vol. 27, no. 1, p. 113, http://dx.doi.org/10.5670/oceanog.2014.14 The dead zone of the Bay of Bengal – Amitav Ghosh and Aaron Savio Lobo," Bay of Bengal: depleted fish stocks and huge dead zone signal tipping point," January 31, 2017, https://www.theguardian.com/environment/2017/jan/31/bay-bengal-depleted-fish-stocks-pollution-climate-change-migration

Acid Seas

Boys called carters are employed in narrow veins of coal in parts of Monmouthshire [England]; their occupation is to drag the carts or skips of coal from the working to the main roads. In this mode of labour the leather girdle passes round the body, and the chain is, between the legs, attached to the cart, and the lads drag on all-fours. – Parliamentary Papers, Great Britain. Parliament. House of Commons. Reports from Commissioners. Children's Employment (Mines). Volume 15, 1842, p. 98.

Breaker boys, Woodward Coal Mines, Kingston, Pennsylvania, ca. 1900., – History of coal mining in the United States, http://www.wikiwand.com/en/History_of_coal_mining_in_the_United_States

October 1919: A man braves the blinding fog to deliver ice around London. Thick smog regularly fell upon the city from the onset of winter in October until the beginning of spring. – Smith, Jennifer, "Foggy London Town: Eerie photographs show the capital in grip of smog during the gloomy winter months in the early 20th Century," Daily Mail, August 31, 2013, http://www.dailymail.co.uk/

news/article-2407768/Eerie-images-London-fog-Grim-mid-winter-pictures-capital-early-20th-century.html

1952 Smog along the Strand, London, which almost completely obscures the midday sun. – Gunn, Angus, Encyclopedia of Disasters - Environmental Catastrophes and Human Tragedies, 2008, p. 426.

Share of global emissions of CO2 in 2014: coal (42%), oil (33%), gas (19%), cement (6%), flaring (1%, not shown) – Global Carbon Budget, International Geosphere-Biosphere Programme, December 7, 2015

CO2 Time Series in the North Pacific from the Mauna Loa observatory in Hawaii showing that as atmospheric CO2 levels have increased, ocean pH has decreased. – "Climate Change and Ocean Impacts," Institute of Climate Studies, USA, http://www.icsusa.org/pages/articles/2016-icsusa-articles/june-2016--climate-change-and-ocean-impacts.php

How carbon dioxide changes ocean chemistry, impacting the ability of animals to build skeletons. – "Ocean acidification," National Oceanic and Atmospheric Administration, https://www.noaa.gov/education/resource-collections/ocean-coasts/ocean-acidification

Glaciers Going Going Gone

The Tibetan Plateau – Ganter, J. Carl, "Tibetan Plateau Water Reserves at Risk," Huffington Post, December 6, 2017, https://www.huffingtonpost.com/j-carl-ganter/tibetan-plateau-water-res_b_100818.html

Dramatic retreat of Ata Glacier, southeast Tibetan Plateau, from 1933 (upper) to 2006 (lower) due to global climate changes. – Yao, Tandong, et al., "Third Pole Environment (TPE)," Environmental Development, April 2012, pp. 52-64, http://dx.doi.org/10.1016/j.envdev.2012.04.002

Shepard Glacier, Glacier National Park, MT, 1913 (top) and 2005 (bottom) – Orlove, Ben, "Glacier Retreat: Reviewing the Limits of Human Adaptation to Climate Change," Environment Science and Policy for Sustainable Development, May 2009.

Flattened Forests

Trees pull water from the ground and release water vapor through their leaves, generating atmospheric rivers of moisture – Pearce, Fred, "Rivers in the Sky: How Deforestation Is Affecting Global Water Cycles," Yale Environment 360, July 24, 2018, https://e360.yale.edu/features/how-deforestation-affecting-global-water-cycles-climate-change

Deforestation through 2015 in Legal Amazonia and the Amazonia biome. The "arc of deforestation" is the heavily impacted crescent-shaped area along the eastern and southern edges of the forest – Fearnside, Phillip, "Deforestation of the Brazilian Amazon," Oxford Research Encyclopedia of Environmental Science, September 2017, DOI: 10.1093/acrefore/9780199389414.013.102

Amazon deforestation from 1970 to 2018 – https://rainforests.mongabay.com/amazon/deforestation_calculations.html

A map of Amazonia 2030, showing drought-damaged, logged and cleared forests assuming the last 10 years of climate are repeated in the future. (PPT = precipitation) – Nepstad, Daniel C., et al., "Interactions among Amazon land use, forests and climate: prospects for a near-term forest tipping point," Philosophical Transactions of the Royal Society B, February 11, 2008, pp. 1737-1746, doi:10.1098/rstb.2007.0

Surface temperature distribution in a mixed landscape with forest – Ellison, David, et al., "Trees, forests and water: Cool insights for a hot world," Global Environmental Change, March 2017, vol. 43, pp. 51-61, https://doi.org/10.1016/j.gloenvcha.2017.01.002

Hotspots of projected forest loss between 2010 and 2030. – Living Planet Report - 2018: Aiming Higher, World Wildlife Fund, 2018, p. 51.

Exterminate

A buffalo hide yard in Dodge City, Kansas, in 1874. Hunters who arrived in the spring of 1874 noted the scarcity of buffalo north of the Arkansas River; vast herds had been decimated in the hunts of 1872 and '73. – Sharon Guynup, State of the Wild 2006: A Global Portrait of Wildlife, Wildlands, and Oceans, 2005, Island Press, p. 101.

1892: bison skulls await industrial processing at Michigan Carbon Works in Rogueville (a suburb of Detroit). Bones were used processed to be used for glue, fertilizer, dye/tint/ink, or were burned to create "bone char" which was an important component for sugar refining. – "Bison Slaughter of the Late 1800s Has Done Lasting Damage to Plains Indian Nations Today, Study Says," February 17, 2019, https://healingmnstories.wordpress.com/2019/02/17/bison-slaughter-of-the-late-1800s-has-done-lasting-damage-to-plains-indian-nations-today-study-says

Pigeon Shoot Historic Illustration. – "The Illustrated Sporting and Dramatic News," July 3, 1875.

Woman with an entire bird in her hat, circa 1890. Late-Victorian and Edwardian fashions led to the deaths of several hundred million birds in the days before state, national, and international laws stepped in to help prevent the extinction of many of them. – Taylor, Stephen, "The 'Bird Bills': A tale of the plume boom," Hoosier State Chronicles, http://blog.newspapers.library.in.gov/the-bird-bills-a-tale-of-the-plume-boom

Grytviken whaling station on South Georgia Island during the First World War. It has long been abandoned. –Cressey, Daniel, "World's whaling slaughter tallied: Commercial hunting wiped out almost three million animals last century," Nature, March 12, 2015, vol. 519, pp. 140-141.

Current and historic distribution of lions: Historically, lions lived across Africa, southern Europe, the Middle East, all the way up to Northwestern India. Today their habitat has been reduced to a few tiny pockets of the original area. – Carrington, Damian, "Earth's sixth mass extinction event under way, scientists warn," The Guardian, July 10, 2017, https://www.theguardian.com/environment/2017/jul/10/earths-sixth-mass-extinction-event-already-underway-scientists-warn

Recent, dramatic declines of the most charismatic animals. Time, but not date, is taken into account, explaining why all trajectories have the same origin. Long, steep lines indicate a large decline at a high rate. Icons represent populations. Wolf is not represented, and 4 subspecies of giraffes are represented. – Courchamp, Franck, et al., "The paradoxical extinction of the most charismatic animals," PLOS Biology, April 12, 2018, https://doi.org/10.1371/journal.pbio.2003997

The percentage of decreasing species classified by IUCN as "endangered" (including "critically endangered," "endangered," "vulnerable," and "near threatened") or "low concern" (including "low concern" and "data-deficient") in terrestrial vertebrates. This figure emphasizes that even species that have not yet been classified as endangered (roughly 30% in the case of all vertebrates) are declining. This situation is exacerbated in the case of birds, for which close to 55% of the decreasing species are still classified as "low concern." – Gerardo Ceballos, Paul R. Ehrlich and Rodolfo Dirzo, "Biological annihilation via the ongoing sixth mass extinction signaled by vertebrate population losses and declines," Proceedings of the National Academy of Sciences (PNAS), July 10, 2017, www.pnas.org/cgi/doi/10.1073/pnas.1704949114

Consumerism Gone Mad

One of the spinners in Whitnel Cotton Mill. She is 51 inches high, and has been in the mill for one year. Sometimes she works at night; runs 4 sides and earns 48 cents a day. When asked how old she was, she hesitated, then said, "I don't remember, then confidentially, I'm not old enough to work, but do just the same." Out of 50 employees, ten children about her size. Whitnel, N.C, December 1908. – Angelova, Kamelia, "The Lives Of Young Workers Before Child Labor Was Abolished," Business Insider, September 10, 2012, http://www.businessinsider.com/the-lives-of-young-workers-before-child-labor-was-abolished-2012-9

Marble Bust of Socrates – "The secret of happiness, you see, is not found in seeking more, but in developing the capacity to enjoy less" - A. Furtwängler and H. H.

Urlichs, Greek & Roman Sculpture, 1914, London J. M. Dent & Sons Ltd., p. 216.
Out of control: The Centers for Disease Control and Prevention (CDC) released

this graphic as part of its The New (Ab)normal campaign. It shows just how meal sizes have increased in the past 60 years. – "How the size of an average restaurant meal has QUADRUPLED since the 1950s - with U.S. burgers now three times as big," Daily Mail, May 23, 2012, http://www.dailymail.co.uk/news/article-2148970/How-size-average-restaurant-meal-QUADRUPLED-1950s--U-S-burgers-times-big.html

Waste products from a garment factory in Dhaka, Bangladesh, spill into a stagnant pond. –"Waste Couture. Environmental Impact of the Clothing Industry," Environmental Health Perspectives, September 2007, vol. 115, no. 9, pp. A449-A454.

Worker at an e-waste recovery site in Guiyu, Guangdong Province, China. – Watson, Ivan, "China: The electronic wastebasket of the world," CNN, May 30, 2013, https://www.cnn.com/2013/05/30/world/asia/china-electronic-waste-e-waste/index.html

1955 LIFE magazine article – "Throwaway Living." The objects flying through the air in this picture would take 40 hours to clean-except that no housewife need bother. They are all meant to be thrown away after use. – Cosgrove, Ben, "'Throwaway Living': When Tossing Out Everything Was All the Rage," Time, May 15, 2014, http://time.com/3879873/throwaway-living-when-tossing-it-all-was-all-the-rage

Toxic World

Raw sewage spills directly into the Yamuna River at the northern edge of New Delhi. The chemical waste from factories manufacturing leather goods, dyes and other goods floats down the river in white blocks of what looks like icebergs of detergent. – McCarthy, Julie, "Can India's Sacred But 'Dead' Yamuna River Be Saved?" NPR, May 11, 2016, https://www.npr.org/2016/05/11/477415686/can-indias-sacred-but-dead-yamuna-river-be-saved

In 2050, more people will be at high risk of water pollution due to increasing BOD, Nitrogen and Phosphorous. – The murky future of global water quality, International Food Policy Research Institute (IFPRI), 2015.

Locations of approximately 275,000 wells that were drilled and likely hydraulically fractured between 2000 and 2013. – Hydraulic Fracturing for Oil and Gas: Impacts from the Hydraulic Fracturing Water Cycle on Drinking Water Resources in the United States: Washington DC, United States Environmental Protection Agency (EPA), December 2016, EPA-600-R-16-236ES, https://www.epa.gov/sites/production/files/2016-12/documents/hfdwa_executive_summary.pdf

Health effects of pollution – "The neglected menace of pollution across world," Gulf Times, March 24, 2018, https://www.gulf-times.com/story/586275/The-neglected-menace-of-pollution-across-world

Air Pollution Catch-22

Violent Volcanoes - The eruption of Mount Tambora in 1815 was the biggest in recorded human history. The diagram shows the cubic miles of ejecta. – Rice, Doyle, "200 years ago, we endured a 'year without a summer'," USA Today, May 26, 2016, https://www.usatoday.com/story/weather/2016/05/26/year-without-a-summer-1816-mount-tambora/84855694

Annual sulphur dioxide (SO2) emissions in million tonnes from 1850-2010. – Hannah Ritchie and Max Roser, "Air Pollution," 2018, published online at OurWorldInData.org. https://ourworldindata.org/air-pollution

The troposphere and stratosphere – "The Stratosphere," UCAR Center for Science Education, https://scied.ucar.edu/learning-zone/atmosphere/stratosphere

Temperature change over global land surfaces from 1958 to 2002 with respect to 1960. While the temperature rise during the period of solar dimming from the 1960s to the 1980s is moderate, temperature rise is more rapid in the last two decades. – Martin Wild, Atsumu Ohmura and Knut Makowski, "Impact of global dimming and brightening on global warming" Geophysical Research Letters, February 2007, vol. 34, doi:10.1029/2006GL028031

Lights Out

Centrally-generated power flows from generation plants through the bulk power system to industrial, commercial, and residential users. – Securing the U.S. Electrical Grid, The Center for the Study of the Presidency and Congress, 2014.

An image of giant magnetic loops on the sun from NASA's Solar Dynamic Observatory (SDO) mission. – National Aeronautics and Space Administration (NASA) Heliophysics Science Division, https://science.gsfc.nasa.gov/sed/index.cfm?fuseAction=highlight_arch.bl&navOrgCode=670

The arc of light heading towards the Earth is a coronal mass ejection, which impacts the Earth's magnetic field, causing magnetic storms. – Steve Tracton, "The scary Halloween solar storm of 2003: a warning for today's space weather," The Washington Post, October 31, 2013, https://www.washingtonpost.com/news/capital-weather-gang/wp/2013/10/31/the-scary-halloween-solar-storm-of-2003-a-warning-for-todays-space-weather

100 Year geomagnetic storm. The regions outlined are susceptible to system collapse due to the effects of the GIC disturbance – Kappenman, John, "Geomagnetic Storms and Their Impacts on the U.S. Power Grid," Metatech Corporation, January 2010.

Resources

Plastic Oceans

Oceans Asia – Investigate and research wildlife crimes, exposing and bringing to justice those destroying and polluting marine ecosystems. To uphold the highest level of integrity, relying exclusively on factual, intelligence-based conservation. – https://oceansasia.org

Trash Free Seas – Part of the organization Ocean Conservancy, Trash Fee Seas has been at the forefront in keeping trash out of our oceans for more than 30 years. With millions of volunteers, they have collected hundreds of millions of pounds of trash from entering the oceans. – https://oceanconservancy.org/trash-free-seas

5 Gyres Institute – The 5 Gyres Institute is a leader in the global movement against plastic pollution with more than 10 years of scientific research expertise and engagement on plastic pollution issues. Since 2009, the team has completed 19 expeditions, bringing more than 300 citizen scientists, corporate executives, brands, and celebrities to the gyres, lakes, and rivers to conduct firsthand research on plastic pollution. – https://www.5gyres.org

Plastic Oceans – Plastic Oceans International is a nonprofit organization whose goal is to end plastic pollution and to foster sustainable communities worldwide. They operate with the belief that we can and must act locally to create change globally through four key pillars of activity: Education, Activism, Advocacy, and Science. – https://plasticoceans.org

Plastic Pollution Coalition – Plastic Pollution Coalition is a growing global alliance of more than 1,200 organizations, businesses, and thought leaders in 75 countries are working toward a world free of plastic pollution and its toxic impact on humans, animals, waterways, the ocean, and the environment. – https://www.plasticpollutioncoalition.org

Surfers Against Sewage – This is a grassroots movement that fights the plastic pollution that blights our beaches and strangles our seashores. Their goal is to change consumer behavior and industry standards when it comes to single-use plastics. They also organize beach cleans and galvanize communities into action – protecting their local beaches. – https://www.sas.org.uk

Fished Out

Oceana – Oceana is an international organization focused solely on oceans, dedicated to achieving measurable change by conducting specific, science-based policy campaigns with fixed deadlines and articulated goals. Oceana seeks to make our oceans as rich, healthy, and abundant as they once were. – https://oceana.org

Ocean Conservancy – Our oceans face many threats like the onslaught of ocean trash, overfishing, and ocean acidification. Ocean Conservancy promotes healthy and diverse ocean ecosystems and opposes practices that threaten marine and human life. Ocean Conservancy advocates for protecting special marine habitats, restoring sustainable fisheries, reducing the human impact on ocean ecosystems, and managing United States ocean resources. – https://oceanconservancy.org

Marine Conservation Society – Our seas are under immense pressure: too many fish are being taken out, too much rubbish is being thrown in, and too little is being done to protect our precious wildlife. Our vision is for seas full of life where nature flourishes and people thrive. The Marine Conservation Society works to ensure that our oceans are healthy, pollution-free, and protected. – https://www.mcsuk.org

The Natural Resources Defense Council (NRDC) – NRDC works to protect our seas from pollution and exploitation. They help implement laws that allow overfished species to rebound and fight to protect coastal communities from offshore drilling. They work to ban destructive fishing practices, conserve ocean treasures, and improve stewardship of the world's shared oceans. – https://www.nrdc.org/issues/oceans

Project AWARE – Project AWARE works to inspire, inform and provide the tools needed to engage and connect individuals, governments, NGOs, and businesses who share our values and vision for a clean, healthy ocean. – https://www.projectaware.org

Seafood Watch – Seafood Watch works to meet the demand for sustainable seafood, working directly with seafood producers, industry leaders, organizations, and governments worldwide who want to improve their fishing and aquaculture practices. – https://www.seafoodwatch.org

Coral Reef Carnage

Coral Reef Alliance (CORAL) – CORAL is a nonprofit environmental NGO that is on a mission to save the world's coral reefs. They work collaboratively with communities to reduce direct threats to reefs in ways that provide long-term benefits to people and wildlife. In parallel, CORAL is actively expanding the scientific understanding of how corals adapt to climate change and is applying this information to give reefs the best chance to thrive for generations to come. – https://coral.org

International Coral Reef Society (ICRS) – ICRS promotes the acquisition and dissemination of scientific knowledge to secure coral reefs for future generations. – http://coralreefs.org

Marine Safe – MarineSafe is a non-governmental organization (NGO). A group of scientific, legal, economic, and policy experts works with MarineSafe to manage individual projects and aspects of its work and provide advice and expertise to the organization, governments, and others. – http://www.marinesafe.org

The Nature Conservancy – The Nature Conservancy is a global environmental nonprofit working to create a world where people and nature can thrive. – https://www.nature.org/en-us/what-we-do/our-insights/coral-reefs

Reef Check – Reef Check is a nonprofit organization leading citizen scientists to promote stewardship of sustainable reef communities worldwide. – https://www.reefcheck.org

Reef Relief – Reef Relief is a nonprofit membership organization dedicated to improving and protecting our coral reef ecosystem. – https://www.reefrelief.org

Dead Zones

The Nature Conservancy – One of the most critical challenges in the Mississippi Basin today is nutrient pollution. Each year massive amounts of nitrogen and phosphorous from sewage treatment plants, farms, and other sources runoff into the river, posing health hazards to people and wildlife, rising water treatment costs, and contributing to the annual Gulf of Mexico dead zone. The Nature Conservancy is working with farmers, agribusiness, policymakers, and others to target science-based solutions in places contributing the highest levels of nutrients. – https://www.nature.org/en-us/about-us/where-we-work/priority-landscapes/mississippi-river-basin

Chesapeake Bay Foundation – Serving as a watchdog, the CBF fights for effective, science-based solutions to the pollution degrading the Chesapeake Bay and its rivers and streams. Their motto, "Save the Bay," is a regional rallying cry for pollution reduction throughout the Chesapeake's six-state, 64,000-square-mile watershed, which is home to more than 18 million people and 3,000 species of plants and animals – https://www.cbf.org/issues/dead-zones

Acid Seas

The Ocean Foundation – The Ocean Foundation's International Ocean Acidification Initiative builds the capacity of scientists, policymakers, and communities to monitor, understand, and respond to ocean acidification both locally and collaboratively on a global scale. – https://oceanfdn.org/projects/ocean-acidification

The Natural Resources Defense Council (NRDC) – NRDC works to protect our seas from pollution and exploitation. They help implement laws that allow overfished species to rebound and fight to protect coastal communities from offshore drilling. They work to ban destructive fishing practices, conserve ocean treasures, and improve stewardship of the world's shared oceans. – https://www.nrdc.org/issues/reduce-ocean-acidification

Ocean Conservancy – Our ocean faces many threats like the onslaught of ocean trash, overfishing, and ocean acidification. Ocean Conservancy promotes healthy and diverse ocean ecosystems and opposes practices that threaten marine and human life. Ocean Conservancy advocates for protecting special marine habitats, restoring sustainable fisheries, reducing the human impact on ocean ecosystems, and managing United States ocean resources. – https://oceanconservancy.org/ocean-acidification.

Glaciers Going Going Gone

World Glacier Monitoring Service (WGMS) – For more than a century, the World Glacier Monitoring Service (WGMS) and its predecessor organizations have compiled and disseminated standardized data on glacier fluctuations. The WGMS annually collects glacier data through its scientific collaboration network active in more than 30 countries. – https://wgms.ch

Stockholm International Water Institute (SIWI) – SIWI is a water institute. It leverages knowledge and our convening power to strengthen water governance for a just, prosperous, and sustainable future. – https://www.siwi.org

Charity: Water – 785 million people live without clean water. That's nearly 1 in 10 people worldwide or twice the United States population without access to life's most basic human need. Charity: Water is a nonprofit organization bringing clean and safe drinking water to people in developing countries. – https://www.charitywater.org

The Water Project – The Water Project, Inc. is an organization unlocking human potential by providing reliable water projects to communities in sub-Saharan Africa. They suffer needlessly from a lack of access to clean water and proper sanitation. This organization helps communities gain access to clean, safe water by providing training, expertise, and financial support for water project construction through our staff and implementing partners. – https://thewaterproject.org

Water.org – Water.org is a global nonprofit organization working to bring water and sanitation to the world. They want to make it safe, accessible, and cost-effective. They help people get access to safe water and sanitation through affordable financings, such as small loans. They give their everything every day to empower people in need with these life-changing resources – giving women hope, children's health, and families a bright future. – https://water.org

Flattened Forests

Rainforest Action Network – Rainforest Action Network preserves forests, protects the climate, and upholds human rights by challenging corporate power and systemic injustice through frontline partnerships and strategic campaigns. – https://www.ran.org

Amazon Watch – Amazon Watch is a nonprofit organization founded to protect the rainforest and advance Indigenous peoples' rights in the Amazon Basin. They partner with Indigenous and environmental organizations in campaigns for human rights, corporate accountability, and preserving the Amazon's ecological systems. – https://amazonwatch.org

Amazon Conservation – Amazon Conservation unites science, innovation, and people to protect the western Amazon - the greatest wild forest on Earth. They have saved over 8.15 million acres of the rainforest, provided cutting-edge tools to government and forest users to protect their lands, empowered hundreds of indigenous communities to develop forest-friendly livelihoods, and hosted thousands of scientists pioneering innovative research at our three conservation hubs. – https://www.amazonconservation.org

Amazon Conservation Team – The Amazon Conservation Team partners with indigenous and other local communities to protect tropical forests and strengthen traditional culture. They see a future where healthy tropical forests and thriving local communities exist in harmonious relationship with each other, contributing to the well-being of the planet. – https://www.amazonteam.org

Rainforest Foundation – The Rainforest Foundation's mission is to support indigenous and traditional peoples of the world's rainforests in their efforts to protect their environment and fulfill their rights by assisting them in securing and controlling the natural resources necessary for their long-term well-being.

– https://rainforestfoundation.org

Rainforest Trust – Rainforest Trust purchases and protects the most threatened tropical forests, saving endangered wildlife through partnerships and community engagement. Every action we take now to safeguard rainforests will have a lasting impact on the future by maintaining our planet's critical ecosystems. – https://www.rainforesttrust.org

Mongabay – Mongabay is a nonprofit environmental science and conservation news platform that produces original reporting in English, Indonesian, Spanish, Hindi, and Brazilian Portuguese by leveraging over 800 correspondents in some 70 countries. They are dedicated to evidence-driven objective journalism. – https://rainforests.mongabay.com

Exterminate

World Wildlife Fund (WWF) – As the world's leading conservation organization, WWF works in nearly 100 countries. At every level, they collaborate with people worldwide to develop and deliver innovative solutions that protect communities, wildlife, and the places in which they live. – https://www.worldwildlife.org/initiatives/wildlife-conservation

International Union for Conservation of Nature (IUCN) – IUCN is a membership Union composed of both government and civil society organizations. It harnesses the experience, resources and reaches more than 1,400 Member organizations and the input of more than 17,000 experts. This diversity and vast expertise make IUCN the global authority on the status of the natural world and the measures needed to safeguard it. – https://www.iucn.org

Save Our Species –Save Our Species supports science-based conservation action that saves animals and plants from extinction. Informed by the IUCN Red List of Threatened Species and in collaboration with thousands of IUCN experts and scientists worldwide, they focus our efforts where they will have the most significant impact. They prioritize the species most in need: those assessed as Vulnerable, Endangered, and Critically Endangered. – https://www.saveourspecies.org

Defenders of Wildlife – Defenders of Wildlife work on the ground, courts, and Capitol Hill to protect and restore imperiled wildlife and habitats across North America. Together, we can ensure a future for the wildlife and wild places we all love. – https://defenders.org

Endangered Species Coalition – The Endangered Species Coalition's mission is to stop the human-caused extinction of our nation's at-risk species, protect and restore their habitats, and guide these fragile populations along the road to recovery. – https://www.endangered.org

Wildlife Conservation Society (WCS) – The WCS is committed to protecting the world's wildlife, focusing on 14 global regions. Their activities are aimed at combining science, conservation, and education. – https://www.wcs.org

Jane Goodall Institute (JGI) – Dr. Jane Goodall discovered that when we put local communities at the heart of conservation, we improve the lives of people, animals, and the environment. JGI advances Dr. Goodall's holistic approach through a tapestry of nine strategies that build on each other and bring the power of community-centered conservation to life. – https://www.janegoodall.org

The Alliance for Zero Extinction – The Alliance for Zero Extinction works to identify and safeguard the most important sites for preventing global extinction and those that have threatened species restricted to just a single location in the world. – https://zeroextinction.org

International Rhino Foundation (IRF) – The IRF works to ensure the survival of rhinos through strategic partnerships, targeted protection, and scientifically sound interventions. – https://rhinos.org

Global Vision International (GVI) – GVI is an organization that focuses on high-impact and high-quality conservation and community development programs. Its mission drives all that it does, "To build a global network of people united by their passion to make a difference." – https://www.gviusa.com

Consumerism Gone Mad

Project Happiness – The World Health Organization has named depression as the most significant cause of suffering worldwide. In the U.S., 1 out of 5 deals with depression or anxiety. For youth, that number increases to 1 in 3. The good news is that 40% of our happiness can be influenced by intentional thoughts and actions, leading to life-changing habits. It's this 40% that Project Happiness Programs help to impact. – https://projecthappiness.org

Becoming Minimalist – Becoming Minimalist is a place that encourages people to embrace minimalism. It does not boldly require anyone to become minimalist overnight—nor does it specifically define the word for you. Instead, it encourages each reader to discover their own journey and the far-reaching benefits that come from owning less. – https://www.becomingminimalist.com

Fair Trade Certified – The Fair Trade Certified™ seal represents thousands of products, improving millions of lives, protecting land and waterways in 45 countries, and counting. – https://www.fairtradecertified.org

ReFed – ReFED is a national nonprofit dedicated to ending food loss and waste across the U.S. food system by advancing data-driven solutions. With their holistic view of the food system, they are working to achieve a 50% food waste reduction following the United Nations' 2030 Sustainable Development Goals. – https://refed.com

Organics International (IFOAM) – Organics International is a membership-based organization working to bring true sustainability to agriculture worldwide. Through their work, they build capacity to facilitate farmers' transition to organic agriculture, raise awareness of the need for sustainable production and consumption, and advocate for a policy environment conducive to agro-ecological farming practices and sustainable development. – https://www.ifoam.bio

Ellen MacArthur Foundation – At the Ellen MacArthur Foundation, they develop and promote the idea of a circular economy. They work with and inspire businesses, academia, policymakers, and institutions to mobilize systems solutions at scale globally. – https://www.ellenmacarthurfoundation.org/our-work/activities/make-fashion-circular

Fashion For Good – At Fashion for Good, they believe that Good Fashion is not only possible, but it is also within reach – what the industry lacks are the resources, tools, and incentives to put it into relentless practice. Their mission is to bring together the entire fashion ecosystem through our Innovation Platform and as a convenor for change. – https://fashionforgood.com

Basel Action Network (BAN) – BAN's mission is to champion global environmental health and justice by ending toxic trade, catalyzing a toxics-free future, and campaigning for everyone's right to a clean environment. BAN currently tackles three toxic waste streams: Electronic waste (e-waste), end-of-life ships, like the cargo ships that carry our goods, and plastic pollution. – https://www.ban.org

e-Stewards – e-Stewards is the globally responsible way to recycle your electronics. Their website lets you find a recycler in your area for e-waste. – http://e-stewards.org

Green Child Magazine – Green Child is a digital magazine devoted to natural parenting and conscious living. They have practical suggestions on how to reduce e-waste. – https://www.greenchildmagazine.com/reduce-ewaste

OfferUp – An online selling platform designed to offer a simple way to buy and sell locally. – https://offerup.com

Ziilch – Provides a way for Australians to give away items they no longer want to other people that may need them. Ziilch's free community-driven platform promotes social responsibility and environmental sustainability by keeping good stuff out of the landfill. – https://au.ziilch.com

GumTree – An online platform for buying and selling spare things. – https://www.gumtree.com

The Global Slavery Index – The Global Slavery Index provides a country by country ranking of the number of people in modern slavery and an analysis of the actions governments are taking to respond and the factors that make people vulnerable. – https://www.globalslaveryindex.org

Toxic World

Environmental Working Group (EWG) – The Environmental Working Group's mission is to empower people to live healthier lives in a healthier environment. With breakthrough research and education, they drive consumer choice and civic action. They have lots of information on cosmetics, pesticides in produce, household cleaning products, endocrine disruptors, and more. – https://www.ewg.org/consumer-guides

The Rainforest Alliance Certified – The Rainforest Alliance seal promotes collective action for people and nature. It amplifies and reinforces the beneficial impacts of responsible choices, from farms and forests all the way to the supermarket check-out. The seal allows you to recognize and choose products that contribute toward a better future for people and the planet. – https://www.rainforest-alliance.org/faqs/what-does-rainforest-alliance-certified-mean

World Fair Trade Organization (WFTO) – The WFTO is a global community and verifier of social enterprises that fully practice Fair Trade. Spread across 76 countries, WFTO members all exist to serve marginalized communities. To be a WFTO member, an enterprise or organization must demonstrate they put people and the planet first in everything they do. They are democratically run by our members, who are part of a broader community of over 1,000 social enterprises and 1,500 shops. – https://wfto.com

The Cornucopia Institute – The Cornucopia Institute engages in educational activities supporting the ecological principles and economic wisdom underlying sustainable and organic agriculture. Through research and investigations on agriculture and food issues, The Cornucopia Institute provides needed information to family farmers, consumers, and other stakeholders in the good food movement. – https://www.cornucopia.org

Food Tank – Food tank is building a global community for safe, healthy, nourished eaters. They aim to educate, inspire, advocate, and create change. They spotlight and support environmentally, socially, and economically sustainable ways of alleviating hunger, obesity, and poverty and create networks of people, organizations, and content to push for food system change. – https://foodtank.com

Alliance for Natural Health (ANH-USA) – The Alliance for Natural Health USA is the largest organization in the US and abroad working to protect your right to utilize safe, effective, and inexpensive healing therapies based on high-tech testing, diet, supplements, and lifestyle changes. They believe a single-mindedly focused system on "treating" sick people with expensive drugs rather than maintaining healthy people is neither practical nor economically sustainable. – https://anh-usa.org

Controlled Substance Public Disposal Locations – Find the location of where to dispose of your medications. – https://www.hhs.gov/opioids/prevention/safely-dispose-drugs

Lights Out

Energy.gov – Planning for a home renewable energy system is a process that includes analyzing your existing electricity use, looking at local codes and requirements, deciding if you want to operate your system on or off of the electric grid, and understanding the technology options you have for your site. –https://www.energy.gov/energysaver/buying-and-making-electricity/planning-home-renewable-energy-systems

Protect Our Power (POP) – Protect Our Power is an independent organization of former electric utility industry, military, government, and regulatory experts focused on the urgent need to make our electric grid more secure and resilient. POP experts work with all stakeholders to advocate for coordination, knowledge-sharing, political cooperation, public support, and funding to ensure needed actions are taken to strengthen the grid. – https://protectourpower.org

Grid Integration Group (GIG) – The Grid Integration Group is part of the Energy Storage & Distributed Resources Division (ESDR) at Lawrence Berkeley

National Laboratory. GIG performs cutting-edge research to make the evolving smart electric grid compatible with the requirements of electric system grid operators and electric utility companies while serving the needs of electricity customers. The GIG team combines multidisciplinary expertise in grid integration, building technologies, data analytics, optimization techniques, electric vehicle modeling, and cybersecurity. – https://gridintegration.lbl.gov

Solar Cookers International (SCI) – Solar Cookers International improves human and environmental health by expanding effective carbon-free solar cooking in world regions of greatest need. SCI leads through advocacy, research, and strengthening the capacity of the global solar cooking movement. – https://www.solarcookers.org

Index

A

acidification 89, 114, 133, 134, 136, 137, 138, 140, 141, 142, 143, 144
aerosol 301, 304, 305, 307, 310
Amazon rainforest 167, 170, 171, 175, 176, 178, 179, 180, 181, 189
Amazon River 167, 168
annihilation trawling 60, 73, 78
aquaculture 56, 69, 70, 71, 72, 73, 74, 77, 115
Arc of Deforestation 175

B

beef 109, 170, 172, 180, 181, 182, 185, 189, 192, 202
Biochemical Oxygen Demand (BOD) 270
biodegradation 18, 278
biofuel 111, 172, 181, 183
biological annihilation 224, 225
biotic pump 169, 179
bottled water 19, 34, 45, 124, 276
BPA 31
Bryozoans 139
buffalo 197, 198, 199, 200, 201, 202, 203, 209, 210, 215, 226
bunker fuel 254, 255
bushmeat 215, 216, 229
bycatch 56, 59

C

cancer 17, 31, 35, 48, 95, 243, 255, 278, 284, 289, 326
carbon 39, 81, 98, 123, 124, 131, 132, 133, 134, 135, 138, 140, 142, 143, 144, 145, 154, 160, 162, 168, 171, 176, 178, 182, 183, 184, 244, 247, 249, 251, 252, 264, 292, 303, 308
carbon dioxide 39, 131, 132, 133, 134, 135, 138, 140, 142, 143, 154, 160, 168, 244, 249, 308
cattle 109, 170, 171, 172, 185, 186, 188, 202, 216, 224, 226, 275
cheetah 213
chemicals 23, 31, 32, 34, 35, 36, 39, 42, 48, 70, 73, 77, 104, 120, 227, 248, 255, 265, 267, 268, 271, 272, 273, 274, 277, 278, 286, 287, 288, 289, 290, 294
cholera 265, 266
cigarette 17, 48
climate change 39, 67, 69, 76, 87, 96, 97, 98, 112, 114, 115, 116, 121, 154,

158, 162, 176, 177, 188, 191, 192, 221, 224, 225, 269, 272, 292
CO2 39, 121, 131, 132, 133, 134, 135, 136, 137, 138, 139, 140, 141, 142, 143, 144, 145, 160, 168, 183, 184, 188, 191, 252, 264, 285, 307, 309, 310
coal 39, 81, 98, 127, 128, 129, 130, 131, 132, 138, 154, 249, 278, 302, 313
cobalt 252, 253
cod 51, 52, 53, 55, 67, 68, 76, 119, 141, 209
consumerism 42, 237, 256
coral 29, 30, 32, 63, 80, 81, 82, 83, 84, 85, 86, 87, 88, 89, 90, 91, 92, 93, 94, 95, 96, 97, 98, 136, 137, 143
coral bleaching 84, 85, 88, 89, 91, 93, 96
coral reef 82, 83, 84, 86, 88, 90, 91, 92, 93, 94, 96, 136, 143
coral reef decline 83, 93
corn 109, 110, 111, 121, 123, 178, 180, 203, 226, 299
Coronal Mass Ejections (CME) 316
cotton 30, 32, 46, 47, 70, 171, 181, 248
cover crops 120, 124
COVID-19 21, 22
crown-of-thorns starfish (CoTS) 88

D

DDE 32
dead zone 91, 106, 107, 109, 110, 111, 112, 115, 116, 117, 118, 119, 120, 142
debt bondage 64, 65, 66
deforestation 84, 154, 170, 171, 172, 174, 175, 176, 177, 178, 180, 181, 182, 183, 184, 185, 186, 188, 189, 190, 191, 192, 193, 228
deoxygenation 113, 114, 115, 122, 135
destructive fishing 87, 89, 97, 98
diapers 35, 36, 48
diarrhea 265, 266, 286
downcycled 42
drought 177, 180, 181, 182, 183, 186, 187, 188
dumpsites 40, 239, 253

E

egret 205, 206, 207
electric grid 313, 314, 318, 320
electricity 131, 264, 281, 282, 293, 302, 312, 313, 314, 317, 320, 322, 325, 327
elephant 155, 214, 215, 226, 228
eutrophication 83, 105, 108
e-waste 251, 252, 253, 254, 259
extinction 56, 62, 75, 87, 96, 122, 141, 142, 196, 202, 205, 206, 207, 208, 210, 212, 213, 214, 215, 216, 220, 221, 222, 223, 224, 225, 226, 227, 228, 229, 272, 275, 308

F

facemasks 21, 22
factory farm 109
farmed fish 72, 73, 74, 78, 171
fast fashion 246
fertilizer 23, 104, 107, 108, 112, 117, 120, 121, 122, 201, 241, 270
fish farm 72
fish stock 55, 63
flaring 132
food waste 144, 242, 243, 245, 258
forest fires 154, 157, 182
fossil fuel 107, 108, 124, 128, 168, 189, 279
fracking 19, 286, 287, 288, 289, 290, 291
freshwater 38, 59, 70, 71, 73, 104, 106, 150, 155, 158, 162, 167, 221, 222, 241, 242, 268, 275, 287
Fukushima 281, 282, 283, 284, 285

G

garbage 13, 15, 24, 41, 43, 238, 239, 247, 293
geomagnetic storm 317, 318, 321, 324
ghost fishing 29, 30, 63, 77
giraffe 211, 212
glaciers 148, 149, 150, 151, 152, 153, 154, 155, 156, 157, 158, 159, 160, 161, 162
global brightening 306
global dimming 304, 305, 309, 310
globalization 191, 254
gold 56, 126, 154, 172, 173, 174, 189, 192, 205, 212, 217, 228, 252
Green revolution 107

H

halibut 54, 55
happiness 196, 234, 235, 236, 237, 238, 257, 258, 260
Himalayan 153, 154, 155
Hindu Kush-Himalaya (HKH) 159
human development 89, 211
hypoxia 104, 105, 106, 108, 118

I

illegal wildlife trade 213, 217
industrial agriculture 98, 111, 122
industrial fishing 48, 60
insect 167, 223, 228, 296

ivory 215, 217, 228

K

kelp 103, 139, 140

L

landfill 35, 42, 241, 247, 249, 253
lion 210, 211, 226
logging 172, 174, 183, 186, 189, 191, 192
longline fishing 62
low-oxygen ocean zone 103

M

Madagascar 185, 186, 218, 292
mangrove forest 73
manta rays 218
marine mammals 27, 30, 59, 61, 72, 95
Mato Grosso 178
Mau Forest 186, 187, 188
meat 46, 55, 109, 111, 172, 181, 183, 192, 199, 200, 202, 203, 204, 209, 213, 215, 216, 218, 271, 296, 299
mercury 72, 173, 174, 248, 253
methane 39, 188, 244
microbead 23
microplastic 22, 23, 25, 28, 29, 32, 33, 34, 36, 43, 256, 276, 277
milliners 204, 205, 206
mining 63, 127, 128, 172, 173, 174, 189, 191, 192, 252
Mount Tambora 300, 301

N

nanoplastic 38
nitrogen pollution 107

O

obesity 31, 243, 271, 292, 326
ocean garbage patch 24
Oceanic Anoxic Event-2 (OEA-2) 122
ocean species collapse 75
overfishing 50, 51, 53, 55, 56, 60, 62, 64, 75, 76, 77, 83, 88, 89, 96, 97, 118, 136, 143, 221, 229
oxybenzone 31, 95, 99
oxygen 90, 103, 104, 105, 106, 109, 112, 113, 114, 115, 116, 117, 118, 119, 121, 122, 123, 131, 142, 168, 184, 190, 270, 279, 282, 303

oxygen minimum zone (OMZ) 113
oyster 30, 140, 141

P

palm oil 171, 172, 185, 189, 192, 228
pangolin 218, 226, 228
passenger pigeons 203, 204
PCBs 31, 32, 94, 273
Permo-Triassic Boundary (PTB) 141
pesticide 228, 270, 271, 272
phthalates 31, 36, 248
phytoplankton 88, 103, 104, 105, 112, 138, 139
plastic 13, 14, 15, 16, 17, 18, 19, 20, 21, 22, 23, 24, 25, 26, 27, 28, 29, 31, 32, 33, 34, 35, 36, 37, 38, 39, 40, 41, 42, 43, 44, 45, 46, 47, 62, 94, 95, 97, 98, 115, 124, 237, 249, 254, 255, 259, 276, 277, 278, 293, 295
plastic bags 13, 14, 17, 20, 27, 40, 46, 124
plastic bottles 19, 22, 45, 47, 249, 277
plastic debris 16, 18, 24, 27, 34, 37, 44
plastic industry 41
plasticizers 31, 277
plastic pollution 16, 18, 21, 24, 27, 33, 35, 36, 40, 97, 295
plastic production 14, 16, 19, 39, 43
plastic straws 20, 46
plastic waste 13, 15, 16, 18, 21, 39, 40, 43, 44, 62, 278
PM2.5 255, 279, 280
pollution 16, 18, 21, 22, 24, 27, 33, 35, 36, 40, 41, 44, 45, 54, 71, 75, 83, 84, 89, 96, 97, 107, 108, 109, 115, 120, 128, 129, 131, 143, 174, 220, 221, 223, 224, 225, 239, 240, 253, 255, 256, 259, 267, 268, 269, 270, 271, 274, 277, 279, 280, 291, 292, 293, 294, 295, 303, 304, 305, 306, 307, 309, 310
polyester 14, 25, 37, 47, 249
polyethylene 14, 19, 20, 34, 276
prescription drugs 273

R

radiation 84, 99, 282, 283, 286, 301, 304, 316
recycling 15, 39, 41, 42, 46, 238, 252
rhino 212, 215, 226, 228

S

salmon 53, 54, 55, 70, 71, 72, 73, 74
sapphire 186
seabirds 26, 27, 28, 30, 33, 47, 61, 62, 74, 77
seafloor 25, 44, 53, 56, 63, 103, 104, 112, 140, 281

sea level 161
seamount 63
sea salt 33
sea surface temperature 69, 91, 183
sewage 23, 37, 38, 72, 84, 90, 108, 115, 117, 267, 268, 269, 293, 322
shark 57, 58, 59, 61, 89, 218, 226, 228
shrimp 25, 27, 73, 74, 105
shrimp farms 73
sixth mass extinction 75, 220, 221, 223, 224, 225
slash-and-burn agriculture 171
slavery 64, 65, 66, 240
smartphone 251, 252
smog 129, 130, 279
SO2 301, 302, 303, 306, 307, 309
solar flare 317, 325
solar storm 318, 319, 322, 323, 325
soy 109, 110, 111, 123, 171, 172, 175, 178, 180, 181, 185, 189, 192, 226
stratosphere 301, 303, 308
sulfur dioxide 142, 301, 302
sunscreen 95, 99, 301

T

tap water 34, 45, 273, 276
Tian Shan 155
Tibet 154
Tibetan Plateau 150, 151, 152, 155
tiger 58, 73, 218
tipping point 116, 176, 177, 178, 182, 191
totoaba 218, 219, 228
transformer 316, 318, 320, 321
trash 13, 17, 24, 27, 44, 59, 238, 239, 242, 359
trawling 53, 56, 60, 62, 63, 73, 77, 78
troposphere 303

U

upwelling 112, 113, 114, 135, 141
urchin 87, 139

V

vaquita 219
volcano 300, 302, 309

W

wastewater 18, 23, 37, 73, 107, 121, 248, 267, 269, 274, 285, 287, 288, 289, 290, 291
wet wipes 35
whale 15, 27, 28, 207, 208, 209, 210, 226

Y

Year Without a Summer 299

Authors

Roman Bystrianyk co-authored with Suzanne Humphries, Dissolving Illusions: Disease, Vaccines, and the Forgotten History. Roman has an extensive background in health and nutrition, a B.S. in engineering, and an M.S. in computer science.

Kathryn Schmutter has a B.S. in Conservation Biology and Environmental Management. Kate owns her own natural medicine clinic where she is a Counsellor and Naturopath with an M.S. in Counselling as well as advanced diplomas in Naturopathy, Nutritional Medicine, Western Herbal Medicine, and Homeopathy.

Printed in Great Britain
by Amazon